SCIENCE AND FUTURE CHOICE

SCIENCE AND FUTURE CHOICE

Volume I
Building on scientific achievement

Edited by

PHILIP W. HEMILY

Deputy Assistant Secretary General for Scientific Affairs of NATO

and

M. N. ÖZDAŞ

Assistant Secretary General for Scientific and Environmental Affairs of NATO

CLARENDON PRESS · OXFORD
1979

Oxford University Press, Walton Street, Oxford OX2 6DP

Oxford London Glasgow
New York Toronto Melbourne Wellington
Kuala Lumpur Singapore Jakarta Hong Kong Tokyo
Delhi Bombay Calcutta Madras Karachi
Ibadan Nairobi Dar Es Salaam Cape Town

© *North Atlantic Treaty Organization 1979*

British Library Cataloguing in Publication Data

Science and future choice.
   Vol. 1: Building on scientific achievement
   1. Science
   I. Hemily, Philip W   II. Özdaş, M N
   III. North Atlantic Treaty Organisation.
   Science Committee
   500   Q158.5   79-40617
   ISBN 0-19-858162-9

Set by Hope Services, Abingdon
and printed in Great Britain by
Lowe & Brydone Printers Ltd.,
Thetford, Norfolk.

Preface

In 1957 the North Atlantic Council accepted a report on the desirability of expanding non-military co-operation that stressed the importance of science and technology for the members of the Atlantic Community, and established its Science Committee. Over the last twenty years, the Committee has sponsored a wide range of scientific programmes designed to raise and strengthen the level of science in the member countries. These programmes have included research grants, fellowships, advanced study institutes, and other scientific meetings and conferences. With no geographical or political restrictions on attendance, the institutes and conferences have included participants from non-member countries, including many from Eastern Europe, and have resulted in the publication of some 700 books; many of these have come to be regarded as standard texts or outstanding descriptions of the current state of knowledge in their fields.

The Committee recently celebrated its twentieth anniversary by holding a Commemoration Conference in April 1978 at the Palais d'Egmont in Brussels. It proved to be a valuable opportunity to take stock of the impact of science and technology on Western societies and was a particularly useful occasion for a critical review of the changing nature and role of science and technology.

This book is the first of a two-volume series. It contains the papers prepared by several scientists who were invited to review the achievements of the last twenty years in the important fields of electronics, mathematics, materials, environmental sciences, biology, systems science, and astrophysics; it ends with a look at the future of the NATO Science Programme. The authors were also asked to project developments in the next twenty years as the industrialized democracies move from predominantly responding to the challenges and limitations of man's natural environment to fulfilling the need for more effective control and management of the technological environment. The companion volume pays particular attention to the future prospects for the interaction between scientific advance and society, and analyses topics such as global energy and demand systems, the impact of technology on standard of living, employment and labour relations, the significance of the approaching 'information society', long-term economic policies, and international relations. As all the authors had the opportunity of reading drafts of each other's papers and in most cases of revising their contributions in the light of the discussion, many interesting links have been made between the two volumes.

As well as scientists, the Conference was attended by national delegations from all the Alliance countries. These were composed of Ministers of Science, senior parliamentarians, Ambassadors, senior officials responsible for science policy, heads of national research councils, and prominent industrialists. Several international organizations were represented, and a number of journalists from the international scientific press and other news media attended the Conference. I believe that this opportunity for association and discussion between those who are responsible for political decisions and those upon whose work these decisions must increasingly rest has been fruitful. I hope that this book will help to extend and improve our understanding of the role of science and technology in modern society.

M. N. Özdaş

Assistant Secretary General
for Scientific and Environmental Affairs
North Atlantic Treaty Organization

Brussels, November 1978

Contents

Volume II: Technological challenges for social change

List of contributors to Volume I

Pierre Aigrain	Secrétaire d'Etat auprès du Premier Ministre chargé de la Recherche, Paris, France
M.F. Ashby	Professor of Engineering Materials, University Engineering Department, Cambridge, UK
J. Brachet	Directeur, Laboratoire de Cytologie et Embryologie moléculaires, Université Libre de Bruxelles, Bruxelles, Belgium
C. West Churchman	Professor, School of Business Administration, University of California at Berkeley, USA
Sir Sam Edwards	John Humphrey Plummer Professor of Physics, Cavendish Laboratory, University of Cambridge, UK
F. Kenneth Hare	Director, Institute of Environmental Studies, University of Toronto, Canada
Mark Kac	Professor of Mathematics, Rockefeller University, New York, USA
J. Maddox	Director, Nuffield Foundation, Regent's Park, London, UK
G. Mathé	Directeur, Institut de Cancérologie et d'Immunogénétique, Hôpital Paul-Brousse, Villejuif, France
Bengt Strömgren	Director, Nordic Institute for Theoretical Atomic Physics, Copenhagen, Denmark

1

Introduction: retrospective and prospective reviews of scientific achievements

John Maddox

Looking back is always a chastening experience — that, no doubt, is why historians have such arresting things to say. One's first reaction to the record of the past is that so much has changed. Can it have been a mere 200 years since the French Revolution — and only a little longer since what some of us in Britain still call the American Revolution? Have all the consequences of the American Civil War, the growth of the most powerful state in the modern world among them, taken place in just over a hundred years? Can the most catastrophic war in the whole of history have dragged to an end only sixty years ago? Is it really possible to contain within what has now become the average human lifespan the Franco-Prussian War, the Boer War, the winning of the universal suffrage in most modern states, the Slump, the Crash and the Depression of the 1920s and 1930s, the rise and then the fall of Hitler? Looking back, we are inevitably surprised by the pace of change in the world we live in.

Our perception of the past, and of the future for that matter, is conditioned by two conflicting habits of thought. On the one hand, we are inclined to attribute to innovations in science and technology (and in other fields as well) an in-built logic, a momentum of their own; we are inclined to think that what is possible will actually come about. This is the spirit in which the development of large central computing systems in the 1950s led all kinds of people to proclaim the arrival of automation — the 'age of leisure' or of chronic unemployment. Yet it is only now, after an interval of twenty years, that microprocessors have given these speculations point. At about the same period, we were all of us inclined to think that the age of nuclear power had arrived, but more recent events have reminded us that all that had really happened was that it had become both technically and economically possible to build nuclear power stations in large numbers — and we all have our different explanations of why this has not happened.

The history of the post-war period is littered with examples of technological developments which have not been as fully or as efficiently exploited as some people say they might have been — telecommunications, high-speed or supersonic flight, genetic manipulation, space travel. By now there has been so much experience of this kind, much of it disappointing, that even those of us who are convinced of the importance and the beneficence of technological change have forced ourselves to acknowledge the propriety of the half-discipline of technology assessment — the analysis of how what is technically possible is modified by economic, social, and even political considerations. The result, however, is that we carry with us a sense of the gulf, some would say a growing gulf, between what the world might be like and what it is. Is it really necessary, we ask, that in this day and age there should be people who starve to death, or die of preventable infectious disease?

The opposing but often coexisting tendency is that of believing that the environment in which we live will remain unchanged. Our recent experience shows that we applaud the development of television, or of wide-bodied civil aircraft, but do not anticipate how profound will be the consequences for people's use of such leisure as they enjoy. We applaud the advances there have been in the use of medical technology, but do not fully appreciate the demographic consequences for advanced societies such as ours. Even now, we welcome the arrival of the microprocessor based on the silicon chip with no more than a hazy and imperfect understanding of how this will change the way we live or the nature of society. The same is true of genetic manipulation. Our present concern, in trying to assess the importance of these new technologies, is to guess which of society's present tasks will be accomplished more efficiently and effectively when the new technologies have matured, yet we know intuitively that in the long run the result may be a radical reshaping of the patterns of our lives, the structure and balance of society.

These two habits of thought — the tendency to believe that what is possible will happen and the tendency to believe that nothing much will change — have in the past few years been widely used by those in the West who doubt the value of science and technology as evidence to support their claim that Western society is not yet mature enough to contain the forces for change which scientific understanding has given it. For reasons that I shall return to, this argument is invalid, not least because it lacks historical perspective. This blend of a sense of change and a sense of no-change has been a part of modern society since the beginning. The Portuguese navigators who first showed us what the surface of the earth is like had every reason to be excited by the enterprises on which they were engaged yet had no means of guessing at the way in which they would in the process enlarge the vision of the whole Western world. Yet which of us, in retrospect, would regret what they accomplished? My present concern with our ambivalence to the prospect of change is for a different reason; inevitably, this ambivalence colours and complicates our understanding of contemporary history which is, after all, the purpose of this volume — the understanding of the history of science and technology in the post-war period and especially in the past twenty years, with the objective of trying to understand what may happen next.

1.1. Post-war science

The distortion of our perspective of the recent past is one of the reasons why we so often forget how rapid has been the pace of change. It is therefore worth reflecting a little on how much, or rather how little, we knew of our environment by the end of the Second World War. Although the war had been a great vindication of the utility of the technological application of scientific understanding, as with the development of radar and nuclear energy, we were woefully ignorant. Here are a few of the conundrums with which people were wrestling in those now far-off days.

The high-energy physicists, in their search for a better understanding of what were the fundamental particles of matter, were still trying to make sense of their observations in the cosmic rays of particles of matter intermediate in mass between electrons and nucleons. Were those 'mesons' the particles that Yukawa had predicted in 1935 as mediators of the forces between nucleons that hold all atomic nuclei together? The answer, we now know, is that for the most part they were not. Most of the mesons observed in the cosmic rays at that time were not Yukawa particles at all, but particles of matter now called mu-mesons that have as much right to be called fundamental constituents of the universe as the electrons with which we are more familiar. The mu-meson, for what it may be worth, is the second, quite-unexpected particle of matter related to the electron whose existence appeared to have been confirmed in 1977. This was the outcome of experiments at two American laboratories with the huge particle-accelerating machines which are themselves a creation of the post-war period.

By now, it is clear that the particles of matter that seemed at the end of the war to be the most deserving of the name 'fundamental' are merely composites: nucleons, protons, and neutrons which constitute the masses of atomic nuclei are made up of other particles called quarks, the products of the imagination of physicists until their existence was confirmed in 1976. Our concept of what matter is has been quite transformed in a period of time so short that we can be confident that historians will regard it as a revolution in our understanding.

Much the same is true of the common concept of how the universe is constructed. In 1945, for example, nobody knew how the chemical elements, the materials from which we fashion all the useful artefacts of the world we live in, had come about. Now it is known that they were not parts of the early universe but, rather, that they have been manufactured within our own galaxy in successive generations of stars − from which it also follows that our own sun and solar system are made from the debris of earlier generations of stars which have long since run their evolutionary course. At that time, at the end of the Second World War, there was a handful of people looking for an explanation of the apparently random radio-noise detected by the early radio telescopes; now it is known that this microwave background is nothing more nor less than a fossil relic of the early universe, the time when the universe was much more densely packed with matter and radiation than it is at present.

Although people such as Tolman had been brooding in the 1940s about the possibility that somewhere in the universe there might be objects that consisted simply of neutrons held together by nuclear forces, neutron stars were discovered only at the end of the 1960s, at much the same time that the current of theoretical speculation carried people to the conclusion that there must be, somewhere in the universe, objects so massive that they have literally collapsed under their own weight. These objects would exert a powerful gravitational influence on the space about them and at the same time provide the seeds of an explanation of why some of the galaxies the astronomers now recognize are more powerful and are more spectacular sources of radiation than conventional explanations will allow.

During this short space of time, astronomers had had put at their disposal a whole battery of new techniques of observation. Indeed, it is sobering to recall that, at the end of the Second World War, the only part of the electromagnetic spectrum used by astronomers in making observations of the stars and the more distant galaxies was that to which, by accident, the human eye is sensitive. Now the prospect is that in less than a further decade, when the first orbiting, large all-spectrum telescope is launched in the United States, our observation and thus our understanding of the universe will be more radically transformed than it ever was by the building of the first 100-inch telescope at Mount Wilson in California, just over half a century ago. It should not be forgotten that before that, the issue was still unresolved whether the spiral nebulae that could be seen with difficulty in the earlier telescopes were objects within our own galaxy, supposedly coextensive with the universe as a whole, or were alternatively galaxies much like our own.

Another of the intellectual upheavals of the past few decades is the change that has come about in the common understanding of the surface of the earth. It is hard fully to appreciate the enormity of the change of attitude forced upon us by the recognition that the surface of the earth, the land masses and the floors of the deep oceans, is continually in motion. At the very least, it implies that what passed in the 1930s for the geology of, say, the formation of mountains must now be counted a pack of lies. The Himalayas are there not because of some mysterious process of upwarping or downwarping – the words the teachers of geology used to use – but because the continent to which India and Australia once belonged was split into two parts about 60 million years ago, and because the Indian part travelled north over the whole length of the Indian Ocean to collide with Asia, and in the process to cause the chain of seismic regions running east into China and west into Iran, Turkey, and Asia Minor.

1.2 Fundamental research and relevance

Nobody will deny the interest of these discoveries, but there may be some who will say that they have no great practical importance. What difference does it make to have recognized that the entities of which matter is made consist of particles like electrons and called quarks, which appear to have the curious property that they cannot exist by themselves? That the objects of which the universe is made are objects of which few people had conceived half a century ago? And that the surface of the earth is continually being refashioned under the influence of the patterns of heat convection still not fully understood? Which of these discoveries, the sceptics will ask, will help to feed a starving community in any undeveloped country? Which of them will throw light on the means by which the prosperity of the Western world, and of the rest of the world, can be further enhanced?

This volume is not, I know, so much concerned with such scholarly inquiries as with the way in which science and technology in the past twenty years have helped to benefit the human condition. That is a quite proper decision for the planners to have taken. Yet it is important that we should not entirely overlook the consequences of these developments in what might be called academic research. The justification of the Western world's support of such research does not rest exclusively or even primarily on the possibility that some of these developments will yield practical benefits. There may indeed be practical benefits, but so long as none of us can guess what they will be or how they will arise, that is a thin excuse for the money that these endeavours cost us, especially at a time when there are so many tangible applications of science and technology crying out to be exploited. Nor is it sufficient to say that so long as such conundrums persist it is right to spend precious resources on finding the answers to them; there is of course something in that argument, but not enough to justify the billion-dollar cost of what now seem to be typical projects in experimental research in the fields that I have listed: a new particle accelerator, an orbiting telescope, or a deep-sea drilling ship and the cost of operating it.

By the same test, it is no justification to say that the improvements of technique – the technological spin-off as the US National Aeronautics and Space Administration used to call it – that research in these esoteric fields will probably bring will ultimately provide benefits that outweight the costs. For although it is probably true that petroleum engineers have a lot to learn from the navigational techniques developed in scientific deep-sea drilling, there is no reason to suppose that they could not have won the same benefits more cheaply by setting out to solve the practical problems with which they are confronted in drilling into the seabed deep in the continental shelves.

The true justification of such activities is quite different and less tangible. It has to do with what our societies are like and with our expectations of them. One of the reasons why we find it pleasurable to live in societies like our own is that it is possible to sense that we are progressively learning more about the universe and the world we live in. All of us would be the poorer if our societies' objectives did not, or could not, include the deepening of our collective understanding of these teasing questions. It is not merely that we thereby satisfy idle curiosity. Rather, we learn to confront the world we live in more confidently – just as 6000 years ago the builders of the huge Megalithic stone circles with which Europe is still scattered must have been able to set about their daily tasks more confidently once they had provided themselves with tangible means for predicting the pattern of the seasons. At the very least, they would have known when to plan their crops. Most probably, they would have rid themselves of other encumbering mumbo-jumbo – superstition and false ideology.

1.3 Science and technology applied

This volume, I acknowledge, is mostly about themes in scientific research that have led to practical applications, but it is proper also to acknowledge the importance of the more esoteric preoccupation of the research community with which I have so far been concerned. As good luck will have it, there is an exception that neatly proves the rule. When Lysenko's doctrine of genetics was at its height in the 1950s, the Soviet Union and even the Soviet Academy of Sciences were much preoccupied with the distinction between 'proletarian' and 'bourgeois' science. One consequence of that was their suspicion of the physical interpretation given to quantum mechanics by Niels Bohr and his associates at the Copenhagen Institute. Bohr's quantum mechanics was suspiciously idealist, the Marxist purists said — and the result was that for a time bright students in the Soviet Union were denied an understanding of what, it is now in retrospect quite plain, is an inescapable part of what the real world is like.

All of us shrink from living with intellectual shackles like these — and it is only proper to acknowledge that, during the same period, Soviet scientists have made many important and stimulating contributions to fields of understanding lying outside the supposedly proper fields of study. It is however plain that in circumstances where freedom of inquiry is inhibited by ideology, the imagination of bright students must be constrained. Conversely, we must count as one of our assets in the West the way in which the ambitions of young men and women are constantly rekindled, and their talents sharpened, by the invitation offered them by the research community to help understand more thoroughly what nature is really like. And if they end not by discovering what quasars really are, but by that equally formidable exercise of the imagination in conceiving of a whole computer on a chip of silicon, can we then count ourselves the losers?

1.4 Molecular biology

The pace of discovery in some academic fields of inquiry has been, since the Second World War, quite without precedent in the previous history of science and yet, for reasons that are entirely proper, none of them features in this book. A few years ago, I suppose, the same might have been said of molecular biology, the theme of Professor Jean Brachet's contribution. He vividly explains how this, too, is a field of inquiry revolutionized since the Second World War. In 1945, some people thought that the repository of genetic information in a cell was nucleic acid, and some thought that it was protein. Now it is known that the nucleic acid is what matters, and for the past quarter of a century there has been a quite breathtaking deepening of our understanding of how nucleic acids function as repositories of genetic information. The genetic code has been deciphered. The ways in which cells make specific proteins in response to instructions embodied in their inherited genetic material has been understood in considerable detail. The nature of many important inherited diseases — haemophilia for example — has been explained. From the start, the molecular biologists have been saying that there would certainly be many important practical applications of this research — and yet, so far, there is nothing much to report.

Why should this be? And what are the prospects for the years ahead? To begin with, it will not be the end of the world if there are never practical applications of molecular biology to boast of. The clinicians will know what they are about more confidently than ever in the past. But there is much more than that to say. Already it is plain that the molecular biologists will soon be in a position to throw light on the origin of a great many diseases of great importance in our societies — the rheumatic diseases for example — because of the understanding of these processes which has been provided in the past few years by their research. Further ahead is the promise of genetic manipulation. As things are, of course, it is too soon to know where these techniques will lead. Their use for the cure of inherited genetic diseases may never be possible. The chances that they will be used for the manufacture on a substantial scale of biological materials that cannot otherwise be cheaply made is almost certain. The ultimate promise lies somewhere in between – and here it is significant that early in 1978 the role of the redundant DNA in the cells of higher organisms has been clarified and in part explained. Professor Brachet describes in his chapter how the genes of higher organisms seem to be mixed up with pieces of DNA that play no obvious part in the functioning of the genes. These extra pieces of material are at first sight simply encumbrances, impediments to the smooth functioning of the cell. But now it turns out that they have a function of their own, and an important one. In ways yet to be understood in detail, these extra pieces of DNA in the genes of higher organisms are the means by which the activity of the genes containing them is controlled – as Professor Brachet suggests. But they are also a means by which the genes of higher organisms can modify their function, most probably an important element in understanding the genetics of evolution and of the differences between cells with very different properties – liver and skin cells for example. And in the long run they are certain to be a means by which the techniques of genetic manipulation, so far applicable with confidence only to the genes of bacteria and other lowly organisms, can be applied to genes from all kinds of complicated organisms. It will not be for much longer that clinicians will be able to show disdain for the benefits of molecular biology.

Molecular biology is, if you like, a field of research in which we can hope to see in the next few years the metamorphosis from understanding to application. With one exception – mathematics – the other topics in this volume have an eminently practical quality. In passing it might be added that no one pretends that they cover the whole of science and technology; the reader can, I am sure, think of gaps that might have been filled. But the fact that there is, for example, no contribution on the progress in the past twenty years in the control of thermonuclear fusion does not lessen the interest of this symposium or suggest that the lessons to be learned from the topics that have been discussed will be any the less valuable. My task is to synthesize and to comment on what has been said in the several discussions that follow, each of them by a distinguished scientist far more expert than me. In the circumstances, the only seemly course I can follow is to draw attention to some of the conclusions that will have been suggested by these discussions.

1.5 Modern electronics

In any case, as I shall try to suggest, synthesis is not feasible – the fabric of science is changing too quickly for that. Take, for example, Professor Pierre Aigrain's topic of electronics. Quite properly, he reminds us that the theoretical understanding of the movement of electrons in solids built up in the 1930s may have been sufficient to have allowed some perceptive physicist then at work to develop the first transistor and so to bring forward the exciting achievements and prospects of the modern electronics industry by a decade or more. Equally, however, he is right to remind us that that hope would always have been a pipedream. For the truth is that the development of modern electronics rests not on a few dramatic inventions, even though milestones such as the microwave generator, the maser (and eventually the laser), and the transistor stand out. This is rather a field in technical and scientific development which has been an unprecedentedly rich field for the investment of the talents of scientists and engineers throughout the world. We are inclined to think of the chief consequences of the electronic devices that now abound as the saving of human labour, as in the application of computers, for example. We tend to forget that each new device is the fruit of development by what may often be a substantial army of highly skilled scientists and engineers. For much of the time, the incentive has not been so much to secure a huge leap forward in technique but, rather, to achieve some limited and practical goal — a data-transmission network light and robust enough to be carried in an earth satellite, a timing device small enough to be fitted in a wristwatch case, a computer whose size is determined not by the electronics but by the size of the smallest keyboard with which human fingers can grapple. None of us can tell where the torrent of electronic innovation that surrounds us will end, or what its economic and social consequences will be. It is also, however, important to acknowledge that this torrent of innovation is not merely the product of science and technology, but of the economic and social environment within which Western industry must sustain itself. It goes without saying, of course, that some of us will be despondent that we are more likely to be the purchasers than the originators of the new electronic devices; but there is nothing we can do about this and of course there is no great credit in being the first to do something new in an environment in which novelty does not naturally flourish.

During the past twenty years, there has been a great deal of discussion within NATO about the reasons for the so-called technological gap — the way in which some countries lag behind others in innovation; and in many of those discussions electronics has been the centre of attention. Two points are therefore worth noting. First, in principle, there need be no shame and despondency in those countries in which the latest electronic devices happen not to have been invented. For any community, NATO or the EEC, it is of course much more efficient that there should be a division of labour between the members, with some developing the electronics industry and others banking services or food production. The measure of what constitutes a technological gap is not the lack of this advanced industry or that, but the lack of adequate economic growth compared with other members of the club. But if there is evidence of a technological gap in the sense in which I have defined it, there is no way of getting rid of it by investing large sums of money in the process of technological innovation. That is like trying to cure measles by the surgical removal of the spots. The only effective course of action, from which some governments may quite legitimately shrink, is to create the economic environment within which innovation will flourish on its own.

1.6 Materials science

The same considerations apply to much of what Professor M. F. Ashby has to say about the development of new engineering materials. This again is a field in which innovation has been spurred on by a diversity of specific needs by the users of these materials — and Professor Ashby shows clearly how the needs of the aero-engine constructors have in the past two decades carried the development of metallic alloys to near the point at which the practical limits of the materials themselves are within sight. We will all accept what he says about the likelihood that the years immediately ahead will see the development of new polymers and ceramic materials that approach the qualities of materials now widely in engineering use. Here is another set of challenges to be faced.

Professor Ashby introduces a line of argument about the balance between the various materials now in use and likely to be used in engineering constructions — questions of scarcity or, more accurately, of relative scarcity. And he is of course quite right to remind us that if it were necessary to extract lead from ordinary rocks, so that the cost of a lead battery for a motor car would be roughly $10 000 (at present prices), people would stop buying car batteries and perhaps even cars. The important question, though, is whether it will ever be like that. I do not believe that society will allow a car battery to cost $10 000. This objective will be accomplished not by draconian price control but by two much more reliable and even familiar devices. If prices threaten to increase that much, then much greater care will be taken of the lead in our batteries just now and some way of using other metals in batteries instead of lead will be developed. Or perhaps somebody will devise a way of starting ordinary cars without a battery. I am not especially unhappy with Professor Ashby's arguments about resource constraints, for he couches them in much more subtle terms than is usually the case. Yet all arguments about resource constraints stem from one important supervening consideration — that the pattern of present practice in our dependence on resources can safely be extrapolated into the future, i.e. things will not change much. Consideration of the recent history of technology shows that the real events are more complicated — and more interesting.

1.7 The environment

Thus there is an argument that bears on another topic in this volume — the emergence of a new science called environmental science in the past twenty years. In my view, Professor Hare has put the case for its continued existence with care — and a nicely judged wryness. The first thing to say is that, on the evidence of his chapter, he is not an environmentalist in the polemical sense. I suspect that we would not find him with a placard in his hand demonstrating at the site of the most recently proposed nuclear power station. His argument is all the stronger for that, and among the many questions that he raises, what most sticks in my mind is Professor Hare's conclusion that the environmental sciences are not a well-ordered set of disciplines, but rather a field in which people of all kinds of disciplines come together for the solution of problems, which almost of necessity are formulated much later than they might have been. It follows that a part of the skill required in this field must be that of recognizing which problems are substantial, and which are merely spurious — and all of us will be aware of how many supposed environmental problems have been thrust on public attention in the past few years, only to have been shown to be illusions, or only to have been tackled from too narrow a point of view.

Two other points stand out from this discussion, one of which is of immediate practical importance. Are the processes by which governments make decisions on the basis of studies in the environmental sciences sufficiently flexible and rational to meet the need? Professor Hare is a good-natured optimist who plainly believes that, while irrational, governments surprisingly often come to sensible conclusions. I wish it were more often the case. I recall that in the United Kingdom the chemicals called cyclamates are still banned from public use for no better reason than that, ten years ago, they were withdrawn from sale in the United States. In 1978 the government of Sweden banned the use of the aerosols to which Professor Hare refers, apparently without reference to the sense of deliberation that is creeping in elsewhere in the world and to the improbability that such a gesture could contribute to the solution of a problem which, if it is a problem at all, is surely an international problem. And we all know of governments that worry about pollution typical of and caused by heavily industrialized countries, without sparing much thought for the ways in which their own populations are still ravaged by infection of one kind or another — what we call in the West the environmental problems of the nineteenth century.

My second question is far from immediate but, rather, is concerned with the long-term future of the environmental sciences. Will they, as Professor Hare suggests, be enabled by their success in dealing with immediate problems to take up some of the global issues of the world we live in: food, population, resources, and the grand issues of human ecology? Certainly there is no reason to suppose that these problems will become less daunting than they have been in recent years. Even if there is no progressive change in the climate of the earth, the pace of technological change itself will help radically to change the forces that determine the future of the human race. I would be the last to deny to anybody with the will the right to contribute what he can to the solution of important problems, but how easy will it be to accommodate within what we now call the environmental sciences some of the changes in the pattern of society that change may bring? What, for example, can the environmental sciences hope to have to say about such massive displacements of human labour as might in principle result from the microprocessor revolution, unless they choose to redefine themselves so as to subsume the social sciences and perhaps political science as well? I wish the dream well, but would prefer to wait to see whether it becomes a reality.

1.8 Systems science

But is this not precisely the set of circumstances in which a prudent population of the world would seek the help of systems analysis, the theme of Professor C. West Churchman's discussion? To be frank, I do not think so, and I am not sure what Professor Churchman himself would say. (Most probably he knows much more than I do about the International Labour Office's large systems analysis of various kinds with the precise intention of throwing light on the causation of unemployment.) My difficulty is partly semantic: I wonder whether the new use of the word 'system' is as accurately cognate as it might be with the usage established more than a century ago in more humdrum contexts: thermodynamics, for example. As students, many of us were drilled in the distinction between open and closed systems, and eventually we learned how to handle them. A closed system is self-contained, an open system can exchange matter or energy, or something else with its surroundings. I take it that 'open' or 'closed' have much the same significance as 'pluralistic' and 'monistic' in Professor Churchman's discussion. Moreover, I take his point that the more general usage of the word implies not merely a collection of parts but an *organized* collection of parts. Thus we speak of the circulatory system when we refer to the heart, the arteries, the capillaries, and the veins. A city is plainly a system in this sense: it is a place where people live and work but with an intricate organization of its own; the busmen have to start early so as to get the office cleaners to their jobs. And a city is a pluralistic system: we cannot look at some aspects of the city system in isolation from the surrounding countryside from which the city-dwellers get their food.

So far, I suspect, Professor Churchman and I are in agreement. Where we begin to part company is with his use of the word 'teleological' to describe the set of systems which are the proper subject of systems analysis. I would suggest that this is an awkward millstone to hang around one's neck, especially if it is the ambition of systems science to extend its analysis to include social or even ecological systems. I do not believe it is anything like as easy as Professor Churchman implies to define the teleological ends even of quite simple physical systems. You might say that the teleological end of a can-opener system is to open cans, but that can be only part of the objective. Questions also arise about the ease with which this task can be accomplished, the safety, the cost, and even the speed. In the eyes of their users, the means by which the teleological objective is accomplished may be an objective in its own right, if only a subsidiary one. With motor-cars, as several manufacturers have learned to their cost, the subsidiary objectives may often be dominant to the teleological end of providing travel over smooth ground. With even as well-defined a social system as a city, the problem of specifying the teleological ends is even more complicated. What is the city *for*? To provide shelter, or income, or culture, or companionship for those who live there? Or to provide a concentration of special skills for the nation state in which it is embedded? And by what yardstick should its utility be measured: the average income of its inhabitants (or the number of times per annum they go to the theatre) or by its contribution to the national GDP? And what is to be said if a strong mayor or city council can somehow convert the city from being a pluralistic system to a monistic system by contemptuously ignoring central government?

I do not wish to make too much fun of difficulties of definition such as these, because I believe them to be serious questions, going to the root of the difficulties in carrying through the most ambitious claims of systems science. So let us agree on what we can agree on: operational research has been shown to be a valuable analytical tool, linear programming is a great help to multiple stores seeking to decide where to site their next establishment, the arrival of large computers has made it possible to construct *models* of real systems much more complicated than any that have previously been constructed. Nobody can doubt the value of exercises such as those now possible: models of the flutter of aircraft wings yield helpful results, so too do models of national economies. The all-important truth is that, in the sense in which it is being used, systems science means the accurate calculation of the behaviour of a model of some real system and not necessarily a calculation of reality. Certainly, in the calculation of the flutter of aircraft wings, we are likely to be on sure ground. In circumstances that are less easily defined, the chief value of a model of some real system may be to provide an understanding of how complicated things really are — and nobody will be surprised that the oil company to which Professor Churchman referred found it necessary to build a model with a billion variables. On other occasions, systems science may help us to understand that things are not what they seem to be. In Britain, for example, the Treasury's model of the economy has helped to reinforce what had always been suspected, that at times of high inflation people save and do not spend (as Adam Smith, if he had been living at such times, would have had them do). I would agree with those who say that in such complicated circumstances the chief function of systems science is to help us understand how the parts of the system are joined to each other: what links are important and what are less important, which bits of our ignorance of how the real world really functions could most profitably be removed. Alas, I do not share Professor Churchman's view that the Club of Rome's disastrous publication *The limits to growth* has helped to spread these lessons wider. What I dispute, within the less flamboyant context of Professor Churchman's discussion, is our capacity — our wit — to build models of real social systems that in our present state of ignorance approximate to the real world, and then to solve the philosophical problem of telling by what number — 'figure of merit' is, I suppose, the more acceptable word — their performance should be measured. Until such questions are answered, I shall be one of those who fear that the use of systems science for the modelling of ethics will have to await a demonstration by somebody that the computers can do better than St. Augustine and people like that.

1.9 Mathematics and science

I have referred briefly to all but one of the contributions to this volume, and I hope that neglect will wherever possible be counted as applause. It would, however, be neither courteous nor possible to attempt some kind of synthesis from what has been written without somehow taking up Professor Mark Kac's account of mathematics in the past twenty years. The connection or the lack thereof between mathematics and empirically based science has of course been given much attention. The paradox that Professor Kac describes is also familiar. Mathematicians are repeatedly being surprised at the discovery that apparently academic exercises in mathematics often turn out to be usable in theoretical physics, following half a century or so behind. And the physicists are often but not always surprised that when they have a problem to solve they can often (but not always) find a mathematical technique to suit their needs. Professor Kac asks whether there is a reason why mathematicians in their pursuit of their peculiar excellence should not pay some attention to the needs of their ultimate users, and to his credit answers in the negative. For my part, I think they have no choice, as I shall try to explain.

First, however, it is important that the fortunate co-emergence of physical problems and the techniques for their mathematical solution is by no means as common as the examples of Einstein and Levi-Civita would suggest. Newton, after all, was compelled to work out his theory of fluxions for himself — and if he had been a better mathematician, he would have devised a better notation. (You may say that he might have saved himself the trouble of inventing the calculus by reading Descartes, but that was before the days of international institutions.) And one of the most successful of the modern theories of physics, called quantum electrodynamics, is still flawed by the perpetuation of a howler that schoolboys are warned against: by now there must be an army, or at least a platoon, of Nobellists who spend their time subtracting infinity from infinity and pretending that the answer is a meaningful finite number. What, I might ask Professor Kac if I were coarse and impolite, are the mathematicians doing to explain why renormalization (subtracting infinity from infinity) works — as it does, given that the magnetic moment of the mu-meson calculated in this illicit fashion is correct to the first eight significant figures? We shall forgive you (I might add) if you have nothing better (which means nothing worth mentioning) to offer us in non-linear differential equations or in a host of other less important fields in which mathematics might have helped to describe the real world.

Second, though, it is in my belief important that we should acknowledge that even mathematicians do not live entirely in a vacuum. Somebody invented *zero* because he thought it would help — and only the Romans could have forgotten the trick. Hilbert did not invent his set of spaces with an infinity of dimension because he had foreknowledge of quantum mechanics, or even because he was simply curious about the behaviour of linear combinations of orthogonal functions (although that may have been a large part of the explanation); it is also, in my view, reasonable to infer that he was also powerfully influenced by the climate of his times and by the recognition of the physical importance of the sets of orthogonal functions which are the solutions of quite ordinary wave equations, as well as by the interest of physicists from the 1880s on in spaces with more than three dimensions. In passing, let me say that in the history of science the history of mathematics has been too much neglected.

The essence of Professor Kac's paper is, however, the conundrum of how mathematicians can reconcile their research interests with their quite proper sense of obligation to science as a whole. Professor Kac is quite right to say that the matter cannot simply be left   to chance; vicarious applications of pure mathematics, however valuable, cannot by themselves satisfy the needs of society and, for that matter, the proper sense of self-esteem of mathematicians themselves. It is, however, easy to overlook what even the purest of the pure mathematicians already do for the scientific enterprise. It is not merely that many of those whose interests in research lie in the more abstruse properties of groups, rings, and ideals nevertheless find themselves teaching beginning calculus from time to time, nor even that their craft informs the practice of modern science at every important turn. It is also the case that fields in which the imaginations of both scientists and mathematicians range overlap at many important points, and that a great many of the problems that stimulate physicists are also problems that intrigue many mathematicians. The mathematicians and the physical scientists have more in common than often meets the eye — an assertion that I hope will not offend too many mathematicians.

1.10 Forecasting the future

So what are the prospects for the next twenty years or so? Philosophically, this is an impossible question. There is no way of telling which unsolved problems will be solved, and how, for science remains what it has always been: an empirical process of inquiry, discovery, and the formulation of new problems. If the pattern of discovery were predictable, research would be by definition unnecessary. So much is fully acknowledged in the contributions to this volume. We are therefore driven back to the acknowledgement that the best we can do is to gauge the present momentum of the scientific enterprise, take account of such threats to this momentum as there may be, and then merely guess where it will all lead.

Some things, however, can be said confidently about the future, for the lead times of many important projects are so long that there is a sense in which the future is almost here. In astronomy, the very large orbiting telescope being planned in the United States for launching in the early 1980s will, if it functions as well as it has been designed, surely contribute as much to empirical knowledge of the universe as did the Mount Wilson telescope more than fifty years ago. In high-energy physics, it is hard to think that the present frenzy of model-building will come to an end without there being some substantial deepening of concepts of what are the inherent attributes of matter. In molecular biology, interest will now turn to the more detailed understanding of the biochemical processes in the nucleus of the cells of higher organisms, with the result that we shall know more about the mechanisms whereby the activity of cells is regulated. It is hard to think that the incentive that has been given to the understanding of the earth's climate in the past few years, in part at least because of anxiety about the consequences of air pollution for the weather, will fail to yield a more adequate understanding than there is at present of the causes of climatic change. We may not yet be sure of the feasibility of controlled thermonuclear fusion, but we can be certain that we are only now at the beginning of the transformation of our lives that will be wrought by the development of the microcomputer industry; habits of work, patterns of communication, and even modes of life will change.

These, unfortunately, are only some of the predictable developments of the years ahead. What we know from our recent history is that most changes are unforeseen — and, in my opinion, no degree of sophistication in what is called technology assessment will radically sharpen, make more specific, our appraisal of the future. Now, as in the past twenty years or, for that matter, in the past 750 000 years, we must go forward hopefully, knowing that the future will be different from the present and that it will increasingly be moulded by developments in science and technology, but fortified in our conviction that we can accommodate our societies to different circumstances by our experience of the recent past.

1.11 The post-war period

It is therefore worth while, even on occasions such as this, to remind ourselves of what has been accomplished in the decades since the Second World War in the industrialized countries of the world. And perhaps the first thing to say is that during that now substantial period of time the industrialized nations have not found it necessary to make war among themselves. Opinions differ about the extent to which this considerable achievement can be counted a benefit of the military research and development. Perhaps the more perceptive view is that we have indeed won stability from military technology, but at the cost of an unwelcome degree of political rigidity, in Europe at least. But one product of the development of military technology in the past few years has been the deepening of intellectual understanding on all sides of the nature of strategy, and of the determinants of stability, in this new age.

The fact that the managers of the Atlantic Alliance have thought fit to sponsor a modest programme of support for science or, perhaps more accurately, for the communality of the scientific community, is one sign of this wider appreciation of the nature of contemporary strategic conflict. A strong alliance needs a strong and a creative scientific community. There are, it must openly be acknowledged, some whose opinions command respect within the scientific community and who are in no sense disaffected from the notion that the West should defend itself who argue that science should be supported either for its own sake or not at all. The other view, which I share myself, is that the industrialized world needs a strong scientific community for all kinds of reasons, that the internal logic of the scientific enterprise ensures that it is virtually immune from perversion and, to be banal, that every little helps.

Our second outstanding achievement since the Second World War has been enormously to improve the well-being of the populations of the industrialized world. Avoidable death is now mostly avoided, and the definition is being broadened all the time. Most industrialized countries have found it possible, even in the recession of the past few years, to be compassionate for the disadvantaged sections of their communities. In general, indeed, people's sense of prosperity has been enormously increased. In Western Europe, for example, it is now hard to find relics of the patterns of life that were all too common in the second half of the 1940s. People no longer live in unheated, unsanitary, or unlit houses. Even in Britain, they no longer queue up for a ration of coal. On the contrary, people now travel and communicate and are educated (and also entertained) in a manner that indicates not merely an improvement of their material standards of life but has also been accompanied by an enhancement of their sense of liberty.

It would be wrong, of course, to suggest that all this largely beneficent change is a simple consequence of developments in recent decades in science and technology. Much of the social transformation that has come about has been fostered by our democratic institutions, or by the enterprise of commercial enterprises, public agencies linked with government (not necessarily at the centre), voluntary organizations, and by people acting as individuals. But science and technology have been necessary, if not sufficient, in the social transformations of the period since the Second World War. What science and technology have made possible is a continuation of the division of labour that is the distinctive feature of all society, and has been a dominant feature of our societies for the past 200 years. We have some reason to be confident that this tendency will continue. We have every reason to hope that it will.

It is only fair that such self-congratulation should be tempered by an acknowledgement that we have not always succeeded as well as we might have done. And indeed we all know how the societies of the West are at present preoccupied with the economic problems stemming from the continuing recession and with their anxiety about the adequacy, even the stability, of our financial institutions. Perhaps the most daunting of the domestic tasks with which we are faced is the continuing high rate of unemployment in several of the industrialized nations of the world; and already there are some who fear that the continuing development of the micro-computer industry can only exacerbate such problems.

Fortunately, there is good reason to believe that the outcome of this and indeed many other technical developments now in train will have quite the opposite effect. Those of us who went to school in Britain were regaled with tales of how, at the beginning of the Industrial Revolution, groups of workers fearful for their jobs set about destroying the new textile machinery and the steam engines then coming into service. As events have shown, however, the consequences of these innovations were quite the opposite of what had been feared at the outset. The new machines created new kinds of jobs, not merely in the processing of their own products but, by the general increase of prosperity, in quite different sectors of the economy. This, after all, was the time when British society first recognized that it could afford to employ the teachers to provide something approximating to universal education.

There is no reason why the consequence of technical developments now in prospect should not be similar in kind. It is true that with what we know — and with our new-fangled habit of trying to guess in advance at each and every aspect of the immediate future — we shall not simply have to trust in good luck or even in Adam Smith's invisible hand. But there are limits to the degree to which we can hope to anticipate what the future will be like, and there are serious dangers in the policies of those who advocate that nothing new should be allowed to happen, in the application of science and in technology, until everybody is assured that the consequences will be universally beneficent.

In my opinion, there is at least as much reason to believe that we only gain from a liberation of the process of technical change from some of the forces which at present constrain it. I am not so much concerned with such things as the steps that have been taken in recent years to protect the environment from industrial development — although some of these are plainly senseless — as with the economic climate in which technology must struggle to survive. In many of our societies, the traditional economic incentives to development have been dangerously eroded. Even within the European Economic Community, presumably one of the most tightly integrated sections of the West, the further economic division of labour that technology promises is inhibited by all kinds of chauvinistic restrictions, the policies of national governments on public purchasing for example. And the Buy America Act is an unhelpful precedent. These, it seems to me, are the directions in which those who are concerned that technology should contribute fully to the development of our societies should concentrate their attention.

It is also proper that we should acknowledge that success falls far short of expectations in the relationship between the industrialized and the developing countries of the world. For the past several decades, it has been acknowledged that in principle science and technology have much to contribute to the process of development, yet many of the developing countries still languish in desperate poverty. Here again, many of the restraints are political and social: there is, for example, no objective reason for believing that the exhortation of the United Nations to the developed nations of the world that they should devote some 0.7 per cent of the resources to development assistance (a target which is achieved in only a few states) is adequate to the need. Yet here it is also clear that we understand only inadequately how best science and technology might contribute to the process of development. And there are few of us who believe that the United Nations conference in 1979 on the subject will turn out to be illuminating. This conundrum, then, is a challenge that should be more vigorously taken up in societies like our own, not merely because it is an intellectual challenge to technology but because the relationship between the industrialized and the developing nations of the world is also a challenge to the political and military stability on which we have set our sights.

I have been concerned to argue that much that is good can continue to flow from the continued development of technology and from the applications of scientific discoveries in the past few years, or those now in prospect — many of which have been described in the contributions to this conference. It is therefore important to know how sure we can be that the exciting torrent of discovery described in this volume will continue unimpeded. There is no reason why it should not but, alas, there are now some alarming signs that the health and the spirit of the scientific community is being harmed by some of the developments of the past few years.

The past few years have seen, in many industrial societies, a decline of the funds available for scientific research in universities and independent research institutes. Even where the funds available have managed to keep pace with inflation or even to grow slightly faster, however, the momentum of the growth of activity in research has been halted. Funds are important, but mere money is not the most important determinant of the quality and the imaginativeness of scientific research. The morale of the research community, and the way in which the pattern of scientific research evolves, are more important.

There are many causes of anxiety. The rapid growth of the research enterprise in the 1960s, followed by the more or less abrupt cessation of growth early in the 1970s, has sharply restricted the recruitment of young men and women to academic research institutions. Throughout our societies, there is a growing concern for the personal plight of such young people, who in normal times would have naturally taken up posts as researchers of potential distinction. As yet, however, we have no means of telling how much their absence will damage the research enterprise as such. But we have every reason to believe that the creativity of that enterprise rests inescapably on the continual renewal of its personnel and thus its strength.

In several of our societies, there are already signs that the research enterprise is faltering. The pattern of research is changing less quickly than it would if it were still practicable for young people to create small research groups. With new problems to solve and the flair for solving them these groups, over the years, can be expected to make a mark. The result is that in the academic sector the research enterprise is more fragmented and thus less fruitful than it should be. And the morale of those concerned is being undermined.

These gloomy developments have been reinforced by the way in which a great many governments — which are of necessity the chief sources of support for scientific research — have in recent years become persuaded that their support of research must be accompanied by attempts to ensure that the benefits of research will become more speedily available to the community at large. There is, of course, a substantial part of the research enterprise in which such intentions are entirely legitimate, and in which mechanisms exist or can be developed for directing the efforts of the research community at definable objectives. But there is also an essential part of the enterprise in which the pattern of research can be constrained only with the risk of damaging its quality. As things are, there is altogether too much evidence that this unhappy process has been set in train.

Let us hope that such problems will be solved before too much damage has been done. To be sure, a failure to find ways of sustaining the research community would not mean that the end of the world as we know it had arrived. Our societies would no doubt be able to carry on. But the whole weight of what we know and understand about the past 20 or even 200 years goes, in my opinion, to show that we shall be impoverished intellectually as well as economically, and able to face the future less confidently, unless steps can be taken to ensure the health of the research community as we have known it these past twenty years.

2

The science of engineering materials

M. F. Ashby

2.1 Introduction: Engineering materials

The great bulk of engineering materials consumed by any society is used for making mechanical devices and structures — uses which subject them to stress. These materials are drawn from four classes of solid.

For the last 100 years, the mechanical engineer has thought first in terms of *metals*; and the range of metallic alloys available to him, and the range of properties they offer, has expanded tremendously during that period. But, although we are heavily committed to metals in mechanical engineering, *polymers* increasingly appear in vehicles, aircraft, and other structures in which they are subjected to stress.

In an industrial nation, the structural engineer, too, is heavily dependent on metals. But on a world scale, the most widely used structural material is a polymer: wood (strictly, it is a composite of two polymers: cellulose and lignin). Polymers tend to have a low modulus — they deflect more than other materials do when they are loaded — and they may decay, losing their strength with time. Partly because of this, the dominance of wood as a structural material is being challenged by structural concrete; indeed in developed nations it has already replaced it.

Concrete is a *ceramic* (which we define as a crystalline inorganic compound, typified by oxides, carbides, and silicates), although it is a very complicated one. The mechanical engineer avoids ceramics, and with them *glasses* (non-crystalline compounds, again mostly oxides and silicates) because they are brittle, and can fail in a sudden and unpredictable way. Stone, brick, and concrete, all of them ceramics, can be used in structures because the stresses to which they are subjected are very small and compressive; and although steel-reinforced concrete is used in tension, the stresses are again, by the standards of the mechanical engineer, very small, and are carried by the steel, not by the concrete. For any high-performance application ceramics are — or have been until recently — completely ruled out.

Fig. 2.1 shows schematically these four basic types of engineering material. To the physicist, they are distinguished by the nature of the bonds which hold their atoms together, and the regular or non-regular pattern of the atoms or molecules within them. Ultimately, it is the strength and directionality of the bonding, and the nature and regularity of the structure that determine the properties which the engineer uses — and which set a fundamental upper limit (which we discuss below) to the value that any of them can reach.

The four basic material types are arranged in a diamond to illustrate that an increasing range of *composite materials* — physical combinations of one or more

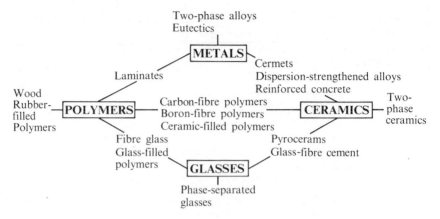

Fig. 2.1. The classes of engineering materials and the composites which can be formed within a class and between classes.

materials — now exists. Natural evolution, and man's ingenuity, have created materials such as wood, and carbon-fibre reinforced polymers, which combine the best of the properties of different material types, or even of two materials of the same type. Some — like wood and ferroconcrete — are cheap and used on a vast scale. Others, like the pyrocerams and the cermets, are expensive, and are used only on a small scale. All offer a combination of properties which surpasses that of the components, used alone.

The modern mechanical and structural engineer thinks, increasingly, in terms of the range of materials illustrated by Fig. 2.1. But this is a recent development. The *pattern of use* of materials, though it is changing, still reflects strongly the more traditional engineering approach, emphasizing, above all, the use of metals.

2.2 The use–pattern of materials

The way in which materials are used in a developed nation — any one of the NATO countries, for instance — is fairly standard. All consume steel, concrete, and wood in construction; steel and aluminium in general engineering; copper in electrical conductors; and so forth; and all use them in roughly the same proportions. Ranked by annual tonnage, or by total value, metals dominate any listing of materials consumed by such a country.

The cost of materials to a developed nation

One of the more instructive ways to rank materials is in terms of the cost to the nation of buying them on world markets. Table 2.1 shows how the total annual import bill of a typical European country — Great Britain — is distributed between five broad categories: *fuels*, mainly oil, coal, and gas; *basic raw materials* other than fuels; *food*, including drink and tobacco; *semi-manufactured goods* like billets of steel, cut wood, paper, etc.; and *finished-manufacture goods* like cars or TV sets. The table shows that in 1974, about 11 per cent of the total import bill was for basic raw materials.[1] To this must be added the segment of the semi-manufactured imports, like steel ingots, or shaped wood, which should be counted as 'engineering materials'. When this is done, the fraction of the import bill which was spent on materials in 1974 was close to 20 per cent.

Table 2.1 *Role of materials in total UK imports*

Material	Proportion of total imports (%)	
	1974	*1964*
Mineral fuels	20	11
Basic materials	11	19
Food, drink, tobacco	16	30
Semi-manufactures	28	24
Finished-manufactures	24	15
Other	1	1

The total import cost for 1974 was £2.3 $\times$ 10^{10}; that for 1964 was £5.7 $\times$ 10^9. Engineering materials include minerals, ores, and scrap which come under the heading of *basic materials* together with partly processed materials such as ingots, slabs, billets, cut timber, rubber, etc., which constitute part of the category headed *semi-manufactures*. Engineering materials accounted for roughly 26 per cent of the total import costs in 1965, and 20 per cent in 1974. The distribution is broadly typical of a developed European nation.
Source: NERC/IGS [1].

Table 2.2 shows how this 20 per cent was distributed among materials. Iron and steel, and the raw materials used to make them, accounted for about a quarter of it. Next is wood and lumber — still widely used in light construction. More than another quarter was spent on the metals copper, silver, aluminium, and nickel. All polymers taken together, and including rubber, account for little more than 10 per cent.

If we include the further metals zinc, lead, tin, tungsten, and mercury, the list accounts for 99 per cent of all the money spent abroad on materials, and we can safely ignore the contribution of materials which do not appear on it.

The table illustrates the dependence of a developed nation on metals: they account for much more than half of the import bill for materials. Indeed, if we leave out wood (much of it which is used for low-stressed structures like furniture) 80 per cent of the remaining materials-import bill is for metals.

For a number of reasons, a slow, long-term evolution away from this emphasis on metals has started. If it is to progress, it will require major advances in the understanding and production of polymers and ceramics, and equally major changes in the design methods employed by the engineer who uses them. One reason for this shift of emphasis is that the rate of development of new metallic alloys, which has been rapid over the last 20 years, is slowing down and is, in some areas, stationary: new advances in technology require, increasingly, the use of completely new materials. Another is economic: metals are expensive, and because their ores are often

---

[1] This was a smaller proportion than in earlier years, mainly because, by 1974, fuels had started to rise steeply in price, while other materials had not.

highly localized, and sometimes genuinely scarce, they may not always be freely available on world markets. We consider this question of availability in the next two sections, returning later to a discussion of the relative merits and future potential of metals, polymers, and ceramics.

Table 2.2 *UK imports of engineering materials, raw and semis*

	Proportion of total (%)
Iron and steel	27
Wood and lumber	21
Copper	13
Plastics	9.7
Silver and platinum	6.5
Aluminium	5.4
Rubber	5.1
Nickel	2.7
Zinc	2.4
Lead	2.2
Tin	1.6
Pulp/paper	1.1
Glass	0.8
Tungsten	0.3
Mercury	0.2
Other	1.0

UK imports of materials in 1974 totalled roughly £4.6 $\times$ 10^9, which constituted 20 per cent of the total import costs for that year. The distribution shown in this table is typical of a developed European nation.
Source: NERC/IGS [1].

2.3 Ubiquitous materials

Let us now shift attention from what we use to what is widely available. Some few engineering materials are synthesized from components found in the earth's oceans and atmospheres: magnesium is an example. But almost all are won by mining their ore from the earth's crust, and concentrating it sufficiently to allow the material to be extracted or synthesized from it. How plentiful and widespread are these materials on which we depend so heavily? How much copper, silver, tungsten, tin, and mercury in useful concentrations does the crust contain? All five are extremely rare: workable deposits of them are so small and highly localized that many governments classify them as of strategic importance, and stockpile them.

The abundance of elements in the earth's crust, oceans, and atmosphere
Not all materials are so thinly spread. Table 2.3 shows the relative abundance of the commoner elements in the earth's crust. The crust is 47 per cent oxygen by weight, or (because oxygen is a big atom) 96 per cent by volume: geologists are fond of saying that the earth's crust is solid oxygen containing a few per cent of impurities. Next in abundance are the elements silicon and aluminium; by far the most plentiful solid materials available to us are silicates and aluminosilicates. A few metals appear

on the list, but iron and aluminium are the only ones which appeared also in the list of widely used materials. I have carried the list as far as carbon because it is the backbone of virtually all polymers, including wood.

Table 2.3 *Abundance of elements* (wt %)

Crust		Oceans		Atmosphere	
Oxygen	47	Oxygen	85	Nitrogen	79
Silicon	27	Hydrogen	10	Oxygen	19
Aluminium	8	Chlorine	2	Argon	2
Iron	5	Sodium	1	Carbon dioxide	0.04
Calcium	4	Magnesium	0.1		
Sodium	3	Sulphur	0.1		
Potassium	3	Calcium	0.04		
Magnesium	2	Potassium	0.04		
Titanium	0.4	Bromine	0.007		
Hydrogen	0.1	Carbon	0.002		
Phosphorus	0.1				
Manganese	0.1				
Fluorine	0.06				
Barium	0.04				
Strontium	0.04				
Sulphur	0.03				
Carbon	0.02				

The total mass of the crust to a depth of 1 km is 3×10^{21} kg; that of the oceans is 19^{20} kg; that of the atmosphere is 5×10^{18} kg.
Source: Verhoogen [2].

The oceans and the atmosphere show a similar picture. Oxygen (and its compounds) are overwhelmingly plentiful: all around we are surrounded by ceramics, o or the raw materials to make them. Some metals are widespread, notably iron and aluminium, but, even for these, the local concentration is almost always small, usually too small to make it economic to extract them. In fact, the raw materials for making polymers are more readily available than those for any of the metals. There are huge deposits of carbon in the earth. On a world scale we extract every month a greater tonnage of carbon than we extract iron in a year, but at present we simply burn it. And the second ingredient of most polymers — hydrogen — is also one of the most plentiful of elements.

Could we, over the next 20 years, shift the emphasis of use of materials away from the expensive and potentially scarce metals, substituting for them ubiquitous materials, such as oxides and silicates, which are almost as common as dirt; or materials like carbon-chain polymers which are as plentiful and widespread as coal? The answer appears to be that we could, although a very substantial research and development effort, spanning the next 20 years; would be required to change the emphasis much. But is it necessary or desirable that we should do so? Are metals really a diminishing resource, and will some of them, like oil, be available on only a very restricted scale in 25 years' time? The resource-bases of most metals appear to be stronger than that of oil, and their projected half-lives longer. But, as discussed in the next section, their future availability is still a matter for concern.

2.4 Potentially scarce materials

There have been many predictions of resource crises in the past. The idea that US oil reserves would not last more than 15 years has prevailed for over 50 years [3]. The Paley Commission [4], reporting in 1952, foresaw exhaustion of the world's lead and zinc reserves by 1970, yet in that year the reserves stood higher than ever before. At least three recent publications [5, 6, 7] quote data showing that the current reserves of some metals are inadequate to last until the end of the century.

Reserves and resources

There is little doubt that, like the projections of the Paley Commission, these recent statements are unduly pessimistic. This is because they are calculated, not from the *resources* of each material, but from the statement of their *reserves*. The reserves are the known ore deposits which can be mined profitably at today's price using today's technology; they bear little relationship to the true magnitude of the resource base; in fact, the two are not even roughly proportional. Data describing the reserves of most minerals are readily available, and to date have always been adequate to cope with demand. Table 2.4 shows that the reserves of copper, lead, and zinc, for instance, have continued to increase as new ore discoveries were made and mining technology improved, despite the exhaustion of earlier reserves. As a result, past forecasts of metal exhaustion based on reserves 'have come to appear almost wildly inaccurate in the light of later production' [8].

Table 2.4 *The changing reserves of metals*

Date	World reserves (tonnes $\times$ 10^6)			
	Copper	Lead	Tin	Zinc
1950 estimate [4]	181	36	5.1	45
Amount left by 1964	130	4	2.5	1
New 1964 estimate [9]	140	29	5.1	64
Amount left by 1974	78	−3	2.9	13
New 1974 estimate [10]	298	108	3.6	101

The resource base includes current reserves. But it includes also all deposits that might become available given adequate prospecting and which, by various extrapolation techniques, can be estimated. And it includes, too, all known and unknown deposits that cannot be mined profitably now, but which — owing to higher prices, better technology, or improved transportation — might reasonably become available in the future. The resource base is the only realistic measure of the total available material. Resources are almost always much larger than reserves. But because the geophysical data and economic projections are so poor, their evaluation is subject to vast uncertainty.

This distinction between reserves and resources is sometimes illustrated by what is called a McKelvy diagram (Fig. 2.2) [11]. It has axes showing the degree of certainty with which the ore is known to exist, and the ease and cost with which it can be mined and the metal extracted from it. The reserves are represented by the shaded area; they are extended downward by improved mining technology or by an increase in price; and they are extended to the right by prospecting.

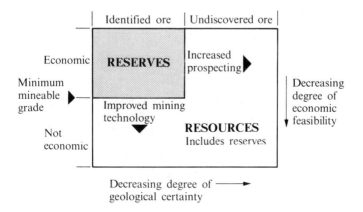

Fig. 2.2. The McKelvey diagram, showing the distinction between reserves and resources. The diagram is conceptually helpful, but cannot be made quantitative.

A more quantitative and informative way of assessing resource magnitude is shown in Fig. 2.3. It shows, for lead, the quantity of metal available as a function of the minimum economically mineable grade. Lead is a good example because its distribution is typical of many of the scarcer metals.

About 15×10^6 tonnes of pure lead is available in stockpiles or is in use and is readily retrievable: it is available at concentration of 100 per cent. Known lead deposits, at a concentration of 4 per cent (the present minimum minable grade), or better, amounted to about 150×10^6 tonnes (which is roughly equal to the total cumulative world production up to that date). If this were all that was available, it would be consumed completely in only 23 years (at the current rate of consumption), and shortages would already be apparent.

Extrapolation suggests a resource distribution as shown by the histogram. This includes as-yet-undiscovered deposits (taken as more than three times the known reserves), as well as the very large amount of lead (albeit at low concentrations) contained in the manganese-rich nodules which lie on the sea bed but which cannot yet be mined economically.

Under current economic conditions, Fig. 2.3(a) suggests that the world resources of lead are over 500×10^6 tonnes. But ordinary rock contains lead at an average level of 15 p.p.m. (0.0015 per cent). Although this is very small, when integrated over the volume of the earth's crust to a depth of 1 km, it represents a tremendous quantity of lead. Most elements occur in this dilute form as well as in concentrated ores. And in all cases, the quantity of contained metal is tremendous: enough to last for several hundred years at current rates of exponential growth. Could we not extract and use it?

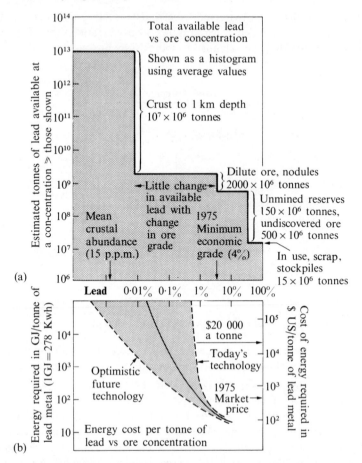

Fig. 2.3. (a) This diagram shows the quantity of lead, in tonnes, contained in the earth's crust, with a concentration (or grade) equal to or greater than that shown on the horizontal axis. The shape of the distribution is obtained from published data for the distribution within currently operating mines; and the absolute value is estimated from the known amounts of lead in countries in which ore bodies have been widely sought, notably the U.S.A. and Australia. Many of the scarcer metals have distribution curves like this one.

(b) The energy cost, per tonne of lead, is shown as a function of the grade. The rising energy cost (together with that of chemicals and plant) sets a limit on the maximum minable grade. It is unlikely to fall below 0.1 per cent, so that the extraction of lead from common rocks (containing 0.0015 per cent) is totally impractical.

The answer is that, with a few exceptions,[1] we cannot. As leaner ores are mined, it becomes necessary to mine, crush, and treat larger and larger quantities of rock in order to produce one tonne of metal. These processes consume energy and chemicals,

---

[1] The exceptions are bromine, chlorine, magnesium, and sodium (and perhaps boron, fluorine, hydrogen, lithium, oxygen, potassium, rubidium, strontium, and sulphur) from sea-water; argon, krypton, oxygen, nitrogen, and xenon from air; silicon (and perhaps aluminium) from wide-spread clays; and perhaps iron from iron-bearing rocks.

and can have enormous environmental impact. As leaner ores are used, the energy consumed per tonne of metal produced rises, slowly at first but then more and more rapidly; if (as is often the case) the ore becomes more complex and hence harder to process, the energy required rises even more rapidly. The lowest usable ore concentration is set by the cost of this energy, the cost of chemicals, and the availability of capital to build new plant.

This is illustrated for lead in Fig. 2.3(b). Bounds on the energy costs are shown (the upper bound corresponds to extraction using today's technology; the lower bound to optimistic future technologies). Using a mean (unbroken line), we find that the cost of extracting lead from common rock is about 10^6 per tonne: at this price a car battery would cost about $10 000. If we assume that lead prices of 100 times the 1975 price of $200 per tonne would price lead products completely out of the market (car battery $300) we find the minimum economic grade of ore to be about 0.1 per cent and the resource base to be about 10^9 tonnes ± 50 per cent.

These results must obviously be treated with caution. But if it is accepted that we can never extract lead economically from common rocks, the exploitable resource base does not depend much on where the cut-off is made. This is because the distribution curve for the ore grades of lead, and of many other metals, is fairly flat over its central portion. Hence, improving extractive technology does not always increase the resource base dramatically; in this example, lowering the minimum mineable grade by a factor of ten, i.e. to 0.01 per cent, increases the amount of lead available by only a factor of two. For such metals, the resource base is effectively constrained, and given enough information to construct a diagram like Fig. 2.3, its magnitude can be estimated.

Consumption rates, present and future

Most materials are being produced and consumed at a rate that is growing exponentially. The current rates of growth of metal production, for instance, vary considerably: from less than 1 per cent per year for mercury to over 10 per cent per year for platinum. But how fast will they be consumed during the next 20 years? The lifetime of a mineral resource obviously depends much more on future demand than on past production. To answer this question we have to examine how each material is used.

It is usually possible to break down the uses into broad categories. Lead, for example, has four broad uses: in storage batteries; in construction, e.g. pipes, sheet, and castings; in alloys, e.g. solder, type-metal, and bearing metal; and in chemical compounds, e.g. tetraethyl lead and lead oxide. Between them, these four uses account for 94 per cent of all lead consumption.

Each use is growing at different rates, some negative, some positive. There are various ways of extrapolating each into the future. Projections on the growth of photography help in extrapolation for silver. Projections on the demand for electric power help in extrapolating for copper. Projections for the growth in demand for cars help in extrapolation for lead because over one-third of all lead produced is contained in car batteries. Against this must be set the effects of legislation which will largely remove lead from gasoline in some countries over the next decade —

12 per cent of today's lead production is used to raise the octane value of gasoline. And so on. Most important of all, the steady increase in the world's population, and their rising expectations, make it virtually certain that demand for all minerals will continue to grow at its present rate, or faster, until A.D. 2000 and probably well beyond.

Resource life

Of course, no material resource will ever be completely expended. As its scarcity increases, its cost will rise and, one by one, substitutes of varying effectiveness will be found. The consumption rate will pass through a maximum and start to decline, as shown schematically for lead in Fig. 2.4; here, any short-term fluctuations are unimportant and have been ignored.

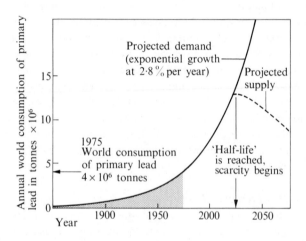

Fig. 2.4. A schematic projection of resource consumption for lead. Shortages are felt when the 'projected demand' curve departs from the 'supply' curve. At an adequate level of accuracy, one can assume that this will occur when roughly half the resource is expended [12].

To project the resource life, a measure is needed which indicates when supply can no longer keep pace with demand, since it is then that scarcity begins to be felt. This point occurs when the exponentially growing demand curve of Fig. 2.4 departs from the curve describing supply. If we assume that this occurs when one-half of the exploitable resource base is exhausted (which may or may not be true, but uncertainty here is negligible compared to uncertainties in estimating the resource base) a time can be calculated at which scarcity due to resource depletion will appear.

By constructing diagrams like that of Fig. 2.3(a) and calculating the energy in the manner of Fig. 2.3(b) estimates can be made of resource 'half-life'. They are often much longer than widely quoted estimates of reserve life [13]. The metals which are in greatest demand are iron, aluminium, copper, lead, mercury, silver, tin, tungsten, and zinc. Their projected resource half-lives are shown in Table 2.5.

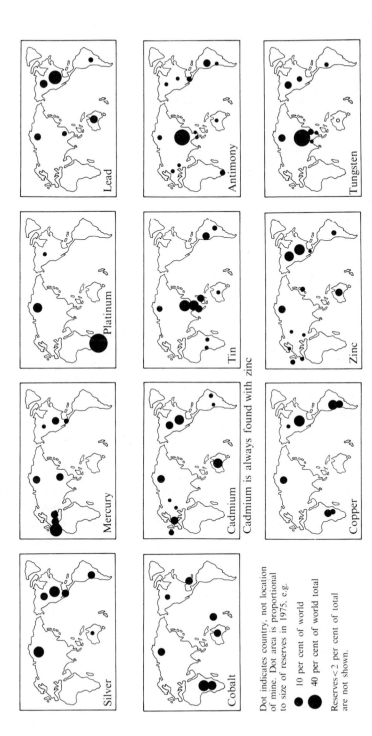

Fig. 2.5. The distribution of the main ore-bodies of eleven of the scarcer metals. In some instances more than half the world reserves are concentrated within a single country.

Dot indicates country, not location of mine. Dot area is proportional to size of reserves in 1975. e.g.

● 10 per cent of world

● 40 per cent of world total

Reserves <2 per cent of total are not shown.

Silver

Cobalt

Mercury

Cadmium

Cadmium is always found with zinc

Platinum

Tin

Lead

Antimony

Tungsten

Zinc

Copper

In general, the resource base appears to be adequate to cover world demand up to A.D. 2000 and well beyond, even allowing for exponential growth. But by about 2025 scarcities due to resource depletion may begin to appear.

Table 2.5 *Projected resource half-lives of metals in great demand*

Metal	Resource half-lives (years) (± uncertainty)
Iron	Over 200
Aluminium	Over 200
Copper	Over 150
Lead	50 (+20, −10)
Mercury	67 (+13, −13)
Silver	50 (+30, −10)
Tin	70 (+50, −20)
Tungsten	60 (+20, −10)
Zinc	50 (+19, −14)

Source: NATO Science Committee Study Group [13].

These times should not make us complacent. Assuming present usage patterns, shortages can be avoided only if research and capital are invested in locating and exploiting new ore bodies, and in improving mining technology. And it must be recognized that the occurrence of some key metals appears to be very localized (Fig. 2.5). Long before deposits are physically exhausted they may become politically inaccessible except on terms dictated by the producing country. Such geopolitical effects have been adequately discussed elsewhere [14, 15].

Many metals, then, are expensive, and potentially scarce; and, at least in some areas, they can no longer supply the demands of increasingly advanced technology. There exist both economic and technological driving forces for examining how polymers, ceramics, and glass might replace metals, or even permit technical advances which have been held back by fundamental limitations of metals and their alloys. To understand the problems associated with this transition we must examine the properties which are required of an engineering material, and which account for the present heavy emphasis on metals.

2.5 The important bulk properties of engineering materials

In selecting a material for use under load, it is the bulk properties which are of first importance; the surface can always be coated, or plated, or hardened, later. And of the bulk properties, there are three which are particularly important.

Resistance to deflection: the modulus

The first is the *modulus*. Low-modulus materials are floppy, and will suffer a large deflection if loaded. Sometimes this is desirable, of course: springs, cushions, vault-

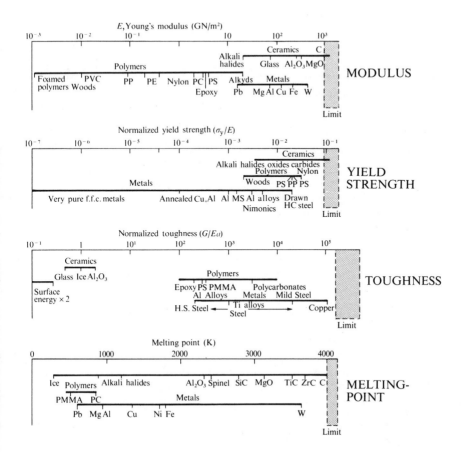

Fig. 2.6. The important bulk properties of an engineering material, showing the fundamental limits (imposed by the electrostatic nature of the forces between atoms, and by the atomic size). Ceramics approach the upper limit in all but toughness; but metals offer the best overall combination, accounting in large part for their present predominance in engineering design. Polymers are now challenging metals in many areas that do not require high-temperature strength, and ceramics may replace metals for very high-temperature applications.

ing poles — these structures are designed to deflect, and the right choice of modulus here may be a low one. But in the great majority of mechanical applications, deflection is undesirable; the engineer seeks a high-modulus material.

Fig. 2.6 illustrates the range of moduli that materials exhibit. In units of giganewtons per square metre (GN/M^2), moduli lie between about 10^{-3} (for very soft foams and rubbers) and 10^3 (for diamond): a range of about 10^6. Metals and ceramics lie near the top of this range. The moduli reflect directly the attractive and repulsive forces which act between atoms in the solid state, and these in turn depend on fundamental quantities such as the charge of an electron (e), and the factors which determine the size (a) and packing arrangement of atoms in the solid;

the moduli of the alkali halides, for instance, scale as e^2/a^4. These quantities cannot be manipulated and they impose a fundamental upper limit on the modulus of solids of about 10^3 GN/M^2. It is shown as a bar in Fig. 2.6. The moduli of many ceramics and metals come fairly close to this upper limit. But common polymers like polyethylene lie nearly four orders of magnitude below it. As we shall see presently, the problem in making polymers useful as structural materials is that of raising their modulus without destroying their other properties.

Resistance to collapse: the yield strength

The second important property is the *yield strength*. The yield strength is the stress level at which a material ceases to be elastic, and deforms, often very extensively and suddenly, in a permanent plastic way: on removing the stress, the deformation does not disappear. Again, except for rather special applications — crash barriers for instance — yielding is thought of as an undesirable property in structural materials. It results in large permanent deflections and usually means that the structure has to be discarded.

Pure, annealed, metals have low yield strengths; typically, they are about 10^{-6} of the modulus. Much of the long history of metallurgy has been a search for ways to increase this strength. It has met with fair success: most structural materials have a strength of at least 10^{-3} of the modulus; and the strongest steels have strengths approaching 10^{-2} of the modulus.

The other classes of materials — polymers, ceramics, and glasses — have yield strengths which are usually greater than 10^{-2} of the modulus. Because the moduli of most polymers are low, the absolute magnitude of their yield strengths may be low too; but ceramics have high moduli, and their yield strengths are enormous.

Resistance to fracture: the toughness

But there is a third vital property, the *toughness*. It is slightly subtler and less familiar than the other two. The name — toughness — conveys roughly the right impression; meat is tough when you cannot sink your teeth through it and politicians are tough when, no matter how you batter them, they fail to fall apart. Glass, which has a high modulus and a high yield strength, is not tough; it breaks only too easily. 'Perspex' has a low modulus and low yield strength but its toughness, too, is low (though it is higher than that of glass). The toughness does not correlate in any simple way with the other two important bulk properties.

In fracturing a solid, new surface is created. One way of measuring the toughness is as the energy required to extend a crack by unit area. For extremely brittle (that is, untough) materials, this energy is little more than twice the surface energy (because, in fracturing a solid, two new surfaces are created). It takes only the lightest tap to cleave a calcite crystal, for example, and the energy absorbed is little more than 1 joule per square metre (J/m^2) of fracture surface — about the same as the surface energy of calcite. But for a tough material, much more energy is absorbed: to fracture copper, for instance, requires almost 10^6 J/m^2 of fracture surface, even though its surface energy is almost the same as that of calcite. This enormous difference appears because the fracture of copper is preceded by extensive plastic flow, while that of calcite is not, and the plasticity absorbs energy.

Like the modulus and the yield strength, the toughness of engineering materials spans a range of about 10^6, from 1 to 10^6 J/m². The toughness of a typical steel is about 10^5 J/m², and is near the top of this range. It is an important quantity for the engineer, who tries to ensure in his design that a sudden load will be absorbed without leading to catastrophic fracture. This he can do if the toughness of the material he selects is high enough.

Fig. 2.6 shows the toughness of engineering materials, displayed on a dimensionless scale which allows the classes of materials to be compared. The toughness G (in units of J/m²) has been divided by Young's modulus E, times the atom size a. This normalization allows the fundamental limits on the toughness to be displayed. On the left-hand side is the lower limit, $G/Ea \sim 0.2$; fracture cannot absorb less than twice the surface energy of the material. On the right-hand side is the upper limit: if the material has unlimited plasticity, it will fracture by deforming until its section has been reduced to zero (involving a strain of about unity); if, in addition, its yield strength is high, of the order of $E/100$, then the energy absorbed in making unit area of fracture is given by $G/Ea \sim 10^6$. The toughness of all materials lies between these limits. Those for metals lie near the upper end of the range, as do those for polymers (though, because their moduli are low, the absolute magnitude of the toughness may be small). Ceramics and glasses lie near the other limit: their toughness barely exceeds their surface energy.

Resistance to softening: the melting-point
Fig. 2.6 includes one further property of a material — its melting-point, T_M. This is not a bulk property in quite the same sense that the other three properties are, and in designing structures for use at room temperature, it is of little importance. But as the temperature is raised, materials become softer. As a general rule the mechanical strength of a material falls off steeply at about 0.5 T_M and it is no longer useful as a structural material. There are exceptions: certain very heavily alloyed nickel-based alloys are used at temperatures as high as 0.7 T_M. But, for fundamental reasons, all materials become very soft, and deform or flow rapidly, if loaded at higher temperatures than this.

I have added the melting-point scale to Fig. 2.6 to emphasize one of the fundamental limitations of the commoner metallic alloys: all have melting-points below 2000 K. Increasingly, energy-conversion technology is seeking higher operating temperatures in order to improve efficiency. This fundamental limitation of metals is constraining further technical advances.

The scale itself has limits. For the same reason that there is an upper limit for the modulus, there is one for the melting-point; it is around 4000 K. A few metals — tungsten, molybdenium, platinum — approach this limit, but all are far too expensive and scarce to use as large-scale structural materials. But many ceramics have melting-points which are in this high range, and they are cheap and plentiful.

The combination of properties required of a successful engineering material
We can now begin to see why metals are so favoured by the engineer when he designs a structure. They have high moduli and a high toughness. It is true that their

yield strengths, when pure, are low, but empiricism and (to a lesser extent) research in alloy development have led to ways of dealing with this which have been so successful that metals have become the prime choice of any designer confronted with a stressed component. But, as materials are used at higher and higher stresses and temperatures, metals are reaching their fundamental limits. For some applications, such as turbines, development of new metallic materials is almost at a stand-still. And metals are expensive and sometimes hard to get. Higher efficiency and innovative design can come about only if radically new materials become available.

Ceramics have certain clear attractions. Their moduli are high; usually higher than those of metals. Their yield strengths are high too; much higher than those of metals, and because they have high melting-points, they retain their strength to higher temperatures. But they lack toughness. This is the class of solid for which the toughness barely exceeds the surface energy; and this is so low that ceramics are generally used in load-bearing structures only in compression. Even then the loads to which they can be subjected safely are small. For reasons discussed later, it seems likely that the development of tough ceramics, and of new ideas in how to design with them, will bring about major changes in the way that brittle materials are used.

Polymers, in general, are much tougher than ceramics. As pointed out earlier, the yield strengths are high — at least when expressed as a fraction of the modulus: they are typically about $E/50$. The problem is that they have low moduli; or at least most of them do. But consider the structure of diamond. It is generally classed as a ceramic, but it can equally well be thought of as the ultimate cross-linked polymer; and its modulus is the highest known. By increasing the degree of cross-linking, or by orienting the chains in it so that the diamond-like carbon–carbon bonds lie parallel to the direction in which the material will be stressed, polymers with very high moduli can be synthesized. Few are yet available to the engineer; but there can be no doubt that the next 20 years will see the appearance of polymers with properties which challenge those of steel.

Let us now examine, in a little more detail, the evolution of metals, polymers, and ceramics as engineering materials.

2.6 The evolution of metals and alloys

The attraction of metals is clear from Fig. 2.6: they combine usefully high values of modulus, yield strength, and toughness. If we were interested in material for use at high temperature, we might consider, too, the scale of melting temperatures, T_M. Materials are useful to about 0.5 T_M; above this temperature they creep rapidly. All large-tonnage metals melt below 2000 K. This is more than adequate for engineering at room temperature, but few metallic alloys have useful strength above 1000 K.

Five thousand years of alloy development

In the past, the introduction of new metallic alloys has produced a series of major, step-like advances in technology. Table 2.6 shows the approximate date at which the alloys first appeared; and when they did they made possible major changes in the design of utensils, of weapons, in transportation, in energy conversion and in many other fields.

Table 2.6 *The evolution of engineering materials*

Period	Metals	Ceramics	Polymers
>10 000 B.C.		Stone, Glass Brick, Cement	Wood
5000	Gold, Silver		
4000	Copper		
3000	Bronze		
2000	Iron		
1000			
0			
1000 A.D.			
1500	Cast iron		
1600			
1700			
1800			
1850	Steel		
1860		Reinforced concrete	
1870	↑		
1880	Alloy steels		
1890	↓		
1900			
1910			
1920			
	Al alloys		
1930	Mg alloys	Fused silica	Nylon
	Ni alloys		Acrylics
1940		Mullite	Polystyrene
1950	Ti V, Cr alloys	Titania	Polyethylene
	Hf, Nb, Mo alloys		Vinyls
1960	Zr, Ta, W alloys	Pyro-cerams	Epoxies
		Spinels	Polysulphones
1970	Glassy metals	Alumina	Urethane
1980	Shape-memory alloys	Silicon carbide Silicon nitride	
1980 ↓ 2000	Rate of development now slow	Tougher ceramics and new design methods sought – if successful rapid development. Cermets and other composites developed.	Evolving rapidly. High modulus polymers entering engineering struc-tures. Rapid develop-ment of composites.

The introduction of copper, and then bronze, about 4000 B.C. gave the first material which could be worked and which combined toughness with a high yield strength – a major advance over stone for knives, axes, and weapons. The next development, the widespread use of iron, is significant not because the iron was a better material; the iron produced in 2000 B.C. was inferior to good bronze. But it was readily available and cheap, and it allowed the desirable combination of properties that metals offer to be used on a wider scale. By A.D. 1400, furnaces had progressed to the point that cast iron, which can be pured to give intricate shapes, became widely and cheaply available; it became one of the mainstays of the industrial revolution. But it was not really until 1856 that steel, a material surpassing bronze in its combination of properties, became available.

For the next 120 years, the history of evolution of materials for use under stress has been entirely one of the development of metallic alloys, and most of this has occurred in the last 40 years. We have seen the development of high-strength steels, aluminium alloys, and titanium alloys which have made modern aircraft design possible; the appearance of iron and nickel-based superalloys which have permitted the development of new generations of turbines; zirconium, haffnium, and vanadium alloys for the reactor technology; and refractory metals like tungsten and molybdenum for rocketry. The understanding of structure and the chemistry of metals has allowed systematic alloying; and the new alloys have often stimulated a whole new technology.

Limits to the further evolution of metallic alloys

But there has been a dramatic slowing down in the rate of development of metallic alloys. That is not to say that no work is in progress — exactly the opposite. There is probably more research under way now than ever before. But its achievements are incremental rather than step-like. A great deal of the work aims at better quality control, at producing existing alloys more cheaply and reproducibly, and in achieving better average properties. Much work, too, aims at understanding and characterizing in more detail the alloys that already exist, and at methods of working them and designing with them. But in most areas, the development of new alloys is almost at a standstill.

A good example is provided by the evolution of the nickel-based superalloys which have allowed the development of modern gas turbines. Between 1950 and 1970, systematic alloy development resulted in a series of alloys for turbine blades which allowed the operating temperature of the turbine to be raised from 1000 K to 1200 K — with large gains in efficiency and cuts in weight. By 1970 the alloys were so strong that many of them could not be worked: they have to be precision-cast to their final shape. Since 1970 there have been further developments. To extract the maximum performance, blades are sometimes cast in a specific way such that all the crystals of the alloy have a common crystal axis along the blade, and in some instances, blades made up of a single large crystal are used, allowing the blade temperature to be raised to almost 1300 K. But here a fundamental limit is reached. The melting-point of nickel is 1726 K, and the blades are now running at 0.75 T_M; nickel-based alloys cannot be used at blade-temperatures significantly higher than this. Of course, the gas-temperature of the engine could be raised by cooling the blades, and since 1974 this has become common practice, allowing a further major increase in efficiency. But this method, too, has now reached a fundamental limit — and great efforts to develop blade materials based on refractory metals such as molybdenum and tungsten have failed.

This is not to say that turbines based on nickel alloys will not be developed further. But it seems likely that improvements will be incremental rather than the large steps that we experienced in the 1960s. If we wish to increase the operating temperature of such turbines by 200 degrees rather than 20 degrees, then we shall have to turn to new classes of materials.

In other applications of metals, other barriers are being encountered. Some are economic: metals are expensive, and because demand sometimes exceeds supply, the price of certain of them can fluctuate widely. The cost of energy, too, is now an important ingredient, both in producing a material and in subsequently shaping, joining, and finishing it to give a product. The use of metals tends to require heavier equipment (and thus capital investment), and is sometimes more labour-intensive than an alternative material might be. For all these reasons, there is growing pressure to examine the potential of polymers, ceramics, and even glasses, where, 10 years ago, metals would have been the automatic choice.

The purely economic factor is not the only one. Because metals have been available for some years, design with them approaches, in many areas, fundamental limits of the kind discussed earlier. Non-metallic materials may permit advances in design that are not possible with metals. I have already cited the example of the turbine, limited at the moment by the melting-point — ultimately, the fundamental nature of the interatomic forces — in nickel and its alloys. In many structures, particularly those associated with aerospace, it is not always the property (modulus E, yield strength σ_y, or toughness G) which is of first importance — it is the property per unit weight. Writing the density as ρ, it is the quantities E/ρ, σ_y/ρ, and G/ρ which matter to the designer. Polymers and ceramics have low densities — much lower than those of metals — so that the ranking of material properties shown in Fig. 2.6 is significantly changed by dividing by it. As new polymers appear, and as engineers learn how to design with brittle ceramics and glasses, the trend to use them where metals were used before seems likely to continue.

2.7 The evolution of polymers

We have seen in Fig. 2.6, polymers show a remarkably attractive combination of properties; only in their elastic properties they are markedly inferior to metals, at least at ordinary temperatures.

Why are the moduli of most polymers so low? The carbon–carbon bond is certainly a stiff one; diamond, which can be thought of as a highly cross-linked polymer (Fig. 2.7(a)) has the highest known modulus: more than 10^3 GN/m^2. It is something to do with the linear chains of which many polymers are built? Fig. 2.7(b), for example, shows the polyethylene chain. Its tensile modulus can be calculated: it is lower than diamond, partly because the zigzag chain straightens out somewhat when it is stretched, and partly because the sheath of hydrogen atoms mean that the density of C–C bonds (when molecules are packed together) is lower than that in diamond. But we might still expect its modulus to be large, and it is: about 3×10^2 GN/m^2 — a little higher than that of steel, and 300 times greater than the polythene with which we are familiar. Why is this?

Fig. 2.7(c) shows a sort of generalized polymer. Some, like polymethylmethyacrylate (PMMA or 'Perspex'), are entirely amorphous. Others, like polyethylene, resemble the figure — they are partly crystalline, partly amorphous, or (in the case of polyethylene) rubbery. Along the chain, the C–C bonding is covalent and strong, but sideways, only Van der Waals forces hold the structure together, and they are weak. The modulus is low for the same reason that cloth, when pulled at an angle to the weave, stretches easily: the fibres slide. But if the fibres are aligned, as they are in woven cloth, and you pull it along the direction of alignment, then it is very much stiffer.

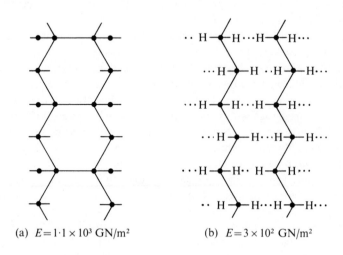

(a) $E = 1 \cdot 1 \times 10^3$ GN/m²   (b) $E = 3 \times 10^2$ GN/m²

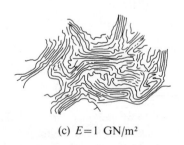

(c) $E = 1$ GN/m²

Fig. 2.7. Schematic drawings of the structure of (a) diamond, (b) polyethylene chains, (c) bulk polyethylene.

Stiffening of polymers

There are some obvious ways of stiffening polymers. Cross-linking — the creation of C–C bonds *between* chains — helps because it prevents the chains sliding over each other. Increasing the volume fraction of crystal helps too, because the bond density in the crystal is higher than in the amorphous phase. And stiffening the backbone of the chain helps: double bonds and benzene-rings are stiffer than single carbon bonds.

These methods go a long way in raising the modulus, and in doing so, the yield strength is increased as well. It is less obvious that the toughness will increase too, but, in some cases — for reasons not yet fully understood — it does. A good example is the polycarbonate which is sold under the trade-name of 'Lexan'; its chemical structure[1] is shown in Fig. 2.8. It is stiff because its backbone contains benzene rings; and it is so tough that it is used for armour-plating windows. A second example is the polyimide sold as 'Kapton', also shown in Fig. 2.8. As its structure might suggest, it is even stiffer than 'Lexan'. And there is further benefit from the tighter bonding which gave the increase in modulus and toughness: the melting-point, too, increases. Kapton is stable at 400 °C, and able to withstand 800 °C for short periods.

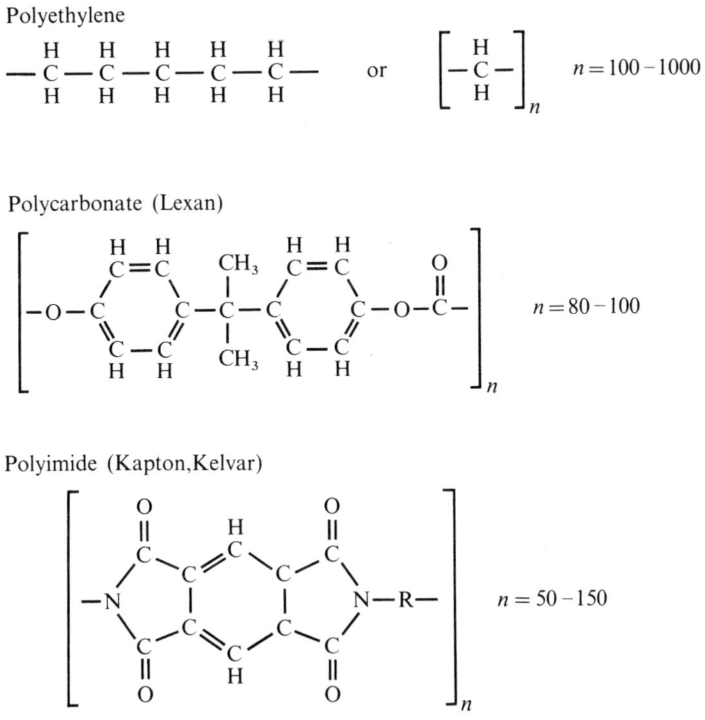

Polyethylene

$n = 100 - 1000$

Polycarbonate (Lexan)

$n = 80 - 100$

Polyimide (Kapton,Kelvar)

$n = 50 - 150$

Fig. 2.8. The chemical structure of three polymers. Polyethylene has a simple C–C backbone and a low modulus. The benzene rings of polycarbonate and polyimide greatly stiffen the polymer chain.

But to obtain the maximum stiffness from a linear polymer, it is necessary to line the chains up so that — like the cloth — they are aligned in the direction in which stress will be applied. The most promising way of doing this is to *draw* the polymer [17]. Drawing orients the chains in the flow direction. Many familiar

[1] For the chemistry of polymers see [16].

polymers are drawn: nylon, for instance. But their moduli, though higher than before, are still low. Recently, there has been a major advance in the technology of polymer drawing; if fibres are extruded under very high pressure, large draw ratios − a factor of 30 or so − can be achieved. The resulting fibres of polyethylenes have a tensile modulus of almost 70 GN/m^2 − equal to that of aluminium. Even more impressive are the oriented polyamides. 'Kevlar', a drawn polymer with a structure very like that of 'Kapton' (Fig. 2.8) has a modulus which is two-thirds that of steel.

Polymer-based fibre composites

There is a quite different way of stiffening polymers. It is by mixing stiffer materials into them, forming *composites*. The most efficient way to do this is to embed fibres of the stiffer material into the polymer, and arrange that the fibres are aligned in the direction in which the material will later be stressed. Wood is an example of this sort of composite: it is made up of fibres of crystalline cellulose embedded in amorphous lignin. Next to wood, the most successful fibre composite, by far, is fibreglass: glass threads in epoxy resin. Not surprisingly, the modulus of a composite increases more or less linearly with the amount of glass in it. The modulus of glass is more than fifty times greater than that of epoxy resin, so the addition of 20 per cent of glass gives a composite which is about ten times stiffer than plain epoxy resin.

Yielding is a little more complicated; but at the simplest level, the yield strength, too, is a linear function of the volume fraction of glass fibres. But it is important to realize that the properties of composites are not always just a linear combination of the properties of the components. Their attraction lies in the fact that frequently something extra is gained.

The toughness is an example. If a crack simply ran through the composite as a whole, one might expect the toughness to be a simple average of the toughnesses of glass and epoxy: and both are low. But, by getting the fibre length right, one can arrange that the fibres do not fracture, but pull out of the epoxy insteady. In pulling out, work is done (Fig. 2.9(a)), and this work contributes to the toughness of the composite. The toughness is greater − often much greater − than the linear combination.

Yet more dramatic stiffening can be obtained by using carbon fibres. These fibres have a modulus ten times greater than that of glass; greater, in fact, than that of any steel, and not far short of diamond. They can be embedded in epoxy to form a sheet or slab with a stiffness close to that of a steel and a density only one half that of aluminium. The particular combination of interface friction and pull-out length which characterizes carbon fibres, however, means that the toughness is low. But this does not mean that carbon-fibre composites themselves are impractical. In some applications stiffness matters more than toughness. And composites are extraordinarily versatile. It may be possible to produce hybrid composites which combine the best features of two sorts of fibre: carbon in the outer layers for stiffness, and glass just below, for toughness.

Such materials have very attractive properties, but they are expensive, and because of this, their uses are limited to high-performance applications. Initially this almost always means military ones, and most applications of carbon fibre and of hybrid composites have so far been in military aircraft. But it is clear that, over the next 20 years, composites like these will appear in commercial aircraft, notably in the wings and tail, where their combination of high stiffness and low density lightens the structure. saving fuel and increasing the payload.

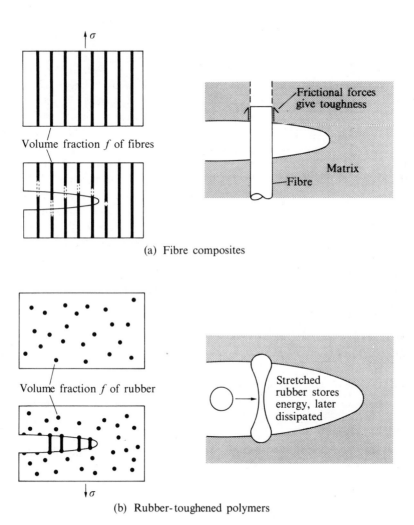

(a) Fibre composites

(b) Rubber-toughened polymers

Fig. 2.9. (a) High-modulus fibres stiffen the composite, raise its yield strength, and increase its toughness because work is done in pulling them out of the matrix. The properties of ceramics like cement and concrete can be improved by similar methods.

(b) Very low modulus rubber particles lower the modulus of the composite. But they act as little springs, linking the faces of a crack in the material; as the crack opens they stretch, storing energy; and when they break, the energy is dissipated as heat. The result is that the rubber gives toughness to the composite.

Filled polymers

Not all new materials are initially restricted, by cost, to military use. When you pick up a telephone, you are holding a remarkable composite material. The plastic used to make most telephones is a rubber-toughened polystyrene. The rubbery spheres of polybutadiene it contains have a very low modulus, and their presence lowers the modulus of the polymer. But above all else, telephones have to be tough − they get dropped, sat on, stepped on; their stiffness is much less important than their resistance to fracture.

When a crack propagates through such a material as this, the rubber particles stick firmly to the polymer, and are stretched, just as if they were little springs linking the crack faces (Fig. 2.9(b)). The fracture strain of rubber is large − about 10 − and in this large stretch, a great deal of energy is stored: that is why catapults are lethal. When the stretched particles finally break, the energy is released and virtually all of it is lost as heat. The rubber spheres provide a new way of dissipating energy during fracture and the material is toughened.

The rubber-filled polymers are an example of a cheap composite, in which a large gain in toughness is achieved with a small sacrifice in modulus. But almost always we want to raise, not lower, the modulus. One method is simply to mix a much stiffer material, such as glass or silica (which is no more than sand) into the polymer. Provided the polymer sticks to the filler, this works well: the modulus and yield strength are both increased and the toughness remains acceptable. Epoxies, stiffened in this way, are increasingly used in engineering applications, and they are cheap, hard, and have good dimensional stability. It is not a bad approach to the problem; we shall return to the reason for its success in the next section.

2.8 The evolution of ceramics

Conventional ceramics − complex silicates which form the basis of pottery, porcelain, and brick − are among the oldest of engineering materials. Research over the last 50 years, much of it in the last 20 years, has added to this list the high-performance ceramics: vitreous silica (really a glass); fully dense alumina and magnesia; high-purity spinels, mullites, and silicates; borides, carbides, and nitrides like SiC and Si_3N_4; and many others. But the mechanical or structural engineer, in considering them for use under stress, is deterred by the deficiency illustrated by Fig. 2.6: they lack toughness.

This low toughness is largely a consequence of the nature of the bonds which hold the atoms of the ceramic together. Localized bonds, particularly the Si-O bonds resist shear. In many ceramics this resistance is so high that − at a crack tip − bonds break before they shear, and the crack propagates without any associated plasticity. Fracture with no plasticity absorbs little more than the energy of the new surfaces: 1 J/m^2 instead of 10^5.

At first sight, there is little we can do: to make ceramics competitive with metals as load-bearing materials in general engineering, we should have to increase their toughness by a factor of about 10^4. But the attractions are immense: they are cheap and plentiful; very stable chemically; have high modulus and yield strength (and are thus very hard); and their densities are low, so that the properties per unit weight are even more attractive. Finally, many of them have melting-points near 3000 K and thus have useful strength above 1500 K: so they offer the large potential increase in operating temperature which would permit a step-like advance in turbine technology. Increasingly, methods are being found for using ceramics under stress. Some of these developments are outlined below.

Ceramic-based composites

Just as polymers can be stiffened by inserting stiff ceramic or glass fibres into them, ceramics can be toughened by the incorporation of rods, wires, or particles of tough metals. The most obvious example is that of reinforced concrete. Concrete, by itself, has almost no strength in tension. But by incorporating steel reinforcement, sometimes pretensioned so that the concrete remains in compression even when the structure is loaded, it can be used in structures which carry tensile forces. More recently, the incorporation of both chopped steel wires, and of alkali-resistant glass fibres, into cement and concrete, has been shown to give useful increases in toughness. Although these developments have attracted less publicity than research on more exotic composites, they are, in many ways, more important. Concrete and cement are very cheap and very widely used. If they could be successfully toughened without great increase in cost, a major advance in building technology — especially that of prefabricated buildings — would follow.

An alternative form of composite: cermets

Cermets, familiar through their common use as cutting tools, are an example of a different sort of composite. They are made of sand-like particles of a ceramic — tungsten carbide is the most common — bonded together by a relatively thin layer of a metal, usually cobalt. Typically, they contain about 70 per cent of the ceramic.

The modulus of a material like this is given roughly by a rule of mixtures. Since it is 70 per cent tungsten carbide, its modulus is high — nearly as high as tungsten carbide itself. But the carbide is discontinuous, and the soft cobalt matrix is continuous, so one might expect that the yield strength would be low — equal to that of soft cobalt. One would be wrong. The cobalt exists as thin films between almost rigid carbide particles. Flow in it is *constructed* for exactly the same reason that flow in a soldered joint is constrained: to get flow, an enormous amount of work has to be done against the yield strength of the material, which might, for instance, shear as shown schematically in Fig. 2.10. The constraint raises the stress required to make the cobalt flow — up to a maximum value equal to the flow (or fracture) stress of the carbide itself.

The beauty of these materials is that they have almost the modulus and yield strength of the ceramic carbide, while still retaining some of the toughness of the metallic cobalt. Fig. 2.10 shows how a crack, running out of a carbide particle, enters the cobalt, where a plastic zone forms at its tip. The size of the zone (and the energy absorbed in making it) is limited by the spacing of the carbides, and this means that the toughness of the cermet increases as the amount of cobalt it contains; a typical cermet has a toughness of about 1000 J/m^2, one hundred times that of pure tungsten carbide. The success of the material depends on the fact that yielding requires *general* plasticity, and this is made difficult by the constraint of the carbide. But the toughness requires only *local* plasticity, at the crack tip; and the cobalt binder permits this.

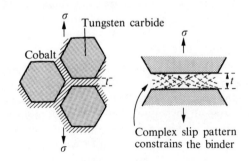

(a) Yielding of cermets

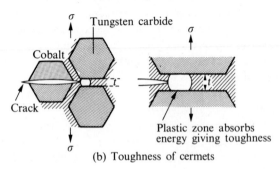

(b) Toughness of cermets

Fig. 2.10. (a) The yield-strength of cermets is high because the binder, shown as a thin layer between the ceramic particles, is constrained in the same way that solder in a soldered joint is constrained. Its yield strength rises as the thickness *t* of the layer decreases.

(b) The toughness of cermets is high because a crack must run into the binder (which is the ceramic particle size). A plastic zone then forms at its tip, absorbing energy. The toughness increases as the thickness *t* of the layer increases.

The properties of the ceramic-filled polymers are closely related to those of cermets.

Cermets appear to be a very attractive area for future research. We have already encountered a cermet-like material in the section on filled polymers: the structure of silica-filled epoxy is rather like that of cobalt-bonded tungsten carbide, and like it, it combines some of the best properties of the ingredients. There is a wide choice of ceramics, many very cheap, which could provide the high stiffness and yield strength, and there is a wide range of metallic and polymeric bonding materials, to give the toughness. Few of the possible combinations have yet been explored.

New design methods

But cermets are expensive, and even the best of them have a toughness far lower than that of a good structural steel; further, the really high-temperature potential of the ceramic has been sacrificed by using a metallic binder. At present, there appears to be no way of getting ceramic toughness up to the level needed for conventional engineering design. We require a new design method.

The chalk with which I write on the board when I teach is a ceramic. It is under stress when I write, and sometimes it breaks. I accept this, if it does not happen too frequently. In normal service, a failure probability of 10 per cent does not cause me undue irritation, though if it were 50 per cent I might change to another brand of chalk. If the component were part of a more expensive structure – a child's toy for instance – one would expect a smaller failure probability – 1 per cent perhaps. If it were the cylinder block of an automobile engine, one might set the acceptable risk at 10^{-4}; if it were the pressure-vessel of a nuclear reactor, one might accept 10^{-7}.

Brittle materials contain cracks, and (because they are brittle) the cracks will propagate easily. The stress at which they do so depends on the size of the largest crack. They resemble a chain with links which have a distribution of strengths. The weakest link (or the largest crack) determines the strength; and the strength decreases as the chain gets longer (or the solid larger) simply because of the larger chance that it will contain a particularly weak link.

Such chains, or the brittle solids they resemble, have a strength which can be described as a survival probability when a given load or stress is applied to it. For many, the survival probability is well described by

$$P_s = \exp - [(\frac{\sigma}{\sigma_0})^m \frac{V}{V_0}],$$

where P_s is the fraction of components of volume V (or chains of length V) which will survive a stress of σ; σ_0 is the mean strength of a large number of samples of volume V_0 (or chains of length V_0); and m is a material property. When m is very large (say 1000), the survival probability is unity while σ is less than σ_0. But it falls abruptly to zero when σ exceeds σ_0, independent of the volume, V, of the component. If, on the other hand, m is small – and for chalk it is about 4 – the strength depends on volume (that is the reason the broken bits of chalk always seem much stronger than the whole stick); and one stick differs very much from the next. Fig. 2.11 shows how the survival probability changes with m.

This approach can be developed into a rational design code for the engineering use of brittle materials. The methods permit the engineer to calculate an allowable design load: one that corresponds to an acceptable probability of failure. The condition of a totally safe working stress (the basis of conventional engineering design) is replaced by that of a stress which will give a certain survival probability. Actual *service* life can be further improved by proof testing the components: subjecting each in turn to a stress equal to the service stress, rejecting those which fail.

These methods are now being seriously applied in engineering designs which use cheap, brittle ceramics such as silicon carbide and silicon nitride in applications which were traditionally filled only by expensive metallic alloys. And they suggest new paths in developing engineering ceramics. Up to this point we have regarded the toughness as the material property which measured the average resistance of a material to fracture – and it is. But, for a given toughness, the quantity m tells us what fraction of a large batch of samples, be they teachers' chalk or turbine discs, will fail at half the average fracture stress. Research directed at developing and understanding ceramics with a large value of *m* is almost as important as that aimed at increasing *G*. Large values of both are desirable in any engineering material.

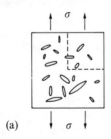

(a)

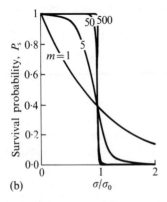

(b)

Fig. 2.11. (a) A brittle solid contains incipient flaws or cracks, with a distribution of lengths as shown. The larger the volume of the sample, the greater the chance of it containing a large flaw: a long stick of chalk has a lower strength, on average, than a short stick.

(b) The survival probability of such a solid can be described by the equation given in the text. It is plotted here. When *m* is high (1000, for example) all samples survive a stress less than σ_0, and all fail if the stress exceeds it. But if *m* is small (for many ceramics it is about 5) many samples will fail at a stress well below σ_0. If σ_0 and *m* are known, a structure can be designed to a given survival probability.

So promising do these new design methods appear, that they are being applied not only to design with brittle ceramics, but to decision-making in the design of large metallic structures like conventional turbines, and even nuclear power plants. Some structures are so complex, and contain so many components (any one of which may fail), that it is unrealistic to hope to design them to be totally without failure. Instead, the consequence of a failure in each section of the structure can be assessed and an acceptable survival probability assigned to it. Each section is built to a design based on its survival probability, to give a statistically designed structure which gives the best combination of economy and safety.

2.9 Conclusions

We are living through an interesting transition. For the last 100 years, mechanical engineering has thrived on a series of remarkable developments of metallic alloys: most notably, steels (for all branches of mechanical and civil engineering); aluminium and titanium alloys (for aircraft); and nickel alloys (for gas turbines). But in a span of only 10 years, a number of obstacles to this continued development have appeared. Some are set by the fundamental properties of metals — their melting-points, moduli, and yield strengths, for instance. Others are economic, geopolitical, or environmental.

This is stimulating the development of a whole range of new materials, based on polymers, stiffened by methods which increase the density of C–C bonds in the direction of loading, or by filling them with a material (like sand or glass) with a much higher modulus. It has also stimulated vast projects to develop new ceramics and explore whether or not they can find general use in load-bearing structures. At the moment, most of these new materials are expensive: often much more expensive than metals. But that is only because any development research of this sort *is* expensive; potentially many of them are very much cheaper and more widely available than the metals they would replace.

The considerations of this chapter point to a number of fields in which long-term fundamental and applied research is needed. Among them are:

(1) Resource assessment;
(2) The analysis of material use-patterns, and the study of substitute materials;
(3) Fundamental limits to further development of metallic alloys;
(4) The science of the strength and toughness of woods;
(5) The production of heavily cross-linked and heavily cold-drawn polymers in industrially useful quantities and shapes;
(6) Forming of heavily cross-linked polymers;
(7) Toughening mechanisms in ceramics, and variability in ceramic strength;
(8) Production of, and design with, fibre-reinforced cement and concrete;
(9) A broad experimental study of both metal- and polymer-bonded cermets with the aims of assessing their long-term potential;
(10) The development of a statistically-based design code for engineering design with brittle solids.

References

[1] NERC/IGS (Natural Environmental Research Council and Institute of Geological Sciences): UK mineral statistics.
[2] Verhoogen, J., *The earth*. Holt, Reinhardt, and Winston, New York (1970).
[3] US National Commission. *Materials policy, material needs and the environment today and tomorrow*, p. 4B–4. US Government Printing Office, Washington, DC (1973).
[4] US House of Representatives Document 527. *Resources for freedom: report of the President's Materials Policy Commission* (Paley Report). Washington, DC (1952).
[5] Brooks, D. B. and Andrews, P. W. *Science* 185, 13 (1974).
[6] Cloud, P. *UN Conference on development and the environment*. Columbia University Press (1972).
[7] US Bureau of Mines, Bulletin 650. *Mineral facts and problems*. Washington, DC (1970).
[8] US Bureau of Mines, Bulletin 667. *Mineral facts and problems*. Washington, DC (1975).
[9] US Bureau of Mines, *Commodity data summaries*. Washington, DC (1964).
[10] US Bureau of Mines, *Commodity data summaries*. Washington, DC (1974).

[11] McKelvey, V. E. *Technology Review*, March/April, p. 13 (1974).

[12] Ashby, M. F. Unpublished work.

[13] NATO Science Committee Study Group. *Rational use of potentially scarce metals*. NATO Scientific Affairs Division, Brussels (1976).

[14] *Conservation of materials*. Proceedings of Harwell Conference, Harwell, UK (1974).

[15] Fried, E. R. International trade review in raw materials — myths and reality. *Science* **191**, 651 (1976).

[16] Brydson, J. A. *Plastic materials*, Butterworths, London (1975).

[17] Ward, I. M. *Physics Bulletin*, February, p. 66 (1977).

3

The environmental sciences

F. Kenneth Hare

3.1 Introduction

In the later 1960s and early 1970s the human environment became the object of widespread public concern. Within two or three years this wave swept every Western industrialized country, taking many forms, but always laying major stress on pollution, air and water quality, and endangered ecosystems. It was highly erratic, in that public attention quickly veered from one worry to another. And it was irrational, in that none of the issues was new, and there was no obvious reason why awareness should have arisen so abruptly.

Within these industrialized countries the environmental movement has since evolved rapidly. At the beginning, in the mid-1960s, it focused on pollution, especially by persistent toxins, as dramatized by Rachel Carson in *Silent spring*. It was seen as the effort to protect the *quality* of the environment from misplaced technological attack. Very quickly, however, the more vocal ecologists enlarged the theme, representing the problems as being the need to find for mankind an ecologically less destructive role. The words 'ecology' and 'ecosystem' became shibboleths used by all those who wanted to side with the angels.

Then, as the world moved into the economic confusion of the late 1970s, the main thrust of public concern moved on to question the adequacy of resources, our ability to feed ourseves, and to find energy for our economy, and to question economic growth as the panacea for social inequities. In a rough sort of way this evolution was a continuum, but it was never well-defined, and always a little muddle-headed.

The 1970s saw the concern spread to the less developed countries, and into the socialist economies. In the former, the environmental movement was seen at first as a rich man's regret at his own folly. Pollution, it was argued, was created by technology, and could be cured by technology. Many Third World politicians said that it was a problem that they would be delighted to have, because it was the consequence of affluence. But in the 1970s environmental degradation within the Third World became inescapably visible: desertification, for example, along the desert margins, and desperate urban squalor. So this sceptical voice was largely silenced.

The orthodox Marxian polemic of the 1960s also saw pollution and environmental stress as the consequence of capitalist exploitation of man and materials. But this, too, was silenced as the socialist countries admitted to themselves that they also had these problems, and must learn to cope with them.

And so, within less than a decade, the world became acutely conscious of a wide range of stresses between man and environment. Things that had been known only to scholars and scientists became matters of commonplace discussion. The environment acquired, and has retained, high political visibility. In North America this may have waned a little as economic problems have worsened. But in Europe the environmental movement remains a potent political force. The anti-nuclear campaign has enlisted the support of most environmental lobbies, and is clearly still gaining momentum world-wide.

I am not primarily concerned in this essay with the political and economic significance of the environmental movement – this will emerge from the reviews in the second volume, which are directed towards 'scientific and technological interactions with respect to socio-political and cultural concerns'. My subject is the impact of the movement on the environmental sciences themselves. They existed, of course, long before the public anxiety of the 1960s, and they will carry on when things quieten down. Meanwhile, they have been heavily influenced by the temper of the times – partly to their advantage, but partly to their detriment.

3.2 What are the environmental sciences?

The long-established basic environmental sciences are those that deal with the various parts of the physical and living complex that surrounds mankind: air, water, ice, soil, rock, biota, heat, radiation, and time. Logically one might include the universe outside the terrestrial system, but we have not yet made that leap; we generally confine ourselves to the planet earth. These sciences are pure, in that they are often studied solely out of curiosity. But they are also applied, in that they depend on the fundamental results of physics, chemistry, and biology. The Royal Society of Canada, for example, treats meteorology, astronomy, and glaciology as interdisciplinary subjects. Only in 1975 did it begin to elect meteorologists as Fellows in physics, where most of them really feel at home.

These disciplines of the earth include the broad fields of meteorology and aeronomy (unhappily separated by scientific history), glaciology, hydrology, soil science, geology, geophysics, oceanography, ecology, and several others, depending on how one cuts the cake. Most of them began long ago, but have been transformed in the past twenty years by the infusion of new ideas, and by huge empirical inputs. All are observational, rather than experimental, in basic method. And all have been touched by the environmental movement. One can find undergraduate courses on environmental issues in academic departments of nearly all these fields, and the earth science journals also show the increased impact of environmental awareness.

The sciences just listed are in fact of two kinds: sectoral on the one hand, systems-based on the other. With the exception of ecology, each science chooses one particular sector of the environment, and develops a methodology suited to its study in isolation from the others. Meteorologists, aeronomers, and physical oceanographers, for example, share a common basis of physics and chemistry; yet they have developed essentially distinct bodies of method, and function to a large extent as separate professions. Air–sea interaction is now so important that there has been a recent convergence of meteorologists and oceanographers. But sectoral specialization, with a high degree of professional isolation, still characterizes these sciences.

Ecology, on the other hand, is obviously an integrative, systems-based science. Biologists have learned that one can treat living organisms in terms of the communities to which they belong, of inter-organism and inter-community relationships, and of their interdependence with the physical world. This tradition is at least as old as von Humboldt, Wallace, Darwin, and Haeckel (who coined the term 'oecology'). In our own century it has blossomed into a sophisticated, comprehensive analysis of the living societies of nature, part-sociological, part-environmental, and part-economic in character. The work of such pioneers as Thienemann, Tansley, Hutchinson, and Lindeman has provided the methodology whereby we can begin to study the human environment as a whole. A significant body of ecological theory exists to influence our choice of method.

Within the past decade we have been groping in this direction. The political situation has compelled us to study specific stresses that man's activities have created. The study of such problems is necessarily interdisciplinary and systems-based. In a broad, ill-defined way, it has been necessary to enlarge ecology so as to include man. And gradually we have learned to look at the human economy in ecological terms. Recent attempts at net energy accountancy, for example, have a marked resemblance to the trophic–dynamic ecology of Raymond Lindeman [1], and books and papers have begun to appear that try to bridge the gap between economics (as a social science) and ecology [2, 3, 4, 5].

To the old-established environmental sciences we have thus been forced to add, in recent years, an ill-defined body of integrative, synthesizing work that one can regard as the beginnings of an effective human ecology. Many have tried this before. American geographers, for example, had a flourishing school of human ecology over fifty years ago [6, 7] which collapsed under its own inconsistencies. So did the sociologists [8]. But these were largely abstract, academic frameworks. What is being attempted now is to generate an ecology that will adequately treat the human condition, and give us an efficient methodology for solving environmental problems. Many commentators would reserve the term 'environmental science' for this second group of studies.

Whereas the basic environmental sciences have clear-cut methodologies derived from the parent disciplines of physics, chemistry, and biology, these environmental sciences of the second kind have to look elsewhere. To a large extent they must depend on the *ad hoc* synthesizing capabilities of their individual practitioners: the art required is the ability to put together evidence from many different specialisms in a useful way. But they also need the help of systems science, of the sort described by Churchman in his parallel review (Chapter 8) and so well practised by Haefele, Sassin, and their colleagues at the International Institute of Applied Systems Analysis (see also Chapter 12).

In order to bring home the character of this new challenge to the scientific method, I offer extensive reviews of three major environmental problems of the 1970s, in all three of which I have been involved personally. They are (1) stratospheric pollution; (2) the phenomenon of desertification; and (3) the disposal of nuclear wastes.

At first sight these problems are utterly disparate, and lack common features. Having worked on each in turn, however, I have come to see resemblances between them. In each case I try to show how various sciences have contributed to the analysis of the problems, and how each of these sciences has thereby been influenced internally. I also discuss briefly the measures that are being taken to combat the problem, and how the environmental sciences can contribute to such solutions.

3.3 Some case-studies

(1) Stratospheric pollution

Concern about air pollution is one of mankind's oldest environmental anxieties; smoky cities, and the proneness of their citizens to bronchial disease, were matters of widespread comment in seventeenth-century Europe. Air pollution continues to be a serious problem of twentieth-century cities and industrial areas. But only within the past decade or so have we begun to worry about global-scale pollution, and its impact on climate, health, and economic productivity.

The stratospheric pollution problem was not identified, in fact, until our own decade. In 1971 H. S. Johnston [9] published a paper suggesting that the flight of supersonic aircraft in the stratosphere might lower the equilibrium amount of atmospheric ozone (O_3), which is concentrated between 15 and 40 mg above the earth's surface. Since the penetration of ultraviolet radiation to the earth's surface is modulated by this stratospheric ozone, such a decrease in concentration would increase the ultraviolet flux to which human beings and other organisms are exposed. This in turn threatened to increase skin cancer in man, and possibly to harm crops, farm animals, and natural ecosystems.

Johnston's paper created an immediate furore, which has not yet subsided. Its impact may have contributed to the decision to phase out supersonic transports of Mach-3 performance, though economic arguments were certainly more influential. His thesis — that oxides of single nitrogen released directly into the stratosphere from aircraft exhausts would dissociate ozone, without affecting its rate of generation by solar ultraviolet — prompted a frantic search for other active reagents that might have similar effects. Other candidates were quickly identified, chiefly among the halocarbons. Molina and Rowland [10] suggested that the photolytic dissociation of small quantities of the chlorofluoromethanes (CFMs) in the stratosphere would release free chlorine, and that this would also destroy ozone. These gases are used as refrigerants, aerosol propellants, and plastic foam inflaters. Calculations suggested that these inert man-made substances might lower ozone concentrations by about a tenth, and hence increase the surface ultraviolet flux by about a fifth.

The response to these suggestions has been extraordinary. They stimulated a large increase in active research into atmospheric chemistry and dynamics. This was accompanied by a hasty re-examination of the epidemiological evidence for a relationship between skin carcinomas or melanomas and ultraviolet irradiance. There was an equally hasty re-examination of the photobiological impacts. Since the conclusion was that there would indeed be adverse effects — though with a wide range of uncertainty— it was also necessary to seek political and legal instruments to phase out the use of the offending compounds, and to re-examine the design of exhaust systems in high-flying aircraft. In addition there has been speculation that the use of agricultural fertilizers, soil fumigants, and insecticides might also affect the issue. Attempts are already in progress — less than seven years after Johnston's original paper — to achieve international agreements governing the manufacture of some of the suspect reagents.

The scare has had, moreover, a marked impact on our approach to the general study of the stratosphere. Before the 1970s ozone was studied mainly by a small group of specialists. One spoke of an 'ozone club' (of which I was briefly a member). It was known that ozone amounts fluctuated with the passage of weather systems, and also showed unexpected latitudinal and seasonal variations. Ozone also had a small impact on the overall radiation balance of the atmosphere, being a strong absorber of solar ultraviolet, and of a narrow band in the terrestrial infrared spectrum. But few suspected that the ozone balance would become the major problem of atmospheric chemistry in the 1970s, or that it would become a political question.

Ozone is created in the middle and upper stratosphere (essentially above 30 km) by the photolysis of oxygen (O_2) molecules [11]. Some of the free oxygen atoms combine with an O_2 molecule to form O_3. This process involves ultraviolet radiation with wavelengths below 242 nanometres (nm), which is abundant outside the atmosphere. The rate of production depends on the intensity of the ultraviolet (which is attenuated as it passes downwards), and on the density of oxygen. It is continuous in daylight. Hence ozone concentrations would increase steadily, unless there existed a compensating dissociation. It now appears that in nature this dissociation is provided by (i) a simple chemical reaction between O and O_3; (ii) catalytic reactions involving O_3 and the hydroxyl radical, HO; and (iii) other catalytic reactions involving oxides of single nitrogen. Process (iii) depends on the slow upward turbulent transfer of nitrous oxide, N_2O, which is released from the soil during ecosystem denitrification processes [12].

This result means that the ozone layer in the stratosphere, which filters out damaging ultraviolet radiation from the solar beam, is actually modulated by a trace gas (the N_2O mole fraction is about 0.3 p.p.m.) that is produced by the decay of organic materials in the soil. The nitrogen cycle at the earth's surface is closely interlocked with the chemistry of the stratosphere. Moreover this third process accounts for over two-thirds of all ozone loss.

Hence we arrive at the conclusion that ozone synthesis by the sun's radiation will be largely constant over time, since ultraviolet irradiance is not thought to vary significantly. But the dominant process of decay — upward transfer of nitrous oxide and other potential reagents from the earth's surface — may well vary, notably as the results of the human economy. If anything accelerates this upward flux of reagents, the ozone layer is likely to be attenuated, and ultraviolet irradiance at the earth's surface increased — possibly to an extent capable of damaging living tissues.

The subsequent research has involved many sciences, of which the chief have been atmospheric chemistry, dynamic and physical meteorology, photobiology, human epidemiology, aeronautical engineering, and multidisciplinary decision theory. One can only briefly sketch the course taken by this research.

(i) *In atmospheric chemistry*, it has been necessary to complete our knowledge of the many reactions possible within the stratosphere and troposphere, especially as regards their kinetics. Most of these reactions involve gases present in the parts per million to parts per trillion[1] range. Over 150 reactions have so far been identified. To predict the possible effects of a pollutant, or of a change in natural components, it is necessary to solve a complex set of simultaneous differential equations, in which some of the reaction constants are only roughly known. Within the past five years, moreover, several new reactions have been discovered – for example, the role of chlorine nitrate, $ClONO_2$, in modulating both nitrogen oxide and chlorine catalytic destruction of ozone. To solve these equations it is necessary to include parameters describing atmospheric transport processes, since many of the inputs and outputs are transported to and from the site of the reaction. To obtain empirical confirmation that the models are valid, it has been necessary to measure the concentration of reagents in the stratosphere and mesosphere at concentrations down to the parts per billion or even trillion range [13, 14]. Some of the reactions, moreover, are photochemical, and many experiments have been needed to establish the optical properties of the species concerned.

(ii) *In dynamic and physical meteorology*, the major problem has been to estimate the rates and nature of vertical transfer processes from the earth's surface, which depend on the structure of atmospheric turbulence or organized wind systems. The modelling experiments carried out by the chemists have so far depended almost wholly on simple one-dimensional diffusion models, in which it is assumed that substances flow from high concentrations to low, at speeds proportional to the gradients. The proportionality coefficient, called the eddy diffusivity, is assumed to be constant in time and location, but variable in height. Estimates of its value have been determined from the observed vertical profiles of such traces as methane (CH_4), carbon-14 (^{14}C) and nitrous oxide (N_2O). This procedure leaves meteorologists uneasy, because they know that, when applied to the travel of more abundant substances such as ozone itself, or water vapour, it gives wrong answers. But their own three-dimensional transport models do not yet offer a satisfactory alternative (though see Vupputuri [15]). In physical meteorology, the main new problem has been to estimate the effects of newly introduced species, such as the CFMs, on the atmospheric radiation balance [16]. It turns out that a wide range of halocarbons and nitrogen compounds are excellent absorbers of infrared radiation in that part of the spectrum where neither water vapour nor carbon dioxide is an absorber. Hence a build-up of these gases should raise surface temperature.

(iii) *In photobiology* the problem has demonstrated how scanty had been the prior research, apart from the work of such pioneers as Caldwell [17]. It had been established that exposure to increased ultraviolet irradiance produced adverse genetic or somatic effects in plants and animals. Such experiments take time,

---

[1] Here and elsewhere 'trillion' is used as 10^{12} and 'billion' as 10^9

however, and sensitivity to relatively small changes in chronic exposure remains controversial. It has been established that photobiological effects due to ultraviolet in the 295-320 nm (UV-B) range (part of which reaches the earth's surface, unlike that below 295 nm, which is absorbed by ozone) include dissociation of the DNA molecule, whose action spectrum is now well-known. Stress-induced aging of cells, rapid killing of certain types of cell, and direct coupling of whole organisms to ultraviolet in this band are probably ascribable to this effect [18].

(iv) *In human epidemiology* there has been a hasty re-examination of the statistical evidence that in white-skinned persons, carcinomas and melanomas increase in incidence southward (i.e., in the direction of decreasing total ozone, which is most abundant in high latitudes). The case seems well established for carcinomas, the incidence being apparently linear with UV-B irradiance (though there is a great lack of synoptic observations of UV-B, as distinct from solar radiation as a whole). There is also a significant relationship for melanomas, with much more scatter [19]. But troublesome unanswered questions remain. What, for example, is the relation of chronic versus acute exposure? Total ozone present in the atmosphere varies with latitude, with season, and from day to day. The effect is to produce, in addition to the well-known latitudinal gradient of UV-B irradiance, a roughly 2 to 1 change from a minimum in spring to a maximum in autumn (for a given solar angle), and large changes from day to day (of order 15 to 30 per cent). Ultraviolet exposure in North America doubles for a southward shift of 8 to 11 degrees of latitude. The ozone reduction expected under some forecasts of future halocarbon manufacture is of the order of 5 to 10 per cent. This corresponds to a shift of one's domicile of less than 5 degrees of latitude − in a continent where people gladly retire from Ontario to Florida, a shift of 15 degrees! Increased ultraviolet irradiance must also have some effect on somatic synthesis of vitamin D, which may be a gain. Hence the epidemiological evidence is still incomplete, and in the eyes of sceptics still inadequate to support strong measures of control.

(v) *In aeronautical engineering* there was a flurry of activity in the wake of Johnston's 1971 paper to re-examine the performance of high-altitude jet aircraft, especially with respect to the engine exhaust outputs and level of operation. First-generation commercial supersonic transports (SSTs) such as Concorde and the TU-144 were designed to operate near 16.5 km, which in most latitudes is well below the levels of active ozone chemistry. But advanced SSTs (such as the abandoned Boeing design) were to fly at greater altitudes (19.5 km) where their exhausts would be discharged into a more active layer [19, 20]. The exhausts contain significant amounts of oxides of single nitrogen, and might therefore accelerate the process of ozone dissociation ((iii) above). Subsequent analysis has shown that Concorde-generation jets contribute very little to ozone wastage, as do subsonic jets. Higher flying SSTs would exert more effect, but the abandonment of the US commercial project reduced the urgency of the question. It remains, however, a matter of some concern as regards military aircraft, which are known to have been operating at these levels for decades. Hence continued research into exhaust design is necessary.

(vi) *In decision theory* and related fields, it proved necessary in at least one major study — that of the US National Academy of Sciences on aircraft impact — to develop a body of new theory in decision analysis, i.e., the theory of choice under uncertainty [21, 22]. The research results from all the above fields were in some degree uncertain, often by a factor of two, sometimes of ten. The Academy panel established chains of consequences, along which it was possible to estimate the most probable results, and also ranges of probability at each stage. They applied the analytical scheme to a series of reasonable scenarios, and attempted to predict the consequences for skin cancer and mean global temperature change. Probability estimates in the presence of objective uncertainty were determined by polls of expert opinion. In the outcome, the most probable estimates had uncertainty factors ranging between 3 and 10. Similar ranges of uncertainty affected later US estimates of the consequences of CFM use; the original National Academy of Sciences forecast gave for the most probable scenario a steady-state reduction in total ozone of six per cent, with an uncertainty range from 2 to 12 per cent. Work of this sort lies somewhere in the limbo between economics, management science, industrial engineering, and common sense, but it is interdisciplinary in intellectual character, and necessary to decision-making in the presence of uncertainty.

By late 1977, only six years after the scare began — the scheme of stratospheric chemistry was thought to be fairly complete, though significant detail needed to be added. The diffusion of CFMs and other inert synthetic substances to the stratosphere has been demonstrated observationally, as has their photolysis by ultraviolet radiation, with consequent release of active chlorine (constituting perhaps half the chlorine in the stratosphere). Significant reductions of ozone concentration have not yet been detected observationally, but they are predicted by modelling to rise to about a tenth to a twentieth of present equilibrium values. Immediate reductions in CFM manufacture will reduce, but not eliminate these impacts, since CFMs already in the troposphere will take many decades to diffuse upwards to the active layer; and no other large sink is known.

There is still controversy as to whether increased use of nitrogeneous fertilizer will increase N_2O release from the soil, and hence increase the nitrogen oxide levels in the stratosphere. Knowledge of the details of the denitrification process in soils is still inadequate, especially concerning the $N_2O : N_2$ ratio in the gases released. The role of the oceans in the overall N_2O balance is also unsatisfactorily known, as is the scale of industrial release.

This list of conclusions has been reached as the result of highly organized, internationally coordinated research programs. The Climatic Impact Assessment Program (CIAP), a project of the US Department of Transportation between 1971 and 1975, involved the work of scientists from all the above disciplines in over twenty countries, and stimulated parallel work in many others. It was followed by a series of US-sponsored investigations of the halocarbon effect. National Academy of Sciences reviews modified, but still basically confirmed, these official studies. The United Kingdom and France carried out independent reviews of the potential impact of aircraft [20] which agreed in most respects with the CIAP review, but were markedly more conservative where potential health or biological impacts were concerned. The same has broadly speaking been true of the United Kingdom's response to the halocarbon issue.

Notwithstanding the degree of uncertainty in the findings of these efforts, several governments (including those of Canada and the United States) have taken initial steps to control and limit the manufacture of the CFMs, especially as regards their use in aerosols (spray-cans). The industry, too, has voluntarily restricted their use to some extent. OECD has undertaken to monitor the production of these compounds.

Efforts are in progress to achieve international action against the excessive use of CFMs. The United Nations Environment Programme (UNEP) has included stratospheric pollution in its Outer Limits Programme. It convened a meeting of experts in March 1977, which adopted a world plan of action on the ozone layer. A Co-ordinating Committee on the Ozone Layer met for the first time in November 1977, and the World Meteorological Organization has rapidly pushed forward a world programme of ozone research and monitoring. In view of the fact that this problem was identified only recently, and the uncertainty that exists about its magnitude, this degree of international action seems remarkable.

(2) Desertification

The desert margin has long been a hazardous place in which to live. Yet much of the history of early civilization unfolded near the edges of the African and Asian deserts, which were also the early homes of Judaism, Christianity, and Islam. This has long been an environment of stress, challenge, and overwhelming achievement. It has also been, on countless occasions, the scene of tragedy and death.

In recent decades, the desert margins of Latin America, Africa, and the Middle East have also been fecund places for mankind. Along the Saharan rim many of the new post-colonial republics have high rates of population increase — as high as 4 per cent per annum in some districts of North Africa. The same pressure has developed in parts of South and Central America. These are, for the most part, regions with subsistence economies, depending on cattle, sheep, or goats, or on simple arable farming where there is enough water. As mouths to feed have grown in number, so also have herds and fields. At the same time there have been two other pressures: breakdown of the ancient systems of pastoral nomadism that fitted these regions so well, and the invasion of the less arid areas by cash-crop agriculture, usually aimed at the export market, so as to earn foreign exchange.

The less developed countries do not, of course, have a monopoly on such environments. Much of Australia's interior has a dry steppe or semi-desert climate like that of the Saharan margin. Mexico's Sonoran Desert extends into California, Arizona, New Mexico, and Texas, and a huge area of semi-arid scrub or grassland stretches through the western United States to British Columbia, Alberta, and Saskatchewan. An even larger subcontinent of steppe, semi-desert, and desert extends across the southern Soviet Union from the Ukraine to the Chinese border, and beyond it into China itself. These richer countries have found it far easier to cope with the dry environments, which have even become the scene of large-scale immigration by people and industries seeking the sun. But here, too, stresses exist, and are mounting.

Between 1968 and 1971 severe drought struck the southern margin of the Sahara — the *Sahel* — from the Atlantic Coast to the Sudan. For part of this period drought extended into Kenya, Ethiopia, and Somalia. Much of this enormous area lay within newly established republics, carved out of the colonies of France and the United Kingdom. None of these new countries had the skills, the money, or the infrastructure to cope with this prolonged disaster. Flocks and herds were halved or worse, and there were many human deaths. There were massive losses of soil to wind or sheet erosion. Supplies of wood for fuel, already badly diminished, were destroyed over large areas. Climatological studies showed that the desiccation actually began over the Sahel as early as 1960. As a result, the entire post-colonial experience of half-a-dozen new republics was clouded by the progressive loss of their most precious resources — water and biological productivity.

International aid was organized in support of the countries affected, though in a haphazard and ineffective fashion. Emergency supplies of food were imported, and in the phase of reconstruction since normal rains resumed in 1974 there has been substantial technical assistance. Some governments have embarked on drastic programmes of social and economic reform. But the most important impact of the Sahelian drought may well have been the questions it raised about the long-term stability of the desert margin. Is the desert spreading so as to engulf human settlements? If so, is changing climate the cause? Or is it human misuse of the land? Or a combination of both?

These questions have become a matter of international politics and have stimulated scientific research in an effort to answer them. In 1974 the UN General Assembly decided to hold a conference on desertification. It took place in August and September 1977, in Nairobi, headquarters of the UN Environment Programme. This conference focused world attention on the problems of the desert margin, and was successful in attracting the support of many natural and social scientists. There has been a large expansion of research into these problems. Once again it has proved necessary hastily to organize a suitable interdisciplinary framework for the research. The fields involved have included ecology, soil science, hydrology, meteorology and climatology, geomorphology, the social sciences (mainly geography, economics, sociology, and political science), the agricultural sciences, and certain aspects of engineering. The background documentation of the conference included four component reviews, on climate, technology, socio-economic matters, and ecological change, together with an overview and an Action Plan (which was adopted with some amendment). I have drawn on these reviews in isolating a series of questions that have been intensively investigated.

(i) *Is the desert advancing?* To this ancient question the answer appears to be 'yes', in many areas. The dust-bowl in the south-west United States in the 1930s was indeed arrested, and much devastated land was reclaimed. But in Africa, flanks of the Sahara provide evidence of both ancient and recent advance. So does the arid land of Afghanistan, Iran, Pakistan, and India. Forty years ago Stebbing [23] wrote **dramatically of the advance of the Sahara across the savannahs and forests of** West Africa. In more recent times, detailed surveys of the Sudanese steppes in 1958 and 1975 [24] have demonstrated a southward shift of the desert by 150–250 km in 18 years, with an average 'advance' of 8 to 14 km per year. But soil and geo-morphological evidence from shifting dunes, plus losses of woody and perennial forests, suggest that the creep is not a linear advance of the desert edge, but the outbreak in the vegetated areas of blain-like desertized areas that gradually coalesce. Desert spread is more like a pox than death in a quicksand. Oliphant described it as the outbreak of many 'little Saharas' around villages, watering-holes, and stand-pipes [25].

(ii) *Is there progressive desiccation of climate in progress?* When the magnitude of the Sahelian drought was first realized – some years after it began – several climatologists raised the possibility that the desiccation was the culmination of a long-term process that would continue, since it was due to fundamental shifts in the position of the subtropical high pressure belt. Bryson speculated that this shift might be due to man's influence [26]. Increasing atmospheric carbon doxide content plus augmented particle loads might combine to force this southward shift, which would probably continue.

Detailed analysis of rainfall and aerological records fails to support this pessimism. On a world scale, there appears to be no progressive desiccation in progress. The records show many instances where the general trend of rainfall is downward for one or two decades (for example, in central Australia and Sahelian Africa). But in all cases so far studied the desiccation finally ends, often with an abrupt return to excessive rains for a few years. The long desiccation at Alice Springs, Australia, for example, was followed in 1973-4 by a series of years of record rainfall. Bryson has reminded us, however, of two important provisos: there have been past droughts in the south-west United States that lasted for two centuries; and that we should not recognize a lasting desiccation until it was long-established, because it would resemble the decadal scale of variation.

(iii) *Do man's economic practices contribute to desertification?* There is general agreement that overstocking, careless cultivation, excessive use of groundwater, destruction of woody species for firewood, use of heavy vehicles, and failure to 'rest' the land weaken the soil, ultimately to the point where it will be deflated by the wind, or eroded by heavy rains. A corresponding loss in actual and potential biological productivity ensues.

Detailed technological and ecological analyses of these processes by Warren and Maizels [27] and Anaya [28] broadly confirm that such effects exist, but call attention to great complexity in the way in which they affect specific regions. To quote Warren and Maizels verbatim: 'There can be little reasonable doubt that many environments have suffered serious damage, and this mostly by cultural practices, but the persistence of the effects is much more debatable. ... Misuse that may go undetected or unremarked over a good period, is accelerated and made evident by drought.'

The major episodes of accelerated desertification clearly arise from unusually prolonged drought, especially when these follow a period of rainy years in which stock numbers increase, and cultivation is pushed too far. Memories are short for such stresses. Where population pressure creates hunger, or the desire for foreign exchange dictates cash cropping, there is a temptation to over-extend the real long-term productivity of the land. Such an over-extended system is vulnerable when the rains diminish, as they inevitably do at intervals of a decade or two.

(iv) *Does human misuse intensify drought?* Otterman [29] called attention to the sharp contrast of surface albedo (i.e., reflectivity to solar radiation), that existed along the fenced Negev–Sinai border. Overgrazed semi-desert in Sinai, with 10 per cent plant cover, was more reflective than the Negev, with 35 per cent cover. It is agreed that overpasturing or overcultivation increases the albedo of the desert margin. If this happens over a large enough area, the effect may be to increase the general atmospheric subsidence that is the root cause of drought [30, 31]. It may well be that such albedo feedback tends to intensify existing drought. Destruction of surface organic litter may have comparable effects [32].

(v) *What are the main socio-economic effects?* About 78 million people live in areas severely affected by such processes, 50 million of whom are directly impoverished by loss of resource base productivity [33]. In spite of this, rural populations are likely to continue to grow by 2–3 per cent per annum – accompanied by further disintegration of the old productive strategies and exchange patterns. To quote Kates *et al.*: 'A long-term chronic decline in productivity of the resource base is the most serious manifestation of desertification, yet its direct social and behavioural effects are poorly understood.' These authors find that health effects due to the desertification are indirect. Overwhelmingly they arise from lack of food, and hence malnutrition. 'The major response of livelihood systems to chronic desertification,' they add, 'is migration and/or change in livelihood patterns.'

Six thousand Somali nomads, for example, were resettled at Brava, a fishing village on the Indian Ocean. I quote the words of one of them, Salawad Warsame Duale, as reported by Graham Hancock [34]: 'For a long time my family and I expected rain every day to give life back to the land again. But no rain came. ... After the rains had failed for a second year my livestock began to die. ... My family had no milk to drink and the meat of the sick animals was impossible to eat. ... I remember thinking – I am going to die, my children will die and this will be the end of our life.' But in fact he was rescued and transported to the coast, where he has become a prosperous fisherman.

The plan of action that emerged from the UN Conference on Desertification takes account of these scientific findings. Probably no previous UN conference had before it such a mass of scientific, technical, and social analysis. Great credit is due to the Conference secretariat for the way in which it drew these materials together and mobilized scientific effort throughout the world.

Yet the actual recommendations seem general and diffuse, and remote from the details of the scientific input. This is inevitable. The stratospheric pollution issue was one of high technology, where the culprit substances and agencies could be readily identified. The action required could be taken by advanced countries, with huge resources and infrastructures. In the case of desertification, the causes and the solutions lie primarily within the actions and life-styles of millions of poor peasants and nomads, whose governments are among the least competent. Something can done within the technical domains by outside bodies – for example, the UN agencies, the World Bank, and the advanced countries that possess arid territories.

But for the most part, this is a problem where those most affected, and those who must protect themselves, are remote from scientific help.

As in the case of stratospheric pollution, there was a significant impact on the sciences themselves. Large-scale meteorological modelling exercises, for example, were designed and carried out to investigate the albedo feedback hypothesis. There were many individual case-studies, some reported to the conference, others in the scientific literature. Overall, the main achievement was to focus scarce scientific resources on the problems of the arid zone. This was formerly a major programme activity of UNESCO, but during the 1960s this effort was to some extent run down.

(3) Nuclear waste disposal

This third problem is of recent origin, though it has been dormant ever since man began to separate fissile elements for use in weapons and nuclear power generation. For over thirty years, the industry has been accumulating wastes of varying degrees of radioactivity, and has contained them (for the most part) successfully. It knew that in the end the longer-lasting wastes would have to be disposed of permanently, so as to isolate them from the biosphere, but nuclear engineers saw this as a minor problem that could wait until more urgent matters could be dealt with. Quite suddenly, however, public anxiety was aroused, mainly by a series of small but embarrassing failures in waste isolation. Environmental activists have taken this up, and in a few years the safety and acceptability of nuclear power generation as a whole have become matters of public debate.

Much of the controversy focuses on the question of the ultimate disposal of highly radioactive wastes from reactors. Since these wastes contain nuclides with half-lives in the range of thousands to millions of years, they must be contained and isolated from organisms for very long periods — longer, in fact, than the history of civilization. Quite properly, this is seen as a unique responsibility resting on modern shoulders. This generation has to protect all subsequent generations of people, plants, and animals.

The debate is acrimonious and partisan. Proponents of nuclear power and weapons see it as a technical problem that will be rather easily solved. This complacency has been so universal that no country is known to have developed an adequate repository for waste disposal, though the Federal Republic of Germany has established a successful experimental site (in the Asse salt mine) for lower-level wastes. On the other side of the argument, the anti-nuclear movement asserts that there is and can be no proof that safe disposal is feasible. From this assertion, the movement argues that further investment in nuclear power should be slowed down or stopped. Some extreme opinion calls for the dismantling of existing facilities.

This question is environmental because the dangerous radionuclides can reach man only along definable environmental pathways, after escaping from confinement. Usually these are waterborne and food-chain pathways. They are similar in character to those whereby other toxic wastes may attack man. They also resemble the routes taken by essential elements in the biogeochemical cycles — notably carbon-14 and tritium, which simply disappear into the large natural reservoirs of carbon and hydrogen. Hence the study of nuclear waste disposal is an exercise in applied ecology, as well as in economics, geology, and mining engineering.

Other aspects of nuclear safety are less clearly environmental, unless that adjective is taken to include occupational and public health (as it really should). These include the safety of uranium mining, milling, refining, fuel fabrication, and reactor operation. In Canada much of the public awareness of this question has arisen from exposure of uranium miners and residents in milling and refining towns to radon-226 emanations from ore bodies and tailings, and the escape of radium salts into surface runoff. There have been no failures of isolation of reactor wastes or spent fuel storage. The public fails to make the distinction, and debate is about all aspects of safety. In what follows, however, I shall confine myself to the problem of true waste *disposal*, i.e., the permanent containment and isolation of the unwanted endproducts of reactor operation.

Such disposal, like the two previous problems, demands a wide range of scientific disciplines for its effective design – and an interdisciplinary machinery to bring results together usefully. The nuclear industry is highly regulated and organized, both nationally and internationally. Even so, it has failed until recently to develop the required machinery for the disposal process. There now exist, however, many detailed studies of the question, and research and development programmes are in progress in several countries. I have drawn on several of these studies in preparing the following summary [35, 36, 37, 38, 39].

Character of the wastes. The term 'waste' may be applied to the following products:

(1) spent fuel from reactors (i.e. irradiated fuel rods stored under water);
(2) reprocessing wastes (i.e. the wastes remaining after extraction of plutonium-239 and uranium-235 from spent fuel); and
(3) reactor wastes (i.e. all radioactive by-products resulting from reactor operation, such as filters, contaminated clothing, etc.).

Characteristically only those materials that are meant to be contained are called waste. Various gaseous and some liquid by-products are allowed to disperse in the atmosphere, or in the hydrosphere, in very great dilution.

After irradiation and removal from the reactor, spent fuel rods contain (a) highly radioactive *fission products*, including tritium and a range of nuclides from the middle of the atomic table, of which strontium-90 is the best-known and leastdesirable example; (b) some unfissioned uranium-235 (small in amount in the case of natural uranium reactors); (c) several *actinides*, which are heavy radionuclides formed by neutron absorption by uranium-238. Chief of these is fissile plutonium-239, whose fission provides much of the heat from the reactor, but which remains in considerable amounts in the spent fuel.

The fission products emit large amounts of heat. Most have fairly short halflives. They are strong sources of beta and gamma radiation. The actinides emit alpha particles, and generally have long half-lives. The radioactivity of the spent fuel comes chiefly from the fission products in the first 600 years of decay. Thereafter the actinides – isotopes of plutonium, americium, curium, neptunium, and uranium –

dominate the radioactivity. For the fuel's radioactivity to fall to levels typical of natural ore bodies requires 10 000–15 000 years. When removed from the reactor, it must be stored in water while the shorter-lived fission products decay. Such storage is feasible up to several decades, and is safe.

Twenty-one countries now operate reactors producing such spent fuel, and all face the problem of disposal. An immediate question is: should the fuel be reprocessed to extract the unused fissionable uranium-235 and plutonium? Except in the case of the Canadian CANDU reactors (which use natural uranium and a heavy water moderator), the nuclear technology has been designed with such reprocessing in mind. Indeed the UK MAGNOX technology requires it.

Advantages of reprocessing are (i) that the extractable plutonium and uranium-235 contain much more binding energy than is extracted during the first cycle (if separated they can be re-burned in thermal reactors, or used in breeder technology) and (ii) that the removal of long-lived plutonium-239 (half-life 24 400 years) greatly reduces the period during which the wastes need to be isolated from the biosphere. The same applies to any uranium-235 removed.

I am not concerned here with the pros and cons of reprocessing, which involve value judgements on the safeguards needed to prevent the diversion of plutonium to illegitimate uses. The important point is that the decision determines the character of the disposal operation:

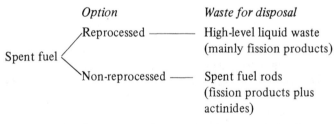

Reprocessing technology creates hot, acid solutions of the fission products plus actinides, from which most (but not all) the plutonium and uranium-235 have been removed. This waste must be immobilized in vitreous or ceramic form before it can be disposed of. Large volumes from military plutonium production are currently stored in surface steel tanks in several countries. This inventory is growing, and is now being added to by civilian wastes in countries where reprocessing is attempted. If reprocessing becomes the norm, such high-level liquid wastes will constitute nearly all the material for disposal. Because they are highly mobile, and produce much heat and radiation while in surface storage, a satisfactory disposal technology is clearly essential.

Spent fuel, on the other hand, is usually made up of ceramic pellets containing unconsumed uranium, plutonium, the fission products, and other actinides. There is also the zirconium alloy cladding (which also contains radionuclides). Being in the solid state, the fuel is not mobile, but it, too, is hot, intensely radioactive, and toxic. If it is to be disposed of, much research will be needed to decide what treatment it will first require.

The two options, then, provide as their end product immobilized high-level wastes *or* spent fuel treated in some way to prevent migration of nuclides. Both will require to be contained and isolated in some suitable repository, where they can be sealed off and forgotten. It is generally agreed that this means effectively for ever, since long-term monitoring is not something that can be imposed on future generations – and at least 600 years of isolation are required even for the fission products.

Several schemes for disposal have been suggested, many of them utterly impractical and unrealistic. The latter include disposal in the Greenland or Antarctic ice (too uncertain), in oceanic subduction zones between tectonic plates (too slow and uncertain), and in the sun or space, to which they would be transferred by rocket (too risky). The serious suggestions now under investigation include burial in the central regions of stable, suboceanic tectonic plates, or in the sedimentary cover of such plates [40], and in deep geological formations – in practice either salt deposits or massive igneous rocks.

A repository must clearly isolate the materials indefinitely. This means, preferably, that it be remote from man's other economic activities, such as mining areas or fishing grounds. It must be immune from disturbance by earthquakes, glaciation, or other cataclysms. It should, as far as possible, exclude circulating waters, though the almost immobile waters of the ocean floor almost meet this requirement. It will have to dispose of large amounts of heat and radioactivity without serious structural change. And it will have to provide, if possible, at least one fail-safe barrier beyond those of the artificial containment of the fuel: if, for example, water unexpectedly does migrate through a deep geological repository, the rock should be capable of chemically retarding the movement of dissolved radionuclides.

In many countries, the option now being examined is that of repositories in deep geological formations. In the United States the main interest has been in salt formations in sedimentary rock, as it is in the Federal Republic of Germany. In Canada small plutonic intrusions are being studied. Whatever the rock-type, however, the effort is to excavate a repository that will meet the requirements listed above, and then be backfilled.

Multidisciplinary research is under way in a few countries to validate some of these concepts. The research includes investigations of immobilization technology for high-level liquid wastes. So far as I know, this has not yet been extended to spent fuel. There has been an extensive search by structural geologists for suitable rock bodies. Recent French studies [41] have confirmed what others have inferred: that the critical quality to be sought in the containing rock is high sorptive capacity, i.e., the capacity to filter out of moving groundwater any dissolved radionuclides that it may contain. In other words, the effectiveness of a repository depends primarily on the geochemistry of the rock body.

What happens if wastes nevertheless escape from a repository? There is a long-standing tradition in the nuclear community for research into environmental pathways. A large body of scientific literature deals with the travel of radionuclides through aquatic systems, and their concentration in certain organisms. The most celebrated case is that of ruthenium-106, a fission product with half-life of about a year that is released into the Irish Sea from the Windscale plant of British Nuclear Fuels Ltd. Against all expectations, it was strongly concentrated by a form of seaweed that is eaten as a delicacy by a small human population in South Wales. If wastes from a repository should ever escape from containment, and travel in moving groundwater to the earth's surface (having escaped sorption *en route*), they may be concentrated in this fashion in aquatic organisms in rivers and lakes, and may thereby reach human consumers. Hence each proposal for an underground repository presupposes a comprehensive pathway analysis for the drainage basin concerned — and an excellent knowledge of the latter's hydrology.

I wish I could report that this problem is being tackled in a comprehensive and satisfactory way, but I cannot. Each of the international regulating bodies has looked at the problem in the past few years. Detailed reports have been published by the Nuclear Energy Agency of OECD and by the International Atomic Energy Agency. In the United States, suitable research is well under way. In 1978-9 expenditures on the civil plan alone will exceed $90 million. In several countries, however, matters are at a standstill because of distrust and dissension, which seems to be spreading. Environmental activists have opposed measures needed to start the necessary programmes of research and development, even though the disposal of existing wastes clearly cannot be avoided. Much of this activism is directed against reprocessing, rather than disposal itself; but it has been effective. Strong local opposition develops in areas being prospected for disposal sites.

Distrust of nuclear power itself seems also to be growing, in spite of the industry's good health and environmental record. In part this must arise from the technology's origin in the Second World War [42]. In most public opinion polls a majority of the population appears to favour the use of nuclear power (though a poll in April 1976, in Canada revealed that less than two-thirds of the population even knew that electricity could be generated in this way). But the minority that is opposed is vocal, passionate, and effective, and it usually grows into a majority in any local community threatened by nuclear development. The arguments used in resisting such development are often fallacious: but they are believed by their proponents. Public perception of nuclear technology often treats it as an environmental and health hazard regardless of statistical evidence to the contrary.

The study of environmental perception has been largely developed by geographers and psychologists, and is a flourishing aspect of the environmental sciences that has not so far been mentioned in this review. The analysis of hazard perception in particular has been taken far, and there is emerging a useful body of method [43, 44, 45]. In particular the concept of acceptable risk [46] has considerably influenced recent approaches to hazard regulation.

In seeking solutions to environmental problems as a whole, and not merely those of nuclear technology, this relatively new body of research and theory will prove very useful. The principles of safety analysis (e.g. [35] pp. 108-20) are now well established in the nuclear industry, and could with profit be extended to other hazardous high-technology industries.

3.4 Dilemmas facing the environmental scientist

The case-studies reviewed above make quite clear a dilemma confronting the environmental scientist. How does he pull together results from many different fields, and translate them into useful advice for the public, or for governments? If he leaves this integration to others, contenting himself with a simple presentation of the various specialized results, the outcome will usually leave him feeling that the real point has been missed, and much of the research wasted. But if he tries it himself he discovers how weak are his synthesizing capabilities, and how unfamiliar he is with the world of decision-making. A scientific training is usually analytical and specialized. The use of science in public action requires the skills of the synthesizing or integrating generalist — once contemptuously described at a meeting of the American Association for the Advancement of Science as 'an unskilled intellectual'.

I offer myself as a horrible example of such an unskilled intellectual. I am Director of the Institute for Environmental Studies at a large Canadian university. My job is to try to mobilize interdisciplinary teams of scientists to study environmental problems. Some of these specialists come from within the university. Others come from government, industry, other universities, or from the general public. I cannot pretend to know much about the various specialities concerned. But I have to know enough to mobilize these groups, and enough to help them translate their results into politically or socially effective forms — though fortunately several of my colleagues are better at this than I am.

I was personally involved, as was the Institute, in each of the three case-studies mentioned above. In the first two — stratospheric pollution and desertification — I started from my own special field of climatology, and at least had the security of feeling that I thoroughly understood part of the problem. In the third — nuclear waste disposal — I did not have even that flimsy support. But I am not alone in facing such situations. All real social problems are interdisciplinary in intellectual character, and multidimensional in political terms. If science is to be effective in tackling such problems, its practitioners have no choice but to go beyond their specialisms. In fact the Peter Principle probably applies: that one rises to one's level of incompetence, and then stays there.

I can also offer personal experience as to how one becomes involved. It is rarely by personal choice. In the case of stratospheric pollution I was appointed by the UN Environment Programme (UNEP) as its observer on the World Meteorological Organization's (WMO) Working Group on Stratospheric and Mesospheric Problems. Later my Institute was commissioned by UNEP to prepare a background paper for the meeting of experts referred to above (p. 57). I could, of course, have declined, but having a background in stratospheric research I was both flattered and intrigued at being invited to participate and contribute. I imagine that every reader of this chapter has been similarly tempted. One must recognize that the mobilization of science in support of such objectives is greatly helped if the scientist enjoys helping.

Much the same circumstances explain my involvement in the desertification issue. My earliest climatological research concerned the climate of the Saharan flanks, and I had subsequently worked rather intermittently on arid zone matters. I was formally approached to undertake the climatological study by UNEP, with

WMO concurrence. I, and my Institute, were seen as suitable for the job by virtue of past experience, and commitment to interdisciplinary approaches.

But my involvement in the nuclear waste disposal issue had a very different basis. I was appointed by Canada's Minister of Energy, Mines, and Resources to chair an interdisciplinary study group whose other members were a nuclear engineer and a geologist, both of whom had previously held senior positions in the nuclear or energy fields. I alone had no visible connection with the official establishment. Moreover I had reasonable credentials as an environmentalist, being a trustee or member of two prominent activist organizations. It was clearly felt that my advice would carry more weight because I had no obvious loyalty to the nuclear industry, but had a long-standing loyalty to the conservation of nature. My lack of knowledge of nuclear technology was seen as a positive qualification for the job.

I apologize for intruding my personal activities into this debate. But I think that they are typical of the present crude state of the environmental sciences. As I said at the outset, no well-developed discipline, or trans-discipline, of scientific synthesis has really emerged. There is no common body of method, no accepted set of critical standards, whereby one can attack these problems. For the present we rely on *ad hoc* synthesizing capabilities, and I can claim no more for my own work.

It is partly to replace such methods by something better that systems science has come into being. In this volume there is an extensive review of this subject by Professor C. West Churchman (Chapter 8). In the second volume, Dr. Wolf Haefele and Dr. Wolfgang Sassin of the International Institute of Applied Systems Analysis discuss the theme of science, technology, and natural resources (Chapter 12). The reports of that Institute are clearly successful attempts at a better level of synthesis than the individual scientist can achieve.

A second dilemma is posed by the sheer scale of the environment. The actual environment of man is made up of natural components (on which I have obviously been concentrating), human artefacts, including a large built environment — bricks and mortar, concrete, and design — a network of neighbours and institutions, and an ocean of information. It also includes pathogens, carcinogens, mutagens, and stresses that endanger mental health. Hence one finds doctors, planners, architects, communicators, economists, and even priests among those who use the term, sometimes imagining that they are alone in the field. In fact it is hard to draw any limit round the idea of the human environment. Hence it is difficult to establish an effective taxonomy of its constituent parts, or any reasonable paradigm for its comprehensive study [47].

Fortunately this dilemma has been recently tackled by the highest international forum for the sciences: the International Council of Scientific Unions (ICSU). This body has a Scientific Committee on Problems of the Environment (SCOPE) [48], founded in 1969, and now chaired by a geographer, Gilbert F. White. SCOPE's eight reports offer an interesting sequence. They began with the need for global environmental monitoring (Reports 1 and 3), now an integral part of UNEP's programmes. There were reviews of man-made lakes as modified ecosystems (Report 2), and of environmental sciences in the developing countries (Report 4). Report 5 discussed principles and procedures of environmental impact assessment, a fast-growing activity in countries with environmental legislation. Environmental pollutants were dealt with in Report 6 and Report 7 covered the biogeochemical cycles. SCOPE-10 (as the reports are labelled) has just appeared, and includes a taxonomy and analysis of environmental issues. It identifies nine *social* issues 'that need to be examined and related rationally to environmental factors':

(i) the need to increase and sustain food production for this and subsequent generations;

(ii) the need to bring 'populations into balance with their environment';

(iii) the need to reduce the disparity between rich and poor in rates *per capita* of converting natural resources into goods and services;

(iv) the need for rational utilization and conservation of natural resources (including energy);

(v) the need to conserve meaningful samples of natural ecosystem as gene banks and national heritages;

(vi) the need for a wiser approach to the design of human settlements;

(vii) the need to transfer scientific and technological manpower from developed to developing countries;

(viii) the need for adequate air, land, and water quality, and to combat present inequalities; and

(ix) the need for 'peace and security in life and poverty'. [44]

The conceptual framework within which these issues are to be examined, in SCOPE's view, embraces the idea of 'the total dependence of life on earth on the sun and on the production of plant matter'. It also insists on 'the overwhelming importance of the natural cycling of nutrients in the environment', and of man's modification of this cycling; it stresses his capacity to control the life-cycles of organisms and to develop productive new strains. It comprehends the role of medicine in protecting and prolonging human life. And it lays special stress on man's rapidly increasing capacity to 'acquire, analyse and present information'.

Recognizing that the social issues listed above are too broad to serve as a basis for the classification of the environmental sciences, SCOPE-10 fell back on the time-honoured device of an identification matrix, into which it could fit SCOPE's investigations in the mid-1970s — and, by inference, those of the scientific community at large. The columns of the matrix (Table 3.1) represent environmental *concerns*, and comprise climate, the biogeochemical cycles, pollutants, ecosystems, non-renewable natural resources, and human ecology (using this term in a narrower sense that on p. 51 above, to mean health, settlements, and direct response to environmental stress or stimulus). The rows represent scientific and technical *responses*. They include problem identification and monitoring, modelling, risk estimation, evaluation and communication, and standard setting and social policy.

As a test of the comprehensiveness of the SCOPE classification, I have indicated by the appropriate letter the squares of the matrix in which significant scientific effort has been expended on the three case-study issues used in this chapter. The matrix does indeed accommodate all the useful effort that I can identify, but it spreads it over most of the squares. In other words the classification does not focus sharply, as complete identification matrices should. These merely confirm what I said at the outset: that a firm framework has not emerged for a comprehensive human ecology,and indeed may never do so. Man's relation to nature and himself is too complex to be readily crystallized. Of the five responses in Table 3.1 I have space to deal with only monitoring, and of the concerns only the biogeochemical cycles.

Table 3.1 *Identification matrix for environmental concerns and responses*

RESPONSES	CONCERNS: Climate	Biogeo-chemical cycles	Pollutants	Eco-systems	Non-renewable natural resources	Human ecology
Problem identification and monitor-ing	SD	SDN	SN	SDN	D	SDN
Modelling	SD	SN	SN	DN	—	SDN
Risk estimation	SD	SN	SN	DN	D	SDN
Evaluation and communication	SD	SDN	SN	DN	D	SDN
Standard setting and social policy	—	N	SN	D	D	SDN

S: stratospheric pollution
D: desertification
N: nuclear waste disposal

Source: SCOPE-10 [48].

3.5 The biogeochemical cycles

One of the most useful ways of looking at the environment (or at ecosystems) is to examine how specific substances and energy cycle into and out of living tissues, soils, rocks, atmosphere, and ocean. We have learned to call these the *biogeochemical cycles*. The idea turns out to be useful in understanding the impact of pollution, and even touches on the chemistry of the stratosphere. Though such cycles were conceived by ecologists, they make sense to most other earth scientists. They are the most genuinely interdisciplinary concepts in the environmental arena.

The chief cycles are those of nitrogen, phosphorus, sulphur, and potassium (the mineral cycles most vital to life); of oxygen and water (the hydrological cycle); and of carbon and energy, the last being often taken together because of their intimate linkages through photosynthesis and respiration. All are in fact linked, for a variety of reasons. Many of the exchange processes, for example, involve changes in energy levels; thus the evaporation of water requires an external heat source, and is hence an energy transaction as well as a phase change for water. The decay of organic debris in the soil, or on the ocean floor, absorbs oxygen but releases carbon dioxide (or monoxide) and heat — the heat that was chemically fixed by photosynthesis. Rivers transport water, minerals, and heat, and convert potential to kinetic energy and thence to further heat. Ecosystem function, a still broader concept, incorporates these exchange processes, phase changes, chemical transformations and linkages to the extent that life is involved. The full cycles, however, usually extend well outside the living cover into ocean, atmosphere, subsoil, and rock, where the processes are purely physical.

The quantitative analysis of such cycles is now a major objective of ecological, geochemical, oceanographic, and meteorological research. It has involved major programmes at such diverse centres as the U.S. Brookhaven National Laboratory (whose main concern is with nuclear science) under G. M. Woodwell, and the International Institute for Meteorology at Stockholm, under B. Bolin. A substantial part of the entire International Biological Programme was aimed at the understanding of these cycles, especially those of carbon and energy. It is generally agreed that future programmes of environmental management will depend on control or manipulation of parts of the cycles.

The nitrogen cycle is among the most important from the standpoint of environmental issues [49, 50, 51, 52]. It is also quite poorly known, many of its reservoirs and transfer pathways being hard to monitor. The natural cycle begins when nitrogen-fixing organisms withdraw nitrogen from the atmospheric reservoir of 3.9×10^{21} tonnes (t) of inert N_2. These organisms synthesize chemically more active and water-soluble compounds, usually of single nitrogen. Fixation by land organisms amounts to about 139×10^6 t per annum (a^{-1}), and aquatic fixation adds something between 30 and 130×10^6 t a^{-1}. Industrial fixation (at 1970 levels) is about 36×10^6 t a^{-1}. Overall fixation is thus of order 250×10^6 t a^{-1}, with human technology contributing about 15 per cent in 1970 — a very marked acceleration of natural overturning.

Research concentrates on discovering the rates of transfer of this active nitrogen between oceanic, atmospheric, soil, fresh-water, and biotic reservoirs, together with chemical cycling between organic, ammoniated nitrite, and nitrate species. The cycle is completed when denitrification, primarily a process involving another suite of micro-organisms, reconverts nitrogen to N_2 and N_2O, which returns to the atmospheric reservoirs. We reproduce Soderlund and Svensson's overall estimates of these flows as Fig. 3.1. The oceanic fluxes in these estimates are especially uncertain.

Many environmental impacts have been ascribed to the human acceleration of the cycle just described. Of these the largest is due to the widespread and rapidly increasing use of nitrogeneous fertilizers on arable land — a key element in the rapid post-war increase in cereal yield, and on the success of the recent 'Green Revolution'. Such increased use, sometimes accompanied by irrigation, tends to leach nitrate or nitrites into streams, lakes, and groundwater, sometimes in toxic concentrations; heavy rains in September 1976, for example, following protracted drought in Britain, washed so much unconsumed fertilizer into streams that rural water supplies became locally non-potable. Eutrophication effects, and other altered ecosystem relations, may ensue.

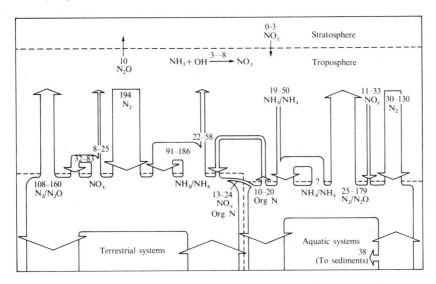

Fig. 3.1. The global nitrogen cycle. The rates are given as Tg nitrogen a^{-1}. The flows of N$_2$/N$_2$O are residuals obtained when balancing the terrestrial and aquatic systems [49].

Widespread release of oxides of single nitrogen (NO, NO$_2$; collectively NO$_x$) or nitrate and nitrite radicals (NO$_3^-$; NO$_2^-$) is a consequence of high-temperature combustion, and hence is emitted from factory chimneys and exhaust pipes. These soluble species tend to be quickly washed back to earth by rain, contributing to the impact on aquatic sysems mentioned above. They are often present in urban smog, where they interact catalytically with hydrocarbons and oxygen to produce ozone and other oxidants — which are responsible for damage to crops and ecosystems, and for adverse health effects [45]. Nitrosamines may also occur, and these have known carcinogenic effects.

Mention was made on pp. 53-4 of the dominant role played by nitrous oxide (N$_2$O) in modulating stratospheric ozone, and hence surface ultraviolet irradiance. The overall annual denitrification product cannot be drastically different from the rate of fixation given above. But it is presumably accelerating. If soil acidity increases, as may now be the case, the ratio of N$_2$O to N$_2$ in denitrification — at present of order one-tenth, imperfectly known — may increase. Thus there are two known processes whereby the flux of N$_2$O to the stratosphere may be substantially accelerated from its present value of perhaps 10^7 t a^{-1}. Unfortunately there is little monitoring of this crucial flux, and estimates of the yields from oceanic dinitrification vary by a factor of four [50].

Clearly, then, man interacts with the nitrogen cycle in a fashion that may injure his health, his economy, and his environment. Yet it is hard to avoid. Extra food, and hence extra fertilizers, are needed. If we can produce self-nitrogenating cereals, to add to the rapidly increasing planting of legumes, the net effect may well be a further increase. And we are a long way from any substitute for the automobile. Most significant of all in the context of this chapter is our own uncertainty about the cycle. We are sufficiently far from being unanimous to guarantee that we shall give conflicting or hesitant advice to the politician.

The carbon cycle is a second fundamental aspect of environmental functions. It is intimately linked to the metabolic energy cycle; the initial photosynthetic creation of hexose carbohydrate from carbon dioxide and water binds solar energy of visible wavelengths into living tissues, which typically yield 4 to 6 megacalories per kilogram on combustion, or during respiration. The entire metabolic energy cycle of the biosphere relies on this source which is, however, less than a hundredth of the terrestrially absorbed solar radiation. The energy required by the evaporation phase of the hydrological cycle is two orders of magnitude greater.

Research concentrates on rates of transfer between reservoirs, of which there are several, and on the chemical interactions of the various carbon speices. These are even more complex than those for nitrogen, and there are again major uncertainties as to how the complete coupled cycle functions.

The largest reservoir, that of carbon in solid rock, is probably of the order of 200×10^{15} t, of which about 10×10^{12} t are fossil fuels (and hence capable of being burned and returned to the active cycle). The atmosphere contains approximately 700×10^9 t, corresponding to a mole fraction (relative concentration by volume) of about 330×10^{-6} (i.e., 330 p.p.m.). Because of fossil fuel consumption (currently approaching 2×10^9 t of CO_2 per annum), the atmospheric fraction is increasing at about 0.7×10^{-6} a^{-1} (0.7 p.p.m. a^{-1}). Large reservoirs of dead carbon exist in the oceans (especially the deep layers), and in the soil. There are wide discrepancies in the estimates of living biomass. Over land the value may be as high as 850×10^9 t, but estimates not much greater than half of this have been put forward [53, 54, 55, 56, 57, 58].

The environmentally most significant questions arising from the carbon cycle are: (i) What is happening to the CO_2 being released industrially or by forest cutting? (ii) What is the rate of exchange of carbon species between the surface ocean layer and the deep reservoir? (iii) Will the atmospheric mole fraction continue to increase? If so, to what level and by what dates? (iv) What impact will such increases have on world climate, and what feedbacks may modify these increases? (v) What impact will the increases have on biological productivity? Only partial answers can yet be given.

The increase of CO_2 since the nineteenth century is firmly established. The gas is well mixed, and its change in concentration can hence be proved by a few suitably distributed monitoring stations. The mole fraction has risen from under 290 p.p.m. in the late nineteenth century to about 333 p.p.m. at present. Since 1970 there has been some hesitation in the upward climb, which still, however, continues. It is known to be due to fossil fuel combustion, and to reduction in the standing biomass of natural vegetation. Half or more of the released CO_2 enters the oceans, but there is uncertainty as to precisely what happens (or will happen) if the ocean surface becomes warmer, or if exchanges with the deep oceanic reservoir become more rapid. The best guess is that CO_2 concentration will continue to climb, and will double, well within a century. Much larger increases are possible within two centuries. Only a major slump in the world economy could avoid such increases.

Climatological modelling of the consequences of such an increase indicates that rising CO_2 mole fractions must mean a rise of surface and low-level air temperatures, of the order of 2 to 3 °C for a doubling of CO_2 [59]. Hence a rise of environmentally effective temperatures should be significant in the next few decades. Preliminary modelling of the various feedbacks that might moderate such a rise is in progress. An increase of cloudiness, for example, might be expected from a general rise of ocean surface temperatures, and this increase would diminish surface absorption of solar radiation — a negative feedback. Modelling suggests, however, that such a moderation of the expected temperature rise would be very limited [60, 61].

Various pollutants now being added to the atmosphere, notably several of the halocarbons, have radiative absorption bands in the so-called 'atmospheric window' band, in which neither carbon dioxide nor water vapour are absorbers. So also does nitrous oxide, which may be increasing due to acceleration of the nitrogen cycle (see p. 71). A synthesis by Kellogg suggests that all these effects are of the same sign [62], so that the CO_2 warming will be exaggerated by other similar effects. He finds that by the mid-twenty-first century, an overall surface warming in the range 1.5 to 6.0 °C is the probable consequence, with 4.5 °C the most likely value. This is a warming greater than any the earth has experienced in the past 10 000 years. Kellogg warns, however, that this crude estimate needs confirmation by comprehensive modelling exercises. He identifies various processes where even the sign of the effect is disputed. This includes the effect of increasing particle load, now believed to be in progress from anthropogenic causes.

Climatologists are wary of announcing yet another potential crisis. Less than five years ago it was widely feared that the earth was on its way to renewed glaciation. They are understandably reluctant to trumpet forth warnings of directly opposite portent. The conservative core of the meteorological profession points to the atmosphere's versatility in adapting to changed boundary conditions, or altered external forcing. They lay stress on the very high natural variability of the system, which permits prolonged departures from 'normal' without any external change. They emphasize (for example in [60]) that the atmosphere–ocean system may behave almost intransitively, i.e., that it possesses no *unique* steady state.

Nevertheless, the effects summarized by Kellogg are so profound, and would have such drastic environmental consequences, that they have to be taken seriously by policy-makers. A rise of world temperature of the magnitude foreseen would probably destabilize or disperse the polar pack-ice sheets (though probably not the continental glaciers). This would produce a positive feedback on world climatic variation. It would also pose dramatic problems of world food supply, general economic well-being, and strategy. These possibilities will be faced by the World Climate Conference, to be organized under UN auspices in 1979.

Thus the perturbed carbon cycle turns out to have major environmental implications, on which this account has barely touched.

3.6 Environmental issues in other cycles

For reasons of space I cannot deal in as much detail with the other biogeochemical cycles, even though they contain numerous important environmental issues.

The *sulphur cycle*, for example, is being profoundly perturbed by man-made releases of sulphur dioxide, SO_2, which is only a trace constituent of the natural atmosphere. Over the industrialized regions of north-eastern North America and Europe, the burning of fossil fuels is releasing steadily increasing amounts of SO_2, and also local releases of hydrogen sulphide, a highly toxic gas. Both gases are very quickly removed from the atmosphere by solution in rain, but ambient levels in large industrial cities are usually raised, occasionally to unacceptable levels. This local air pollution effect is, of course, of long standing.

Much more recent has been the realization that SO_2 is rather quickly converted into sulphate (SO_4^{2-}) aerosols, which may travel long distances before being washed out in rain, which is thereby rendered acid. The Scandinavian countries, being downwind from Europe's major industrial belt, early discovered this effect. Various adverse consequences have been ascribed to the acid rainfall, notably falling pH in streams and lakes, and effects on forest vegetation. Sulphate aerosols have also been thought to penetrate the human lung rather deeply, with adverse health effects. Swedish initiatives on this question, led indirectly to the convening of the 1972 UN Conference on the Human Environment, a landmark for the environmental sciences. OECD organized a study of the long-range travel of air pollutants, and its results have already been reported [63, 64]. The Economic Commission for Europe (ECE) has now initiated a programme that will cover both sides of the Iron Curtain. In North America, realization of the scale of the acid rain effect was slower, but is now well recognized [65]. NATO organized an Advanced Research Institute on acid rain effects in Toronto in May 1978.

The *phosphorus cycle* is also perturbed. Ordinarily it is a slow cycle. Within terrestrial ecosystems, exchanges between vegetation and soil are largely closed, with little leaching. In ocean ecosystems, available phosphorus is kept low by the slow overturning (of the order of 1200 years) between the deep reservoir and the euphotic zone. Where local accelerations of this overturning occur (for example in the cold upwellings off continental tropical west coasts), life is much more abundant. There is little atmospheric exchange of phosphorus, and stream content is low in natural ecosystems.

Widespread use of mined phosphates on agricultural land in the past century — and especially in the past two or three decades — has obviously changed this situation. Substantial releases of soluble phosphate into municipal sewerage systems, for example in detergents, has done the same. Hence many fresh-water bodies and some estuaries have seen spectacular rises in dissolved phosphates. The result has been eutrophication, a chronic condition of many lakes in North America and Europe.

3.7 Environmental monitoring and theory

It is natural to ask the question: is our environment changing? We can answer it only if we monitor the system adequately. And at present we cannot do so — in part because of lack of resources, but mostly because we do not know *how*. We lack a body of theory to suggest the right way.

As I said at the outset, the environmental sciences are at root observational, rather than experimental. They depend on systematic, quantitative, and replicable observations of chosen parameters. These observations have to be distributed rationally over space and time. Meteorologists, who have the best-developed observational system, use the term 'synoptic' for regularly spaced, indefinitely repeated observations of specified parameters. Oceanographers accept the same principle, but do not have the resources to collect data as regularly as the meteorologists. Few of the remaining sciences are as far forward. Monitoring of the environment is hence in a crude state. (Munn, in SCOPE-3 [48] ; Goodman, in SCOPE-10 [48] .)

This is one of the root differences between the environmental sciences and their more fortunate neighbours. The solid-state physicist, the molecular biologist, and the nuclear engineer can all rely on the principle of replicability. Once a given relationship has been tested by experiment, the result can be replicated when others repeat the experiment several times. If this is achieved, the relationship need never be tested again, nor need one monitor it: only at the level of statistical mechanics does the certainty of experimentally established relations vanish. Moreover, each field can draw on theory to a formidable extent in choosing its experimental objectives.

But the atmosphere, ocean, soil, crust, and ecosystems all function within laws that permit unending variability. For the atmosphere and ocean, for example, one can write down the set of equations within which all behaviour must be constrained. One can also define domains which are for all practical purposes forbidden. But the remaining degrees of freedom are so large that the set of solutions to the equations is effectively infinite. We know the laws that govern the behaviour, but they do not specify what that behaviour will be at any given instant.

Certain solutions, however, are more probable than others. Hence the atmosphere and oceans generate a set of frequency distributions, whose statistical moments we can calculate. These distributions seem to be derived from deterministic systems. Many of the generalizations derived from kinetic theory — the study of molecular chaos — are readily transferable to the study of the gross behaviour of atmosphere and oceans, i.e., to their 'general circulations'. So are many of those of statistical mechanics. Yet others are not. So there is no easy short-cut. The slow progress of meteorology and oceanography as predictive sciences is understandable.

If one then turns to the still more complex systems containing life, with its hoard of genetic information, predictability becomes even less likely, and one's dependence on statistics even greater [66]. We are unlikely ever to be able to deduce much about ecosystem function and dynamics, except as regards their gross characteristics. The substantial body of ecological theory now existing is heavily dependent on the prior existence of masses of observational data [1, 67].

If, finally, one adds human freewill and decisions, the system becomes infinitely complex. None of the social sciences has a body of theory in any way comparable with that of the physicist. Certainly one cannot, to quote my previous words, 'write down the set of equations within which all behaviour must be constrained'. Economic theory offers some insight into the mass, statistical behaviour of humanity in search of a living: but it has little predictive power. Where environmental relationships are concerned, the theory is in any case primitive [68].

Hence we have little choice but to rely on observational data. We cannot predict with any accuracy what will be the effects of a given step in environmental management, nor can we deduce what the past effects of pollution have been. We have to measure them — and measure them systematically. SCOPE-10 defines monitoring as:

the collection, for a predetermined purpose, of systematic, inter-comparable measurements or observations in a space–time series, of any environmental variables or attributes which provide a synoptic view or a representative angle of the environment. ... Monitoring is thus a systematic method of collecting data needed for environmental problem solving.

I would personally omit the last sentence, since the environment is worth observing for its own sake. But the rest will stand.

The SCOPE analysis adds that monitoring is essentially of four kinds:

  (i) measuring *levels* of potentially harmful or beneficial substances in various media;
 (ii) measuring physical *attributes* of the media;
(iii) measuring beneficial or harmful *effects* on living things of specified substances or physical states, notably in terms of dose–response relationships; and
 (iv) taking inventories, for example those which offer insight into climatic change (e.g., glacial area and mass); or those quantifying human impact on nature (e.g., deforestation, desertification); or those representing ecosytem change.

Such monitoring needs to be prolonged, to be consistent as to method over long periods, and to be geographically extensive. This is especially true in classes (i) and (ii). Class (iii) offers the possibility of conclusive answers (e.g., dose–response relationships for ionizing radiation).Class (iv) is likely to be chiefly of value if the inventories are repeated at regular intervals, as are censuses of human population.

Even in the physical sciences of the environment, these conditions are inadequately met. Thus it takes immense effort for the climatologist to answer the question 'is the world getting warmer or colder?' The difficulties include lack of observations from remote areas, from the oceans, and before about 1950, from the layers above 3 km. There have been failures to standardize instruments; failures to agree on the best parameters; and failures to use the available observations to best advantage. Long-established observing stations are often closed for administrative reasons. Monitoring is neither understood nor valued by many meteorologists, especially by those concerned with weather forecasting.

There are also myopic gaps in what is observed. The various radiative fluxes, for example, have been systematically observed in a very haphazard and inconsistent fashion. We do not even know precisely the value of the solar constant, much less its secular variation. Few chemical characteristics have been adequately monitored. The oldest continuous infrared gas analyser data on CO_2 date from 1957 (at Mauna Loa), and spectrophotometric observations of ozone from the late 1920s. Still more depressing is the fact that the present flood of new observations of cloudiness from orbiting satellites is not being systematically stored. Changes in techniques, or in vehicle, take place so often that such storage would in any case be of inhomogeneous data.

Oceanographic monitoring is similarly imperfect. The bottom sediments of the deep oceans do, of course, accumulate a record of events above them, and this has been widely exploited during the past two decades to give a long-term, low-resolution history of marine climate. But the question of marine pollution poses critically difficult problems. In a UNESCO-sponsored review of such pollution, Edward Goldberg put it much more pungently than I could:

The many proposed global marine monitoring programmes are characterized by their vastness and complexity which lead to their doom on paper.... While such documents pass for review from one international organization to another, the world ocean continues to receive man's wastes, and there is no systematic attempt to measure the exposure levels of identified pollutants in the various parts of the ocean. [69]

Goldberg's response to this is to use indicator or 'sentinel' organisms that concentrate pollutants in their hard or soft tissues. Thus the common mussel *Mytilus edulis* concentrates the heavy metals (including plutonium and mercury), halogenated hydrocarbons, alkanes, aromatics, and simple hydrocarbons. It has already been used to monitor total hydrocarbon burden. Being widely distributed in the polluted northern hemisphere, and being readily transplanted, it thus offers itself for calibration as a sentinel. The goose barnacles (*Lepas* spp) may also be suitable for such a purpose.

Immediately upon its creation the UN Environment Programme saw this monitoring *impasse* as a critical job for it to tackle. With the aid of ICSU (SCOPE-3 [48]) it designed, and has since begun to develop a Global Environment Monitoring System (GEMS). This system takes account of existing monitoring schemes, and tries to extend, consolidate, and diversify the effort. A Monitoring and Assessment Research Centre (MARC) has been established at Chelsea College, University of London, and was until 1977 under the direction of Gordon Goodman.

The difficulties in the way of effective monitoring are obviously enormous. Yet without it we are largely crippled.

3.8 International teamwork: acronyms in action

A welcome outcome of the environmental movement has been a strengthening of international team efforts. Most of the major problems can only be solved if some kind of coordinated effort can be achieved. Scientists have known this for decades, but politicians have been slow to recognize that local or national programmes are rarely enough. Funds have thus been hard to get.

The UN specialized agencies have long made such efforts. WMO goes back (as the International Meteorological Organization, or Committee) to the nineteenth century; it is thus older than its parent. The World Health Organization (WHO) and Food and Agricultural Organization (FAO) are younger, but both have attempted bold schemes in the environment and resource areas. UNESCO has been a very significant force. Its Arid Zone Programme of the post-war decades arose from a conviction that problems of the desert margin would become critical – a remarkable example of prescience. These and several other UN-sponsored efforts helped environmental understanding at a time when it had little political appeal.

More recently ICSU and its committees have added non-governmental support to many world programmes. The International Geophysical Year in 1957–8 was extended by other, lower-key ventures. Cooperation between ICSU and UN agencies became a common mode. Longer periods were seen as necessary to the solution of major problems. The International Biological Programme (IBP), for example, and the International Hydrological Decade (IHD) were very extended efforts. Studies of biological productivity and of the world's water resources, both critical to human welfare, can hardly be completed in a year or two's sustained effort. IBP's UNESCO-sponsored successor, Man and the Biosphere (MAB), is directly environmental in its thrust, but has received less political support than it needs. The WMO–ICSU Global Atmospheric Research Programme (GARP) has had substantially better luck, as have the various support programmes it depends on.

The 1972 Stockholm Conference on the Human Environment led to the creation of UNEP, and to several of the programmes and conferences discussed in this paper. Since 1973 leadership and funding for international efforts have increasingly come from Nairobi. There is still jealousy, fence-mending, and friction between members of the UN agency family, but these are trivial by comparison with what has been achieved, and what may still come. It is fashionable to snipe at the UN and its programmes, especially now that the political majority in the organization has swung away from the Western industrialized countries. But the environmental sciences need global programmes, and there are no alternatives to the UN, or the ICSU family of non-governmental organizations, to mount these programmes.

There are also local and regional problems, where other sponsors are possible. OECD, the so-called 'rich nations club', has done an excellent job in supporting environmental programmes in Europe, with some spillovers in North America. Some of these have been passed on to the European Economic Community, or (when Eastern European participation is needed) to the UN Economic Commission for Europe (ECE).

Here I should like to pay tribute to the North Atlantic Treaty Organization. Its resources and its mission make it a much less obvious sponsor for international scientific programmes. Nevertheless, since the Committee of Three's report in 1956 the scientific and political wings of the organization have chosen to play a growing role in environmental and resource studies.

The Science Committee, for example, has sponsored special programme panels, of which several are related to the environmental objective. The Ecosciences Panel, which I had the honour briefly to chair, has chosen to support work among the member states that fills important gaps in environmental research. Several Science Committee or Panel conferences have dealt with key issues in this area. Some Advanced Study Institutes and Advanced Research Institutes have produced distinguished publications. The Conference on the Rehabilitation of Severely Damaged Land and Fresh Water Ecosystems in Temperate Zones held in Iceland in 1976 broke much new ground.

The Pilot Studies of the Committee for the Challenges of a Modern Society have also been effective. Of eight completed studies, five were concerned with environmental quality, or human response to hazard. These included studies on coastal water pollution and oil spills (Belgium was 'pilot' member); disaster assistance (USA); environmental and regional planning (France); air pollution (USA, Germany, and Turkey); and inland water pollution (Canada). Each study involved several other assisting members, so that each was a significant essay in international cooperation. Studies still in progress include advanced waste water treatment, disposal of hazardous wastes, air pollution assessment methodology and modelling, control of pollution of the seas, flue-gas desulphurization, and a number of interlocking energy studies.

Thus the organization can claim to have supported environmental programmes very thoroughly. I hope that this will continue. There may be some room for rapprochement between the Science Committee and the Committee for the Challenges of a Modern Society in these questions. The purposes and methods of these two committees are rather different, but they may well prove complementary in tackling future issues of an environmental nature.

3.9 The future

As I look back on this long, rambling chapter, I am disturbed by what I have left out, as well as dissatisfied with what I have put in. Choosing to work in the environmental sciences is bad for one's self-esteem — and for one's style.

I have said very little, for example, about the huge question of toxic substances in the environment. As my friend Ross Hall is fond of saying, we are afloat in a chemical sea. We are bathed in a fantastic wash of chemicals whose potential malignancy we cannot predict — and do not have the resources to analyse. Thalidomide, methyl mercury, the polychlorinated biphenyls, and dioxin have already shown their colours. Most have not. And I have said very little about the related questions of environmental health — even though my closest colleagues have been steeped in both questions for many years. I have neither the time nor the skill to do justice to these enormous questions.

The same applies to the large body of environmental economics and law that has grown out of the situation. I have not covered the adaptation of technology to the new challenge, and I have not even mentioned technology assessment.

I want to conclude, in spite of these deficiencies, on some optimistic notes. There seem to me to be many grounds for qualified cheerfulness. Ashby, for example, has stressed the irrationality of the decision-making process in this, as in all other domains [70]. He appears little impressed by the kind of decision theory that I described above on pp. 55–6. Politicians judge these complex issues, he says, by 'hunch'; and little or nothing is known about the psychology of hunch. Yet in the United Kingdom, as elsewhere, political action for a cleaner, healthier, and more productive environment has been supported in many ways. In spite of the shocking impact of the Industrial Revolution, Britain's landscapes, streams and air are actually being restored. Political hunch has worked *for*, and not against, the environment.

In Canada, my own country, this has also happened. Recently, for example, we have been arguing at enormous length – for over seven years, in fact – about the possibility of pipeline construction along the Mackenzie Valley, a nearly pristine Boreal Forest and tundra world with a thin scatter of Inuit and Indian population. After the discovery of oil and natural gas on the Alaskan north slope, in the Mackenzie Delta and on the floor of the Beaufort Sea, the Federal Government quickly moved towards the idea of pipeline construction, and consortia of private interests appeared with this in mind. All the previous history of Canada's economy suggested that these pressures would steamroller matters towards construction permits. The question of environmental impact, and of the rights of the native peoples to land and resources, would be ruled irrelevant.

But miraculously they were not. After mounting public concern the federal government appointed in March 1974 a one-man commission of inquiry, Justice Thomas R. Berger, to consider the social, environmental, and economic impact of the proposals. In his report, Mr. Berger recommended a ten-year delay in construction of a line along the Mackenzie Valley, and a permanent ban on a line across the environmentally sensitive Northern Yukon, where some of North America's most hard-pressed wildlife finds its habitat. His recommendations proved irresistible. Licences were not issued to either proponent, whose large investment in environmental impact research was unavailing. The governments of Canada and the United States have preferred a more southerly route, along the alignment of the Alaska Highway [71].

For the environmental scientist, the Berger inquiry and others like it have imposed enormous strains. The field scientists and the engineering community have been hard pressed to find enough skilled people to carry out the background studies in time. And time is of the essence. In a huge country like Canada there is rarely enough general background knowledge of local environments on which to base sound judgements. It has to be generated as issues surface.

The new concern has thus given a shot in the arm to Canada's environmental scientists, which for the most part they have thoroughly enjoyed. Much the same is true world-wide, as public attention – and hence funds – have been focused on the successive issues dealt with in this paper.

But there have also been losses. Few governments have been willing simply to increase the overall funding of these sciences. To use my own country as an example, the new environmental concerns have had to be met by a transfer of resources and not the creation of new ones. After the announcement by Canada of a 200-mile economic zone at sea, for example, the major new scientific commitments involved were met by a selective reduction of resources devoted to air, land, forest, fresh water, and wildlife inland. Again the process has been world-wide. Some old environmental themes have been neglected as new ones have appeared. And it may well be that some fundamental work has gone undone because of the exigencies of new, *ad hoc* programmes.

I expect the next two or three decades to be dominated by issues of world security. These will regrettably include the continuance of power blocs, and of the need for endless vigilance against military threats, especially of terrorism and nuclear weapons proliferation. Supposing, however, that we somehow preserve world peace, and prevent the nuclear holocaust that make concerns like mine preposterous, we shall still have problems. Essentially these will relate to the adequacy of the earth to sustain the explosive growth in numbers that we seem unable to avoid.

It is quite likely that the Western world will learn to curb its dependence on economic growth. There are many signs that the richest communities have already outrun their appetite for further material satisfactions. Two cars per family tends to make the third a marginal need. Such communities may well learn, as Maurice Strong says they must, to prefer growth in quality to quantity: conceivably, to prefer enrichment of their spirits to expansion of their bellies [72]. But such communities are few. For the rest of the world, or at least for two-thirds of it, there is still hunger for the many things that they cannot now get.

And so I expect the energies of the scientific community to be increasingly concerned with such questions as the adequacy of world food supply, and of the productivity of all the life-support systems that we cannot do without. Environmental quality and pollution are Western ideas, appropriate to the advanced industrial societies. The productivity of renewable resources, the efficient use and renewability of energy, the stabilization of population growth are vastly more universal. I am quite sure that the peoples of what we call the Third World will demand that we shift our attention in that direction.

All of which makes me confident that the environmental sciences will still have a big job to do long after pollution has been banished, and technology fully humanized. It will be the needs of the vast numbers of Asian, African, and Latin American peasants and nomads that dictate the character of our work, whether we like it or not. It is no accident that the Executive Director of the UN Environment Programme is an Egyptian, nor that under his guidance the Programme should focus its attention on such questions as desertification and ecodevelopment. These are indeed, for the majority of the world's peoples, the imperatives of the future.

This looks to me like a bigger and better job than clearing up the mistakes of the Industrial Revolution. If my predictions are right, the environmental scientist will have his hands full in the next few decades. But he will enjoy himself!

References

[1] Lindeman, R. L. The trophic-dynamic aspects of ecology. *Ecology* 23, 399–418 (1942).
[2] Odum, H. T. *Environment, power and society*. John Wiley, Toronto (1971).

[3] Odum, H. T. and Odum, E. C. *Energy basis for man and nature.* McGraw-Hill, New York (1976).

[4] Rapport, D. J. and Turner, J. E. Economic models in ecology. *Science* **195**, 367–73 (1977).

[5] Westman, W. E. How much are nature's services worth? *Science* **197**, 960–4 (1977).

[6] Barrows, H. H. Geography as human ecology. *Annals of the Association of American Geographers* **13**, 1–14 (1923).

[7] Hartshorne, R. *The nature of geography.* Association of American Geographers, Lancaster, Pa. (1939).

[8] Hawley, A. *Human ecology: a theory of community structure.* Ronald Press, New York (1950).

[9] Johnston, H. S. Reduction of stratospheric ozone by nitrogen oxide catalysts from supersonic transport exhaust. *Science* **173**, 417–22 (1971).

[10] Molina, M. J. and Rowland, F. S. Stratospheric sink for chlorofluoromethanes; chlorine atom-catalysed destruction of ozone. *Nature* **249**, 810 (1974).

[11] Chapman, S. A theory of upper atmosphere ozone. *Memoirs of the Royal Meteorological Society* **3**, 103–25 (1930).

[12] Crutzen, P. J. The influence of nitrogen oxides on the atmospheric ozone content. *Quarterly Journal of the Royal Meteorological Society* **96**, 320–5 (1970).

[13] Evans, W. F. J., Kerr, J. B., Wardle, D. I., McConnell, J. C., Ridley, B. A., and Schiff, H. I. Intercomparison of NO, NO_2, and HNO_3 measurements with photochemical theory. *Atmosphere* **14**, 189–98 (1976).

[14] Lovelock, J. E. Atmospheric halocarbons and stratospheric ozone. *Nature* **252**, 292 (1974).

[15] Vupputuri, R. K. R. The steady state structure of the natural stratosphere and ozone distribution in a 2-D model incorporating radiation and D–H–N photochemistry and the effects of stratospheric pollution. *Atmosphere* **14**, 214–36 (1976).

[16] Ramanathan, V. Greenhouse effect due to chlorofluoromethanes: climate implications. *Science* **190**, 50–1 (1975).

[17] Caldwell, M. Introduction (Chapter I). In *Impacts of climatic change on the biosphere.* CIAP Program pp. 1–1 to 1–5; plus remainder of report, of which Caldwell was chairman and editor (1974).

[18] CIAP (Climatic Impact Assessment Program). *The effects of stratospheric pollution by aircraft* (by A. J. Grobecker, S. C Coronti, and R. H. Cannon, Jr.) (1974).

[19] NAS (US National Academy of Sciences). *Environmental impact of stratospheric flight.* Washington, DC (1975*a*).

[20] COVOS–COMESA. *Meteorological effects of stratospheric aircraft.* Anglo-French Symposium, 2 vols., Oxford (1974).

[21] Raiffa, H. *Decision analysis: introductory lectures on choices under uncertainty.* Addison-Wesley, Reading, Pa. (1968).

[22] Arrow, K. (Chairman). Uncertainty analysis. Appendix K in [19](1977).

[23] Stebbing, O. The encroaching Sahara: the threat to the West African colonies. *Geographical Journal* **85**, 506–24 (1935).

[24] Rapp, A. Sudan. In *Can desert encroachment be stopped?* pp. 155–164, Ecological Bulletins/NFR 24, Stockholm (1976).

[25] Oliphant, F. M. Discussion of paper by Stebbing [23], pp. 520–2 (1935).

[26] Bryson, R. A. The lessons of climatic history. *Environmental Conservation* **2**, 163–70 (1975).

[27] Warren, A. and Maizels, J. K. *Ecological change and desertification*. Component review, U.N. Conference on Desertification (1977). To be published in *Desertification: its causes and consequences*. Pergamon Press, London.

[28] Anaya, M. *Technology and desertification*. Component review, UN Conference on Desertification (1977). To be published in *Desertification: its causes and consequences*. Pergamon Press, London.

[29] Otterman, J. Baring high-albedo soils by over grazing: a hypothesized desertification method. *Science* **186**, 531–3 (1974).

[30] Charney, J. Dynamics of deserts and drought in the Sahel. *Quarterly Journal of the Royal Meteorological Society* **101**, 193–202 (1975).

[31] Charney, J., Quirk, W. J., Chow, S.-H., and Kornfield, J. A comparative study of the effects of albedo change on drought in the semi-arid regions. *Journal of the Atmospheric Sciences* **34**, 1366–85 (1977).

[32] Schnell, R.G. and Vali, G. Biogenic ice nuclei. Part I: Terrestrial and marine sources. *Journal of the Atmospheric Sciences* **33**, 1554–64 (1975).

[33] Kates, R. W., Johnson, D. L., and Johnson Haring, K. *Population, society and desertification*. Component review, UN Conference on Desertification (1977).

[34] Hancock, G. The drought with the long tail. *Mazingira* **2**, 31–5 (1977).

[35] Flowers, Sir Brian (Chairman). *Nuclear power and the environment*. Royal Commission on Environmental Pollution: Sixth Report. HMSO, London (1976).

[36] Hebel, C. (Chairman). *Nuclear fuel cycles and waste management*. Report of the American Physical Society, New York (1977). [To be published later in *Review of Modern Physics*.]

[37] ERDA (Energy Research and Development Administration). *Alternatives for managing wastes from reactors and post-fission operations in the light water reactors*. 5 Vols. (1976).

[38] Aiken, A. M., Harrison, J. M., and Hare, F. K. *The management of Canada's nuclear wastes*. Report EP 77-6F, Department of Energy, Mines and Resources, Ottawa (1977).

[39] Lieberman, J. A., Rodger, W. A., and Baranowski, F. P. *High level waste management*. Presentation to Energy Resources Conservation and Development Commission, Sacramento (1977).

[40] Frosch, R. A. Disposing of high-level radioactive waste. *Oceanus* **20**, 5–17 (1977).

[41] de Marsily, G., Ledoux, E., Barbreau, A., and Margot, J. Can the geologist guarantee isolation? *Science* **197**, 519–28 (1977).

[42] Hohenhemser, C., Kasperson, R., and Kates, R. The distrust of nuclear power. *Science* **196**, 25–34 (1977).

[43] See [35].

[44] Holdgate, M. and White, G. F. (1977) See [48].

[45] Science Council of Canada. *Policies and poisons*. Report no. 28, Ottawa (1977).

[46] Lowrance, W. W. *Of acceptable risk – science and the determination of safety*. W. Kaufman, California (1976).

[47] Hare, F. K. How should we treat environment? *Science* **167**, 352–5 (1970).

[48] SCOPE (Scientific Committee on Problems of the Environment). Reports are as follows:
SCOPE–3 *Global environmental monitoring system: action plan for phase I*. (1973) (R. E. Munn)

SCOPE-5 *Environmental impact assessment: principles and procedures.* (1975) (R. E, Munn)

SCOPE-6 *Environmental pollutants, selected analytical methods.* (A. Gallay, ed.) Butterworth, Sevenoaks, U.K. (1976).

SCOPE-7 *Nitrogen, phosphorus and sulfur – global cycles.* (B. H. Svensson and R. Soderlund, eds.) Ecological Bulletins Vol. 22, Stockholm, (1976).

SCOPE-10 *Environmental issues.* (M. W. Holdgate and G. F. White, eds.) Wileys, London (1977).

[49] Soderlund, R. and Svensson, B.H. (eds.) (1976) See [48].

[50] Hahn, J. The North Atlantic Ocean as a source of atmospheric N_2O. *Tellus* **26**, 160–8 (1974).

[51] Delwiche, C. C. The nitrogen cycle. *Scientific American* **223**, 137–46 (1970).

[52] Crutzen, P. J. and Fishman, J. Average concentrations of OH in the troposphere, and the budgets of CH_4, CO, H_2 and CH_3CCl_3. *Geophysical Research Letters* **4**, 321–4 (1977).

[53] Woodwell, G. M. The energy cycle of the biosphere. In *The biosphere*, pp. 26–36. Scientific American Books, Freeman, San Francisco (1970).

[54] Bolin, B. The carbon cycle. In *The biosphere*, pp. 47–56. Scientific American Books, Freeman, San Francisco (1970).

[55] Bolin, B. *Energy and climate.* University of Stockholm, Secretariat for Future Studies, Stockholm (1975).

[56] Rotty, R. M. A note updating CO_2 production from fossil fuels and cement. IEA(M)-75-4, Institute for Energy Analysis, Oak Ridge, Tenn. (1975).

[57] Munn, R. E. (1976). CO_2 'Greenhouse' warming of the atmosphere; MS. [See also: The greenhouse effect. *Mazingira* **2**, 78–85, 1977.]

[58] Keeling, C. Industrial production of carbon dioxide from fossil fuels and limestone. *Tellus* **25**, 174–98 (1973).

[59] Manabe, S. and Wetherald, R. The effects of doubling the CO_2 concentration on the climate of a general circulation model. *Journal of the Atmospheric Sciences* **32**, 3–15 (1975).

[60] NAS (US National Academy of Sciences). *Understanding climatic change.* Washington, DC (1975*b*).

[61] WMO–ICSU (World Meteorological Organization – International Council of Scientific Unions). *The physical basis of climate and climate modelling.* GARP Report 16, Geneva (1975).

[62] Kellogg, W. W. *Effect of human activities on global climate.* WMO Geneva (1976).

[63] AMBIO. *Journal of the Human Environment.* Vol. 5 (2) The sulphur cycle; (5–6) Acid precipitation; vol. 6, (2) The nitrogen cycle (1976–7).

[64] OECD (Organization for Economic Cooperation and Development). *The OECD programme on long-range transport of air pollutants (measurements and findings).* OECD, Paris (1977).

[65] Cogbill, C. V. and Likens, G. E. Acid precipitation in the Northeastern United States. *Water Resources Research* **10**, 1133–7 (1974).

[66] Monod, J. *Le hasard et la necessité.* Editions de Seuil, Paris (1970).

[67] Margalef, R. *Perspectives in ecological theory.* University of Chicago Press (1968).

[68] Mäler, K.-G. *Environmental economics: a theoretical enquiry.* The Johns Hopkins University Press, Baltimore (1974).

[69] Goldberg, E. D. *The health of the oceans.* The Unesco Press, Paris (1976).

[70] Ashby of Brandon, Lord. *Environment and politicians*. Sesquicentennial seminar, University of Toronto, 17 November 1977 (verbal communication).

[71] Berger, T. R. *Northern frontier, northern homeland*. Report of the Mackenzie Valley Pipeline Enquiry: Volume One. Ministry of Supply and Services, Ottawa (1977).

[72] Strong, M. R. The international community and the environment. *Environmental Conservation* **4**, 165–72 (1977).

4

Molecular biology

Jean Brachet

The main characteristics of living organisms are *reproduction* and *heredity*. These two properties distinguish them from machines; for instance, a car can move, accelerate, or reduce its speed, but it cannot grow and divide into two identical cars (unfortunately for car-users and luckily for car-makers). The car 'respires' as we do; it oxidizes hydrocarbons, while we prefer carbohydrates, fats, and proteins, for energy production. A crystal can grow and can even regenerate if it has been broken into two halves. However, neither cars nor crystals are alive; they are unable, unlike all living organisms, to perpetuate their own species through reproduction and heredity.

There have been, in the past, many philosophical attempts to explain life: these range from the oversimplified mechanistic approach of Descartes – the heart is a pump, the kidney a filter, etc. [1] – to the vitalistic theories, the ultimate form of which is the existence of a principle (H. Driesch's entelechy [2]) which is distinct from the soul and which regulates embryonic differentiation. Entelechy is comparable to the driver of a car, who can either set it into motion, accelerate, slow down, or stop the car; however, according to Driesch's conception, it does not exist in space and, by definition, escapes experimental investigation. This philosophical conception is the opposite of the basic ideas of molecular biology.

Molecular biology attempts to explain life by the physical and chemical properties of the molecules, especially the macromolecules nucleic acids and proteins, which build up living organisms. The combination of genetic, biochemical, and biophysical methods during the last three decades has led to a tremendous and completely unexpected progress in our understanding of heredity: however, the basic mechanisms of reproduction are still the subject of close scrutiny. The purpose of this chapter is to summarize the contribution of molecular biology to our understanding of heredity and reproduction.

4.1 Cells, bacteria, viruses

Cells
Since the days of the botanist Schleiden and the zoologist Schwann [3, 4], we have known that all animals and plants are made of cells and that these are formed of a central nucleus and a surrounding cytoplasm. In view of the morphological and biochemical complexity of their nuclei, animals and plants are called eukaryotes (in contrast to bacteria, which are prokaryotes, because their nucleus is extremely simple).

Many methods are now available for the study of cells. Electron microscopy is used to visualize the fine structure of the morphological constituents of the cell (the cell organelles); cytophotometry and autoradiography can be used to identify the chemical composition of the organelles and to follow their synthesis. The cell organelles can also be isolated from broken cells and their chemical properties studied [5].

Fig. 4.1 shows a schematic, grossly over-simplified, representation of a 'typical' cell; however it should be pointed out that there is no such thing in nature as a 'typical' cell. Cells belong to different families called tissues (a muscle cell is very different from a nerve cell): they are *differentiated*. Cell differentiation will be discussed at the end of this chapter.

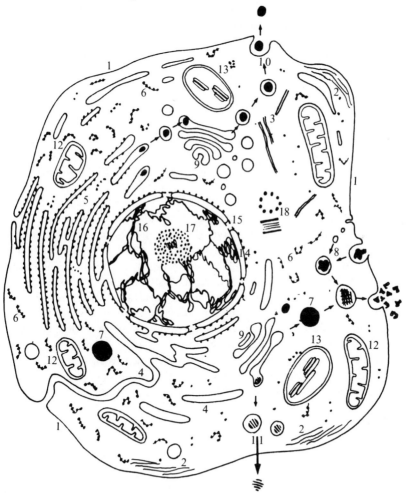

Fig. 4.1. Schematic representation of a cell. 1, plasma membrane; 2, microfilaments; 3, microtubules; 4, smooth endoplasmic reticulum; 5, rough endoplasmic reticulum; 6, polyribosomes; 7, lysosomes; 8, endocytosis vacuoles; 9, Golgi bodies; 10, protein secretion; 11, glycoprotein secretion; 12, mitochondria; 13, chloroplasts; 14, nuclear envelope; 15, nuclear pores; 16, chromatin, 17, nucleolus; 18, centrioles.

Returning to the theoretical cell of Fig. 4.1, it can be seen that it is surrounded by a *plasma membrane*, formed by a double layer of lipids in which protein molecules are embedded. The fluidity of this lipid bilayer varies from cell to cell and allows easy or restricted lateral movement of intra-membrane proteins. Outside the membrane is often a layer of glycoproteins (i.e. proteins with a high sugar content) forming a matrix which holds the cells together; destruction of this intercellular matrix leads to dissociation of the tissue into its individual cells. The plasma membrane controls all the exchanges between the cell and the surrounding medium. The latter is generally rich in sodium (Na^+) and poor in potassium (K^+) ions, and a good deal of energy is spent by the cell in order to keep the 'sodium pump' going (expulsion of excess Na^+, intake of K^+). Deformations of the membrane allow the cells to catch by 'endocytosis' large molecules or small solid bodies (bacteria for instance) which cannot penetrate by permeability alone. The specific chemical composition of the membrane proteins and glycoproteins allow cells to recognize each other. Cells are very xenophobic; they aggregate with cells from the same tissue of the same species and they reject all the others. However, cells are also well-bred; when cells, even from the same tissue, come into direct contact, they stop moving and dividing (contact inhibition). In contrast, malignant cancer cells do not display these qualities; they stick together, continue dividing, and form a solid mass instead of a single layer of cells.

Thus cells can respond to *signals*, originating from either the external medium or neighbouring cells which are perceived at the plasma membrane level. Their main responses are of three different kinds: changes in locomotive activity, changes in cell shape (cell morphology), and induction of cell division.

Individual cells are capable of locomotion. The molecular bases of their motility are very similar to those of our own muscles, since the cells possess bundles of *microfilaments*, made of contractile proteins (actin, myosin) which are almost identical to those present, in much larger amounts, in muscles. Disruption of the microfilament bundles by the drug cytochalasin B leads to the arrest of cell motility.

A spherical cell can elongate and come back to its initial shape; cells contain a 'cytoskeleton' made of *microtubules*. If these microtubules are long and parallel, the cell will assume an elongate shape; it will, however, remain spherical if the assembly of the microtubules is prevented by treatment with the drug colchicine. Microtubules result from the linear polymerization of a protein, tubulin. Thanks to molecular biology we know a good deal about the mechanisms of actin and tubulin polymerization, which are responsible for cell contractility and morphology respectively.

The division of cells into two daughter cells in response to signals received by receptors located in the cell membrane is another important feature of the cell membrane: induction of cell division by addition of substances that bind to the carbohydrate groups of the cell membrane proteins is a frequent event, especially in lymphocytes.

Electron microscopy shows that the cells contain a network of membranes, the *endoplasmic reticulum* which consists of small granules attached to membranes made of proteins and lipids. These small granules are the *ribosomes*, which play such a fundamental role in protein synthesis and which will be discussed further below. The main functions of the endoplasmic reticulum are protein synthesis and the transport of proteins that the cell might need to export. For instance, the endoplasmic reticulum plays a key role in both the synthesis of trypsin in pancreatic cells and its transport into the pancreatic ducts [6, 7].

Connected in some way with the endoplasmic reticulum are the *lysosomes* [8] and the *Golgi bodies*. The lysosomes are an intracellular digestive apparatus. Large molecules taken up by endocytosis become surrounded by a membrane, and the lysosome which contains digestive enzymes fuses with this and breaks down the ingested molecules. Golgi bodies are a concentric accumulation of membranes, and are believed to be storehouses for large molecules (proteins and others) synthesized by the cell. The addition of sugars to proteins in order to synthesize glycoproteins, which will leave the cell and form the intercellular matrix, is another function of the Golgi bodies.

The *mitochondria* are also very important as they are the sites of energy production in the cells. The energy comes from the oxidation of small molecules by a whole array of enzymes localized in the inner membrane of the mitochondria (which possess a double membrane), and is stored in the form of the energy-rich phosphate bonds of adenosine triphosphate (ATP), a nucleotide made of a purine base (adenine), a sugar (ribose), and three phosphate bonds (adenine-ribose-P~P~P, where ~ indicates energy-rich chemical bonds). When a cell synthesizes a protein or a nucleic acid, it must draw energy from its ATP store, and ATP is transformed into ADP (adenine-ribose-P~P). Thus the oxidation of small molecules and the phosphorylation of ADP into ATP (oxidative phosphorylation) is the main function of the mitochondria and the two phenomena are intimately coupled. Consequently, poisons which uncouple the two processes (e.g. dinitrophenol) deprive the cells of their energy supply. Mitochondria are interesting in another respect – they contain nucleic acids (DNA and RNA). Their role is discussed below.

Green plants also possess a specific cell organelle, the *chloroplast*, which is capable, by the complicated process of photosynthesis, of transforming energy from light into chemical energy (synthesis of glucose and starch and also phosphorylation of ADP into ATP by photophosphorylation). Life would soon disappear from the earth in the absence of green plants, and we cannot imagine the possibility of life on another planet without the presence of photosynthetic organisms. Chloroplasts, like mitochondria, contain their own nucleic acids, which are different from those present elsewhere in the cell.

The cell *nucleus* in eukaryotes, has a complex structure shown diagrammatically in Fig. 4.1. It is surrounded by a *nuclear membrane* (or nuclear envelope), which consists of an external and an internal layer pierced by thousands of small pores; the latter allow an exchange of materials between the nucleus and the cytoplasm. If, as many believe, there is a selective control of these nucleocytoplasmic exchanges, the nuclear pores (which have a complex ultrastructure) would play a role comparable to police and customs at the border between two countries. Inside the nuclei, we find *chromatin* and *nucleoli*, both of which are discussed in more detail below. Chromatin is made of fibrils, which are coiled to various extents. When the fibrils are loosely organized, one speaks of *euchromatin*; if they are so strongly coiled that the chromatin becomes highly condensed, one is dealing with *heterochromatin*. Chromatin, as we shall see, is made of deoxyribonucleic acid (DNA) associated with various proteins. The nucleoli are spherules (often two per cell nucleus) made of ribonucleic acid (RNA) and proteins which are different from those present in chromatin; they are formed of two parts, a central fibrillar core, and a granular outer layer. The granules which form this external layer bear many similarities to the cytoplasmic ribosomes.

When they have reached a given size, cells divide by *mitosis* (Fig. 4.2). At *prophase*, the nuclear membrane breaks down and the apparently homogeneous chromatin of the 'resting' (interphase) nucleus is transformed into filaments, the *chromosomes*. The number of chromosomes is constant in a given species. Since they all derive from a fertilized egg, somatic cells are diploid (2*n* chromosomes); the egg and the sperm, however, have only *n* chromosomes and are said to be haploid. At *metaphase*, the chromosomes split into two chromatids helds together by the centromere (also called the kinetochore). The centromeres are attached to the

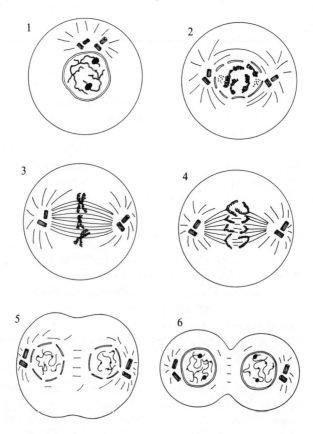

Fig. 4.2. Mitosis. 1, beginning of prophase; 2, end of prophase; 3, metaphase; 4, anaphase; 5, telophase; 6, cytokinesis.

mitotic apparatus, which consists of a spindle and, at the two poles, of asters. Each aster is centered around a tiny, but complex granule, the *centriole*. The spindle and asters are made of microtubules which are almost identical to those of the cryto skeleton and made of polymerized tubulin molecules. Their polymerization (and thus mitotic-apparatus formation) is inhibited by colchicine, a potent antimitotic drug.

At *anaphase*, the centromeres divide and the separated chromatids, which are now daughter-chromosomes, move towards each pole along the spindle fibres. There is increasing evidence for the view that chromosomes movement at anaphase is due to the presence of contractile molecules (actin) in the spindle. Finally, at *telophase*, the mitotic apparatus regresses, while a furrow separates the cell into two daughter-cells. Furrowing is the result of the contraction of an actin microfilament ring present under the cell membrane concomitantly the chromosomes become indistinct and the nuclear membrane and nucleoli form. The two daughter-cells have a typical interphase nucleus, containing the diploid number of chromosomes.

Mitosis is often a delicate step in the life of the cell; unequal distribution of the chromosomes by accident or by chromosomal breakage can occur, with unfavourable or even disastrous consequences for the cell. It is of interest that chromosomal abnormalities are frequent in both senescent and cancer cells.

Bacteria

Bacteria (Fig. 4.3) and primitive unicellular algae have been called prokaryotes because their 'nucleus' is of the utmost simplicity compared to that of the eukaryotes.

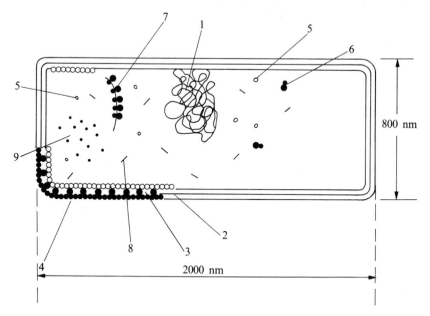

Fig. 4.3. Schematic representation of the bacterium *Escherichia coli*. (Modified from J. D. Watson 1969). 1, DNA thread attached to plasma membrane; 2, respiratory chain; 3, plasma membrane; 4, cell wall; 5, free enzyme; 6, free ribosome; 7, polyribosome; 8, transfer RNA; 9, small molecules.

It is a single circular chromosome and consists of a continuous thread of DNA. Bacteria are surrounded by a membrane which regulates the exchanges with the outer medium; but, in contrast to eukaryotes, which possess specialized mitochondria, this membrane also contains the enzymes required for energy production (oxidative phosphorylations). A bacterium possesses about 15 000 ribosomes, which represent as much as 25 per cent of the total cell mass. These ribosomes, as we shall see, are smaller than those of the eukaryotes.

Viruses

Viruses (Fig. 4.4) are not true living organisms, in the sense that they cannot multiply unless they are introduced into a prokaryotic or eukaryotic cell. They require the machinery present in these host cells in order to exhibit the two main properties of all living organisms: reproduction and heredity. Viruses are parasites responsible for many diseases in men, animals, and plants. They are made of a nucleic acid (DNA or RNA) and one or several specific proteins. Some of the largest

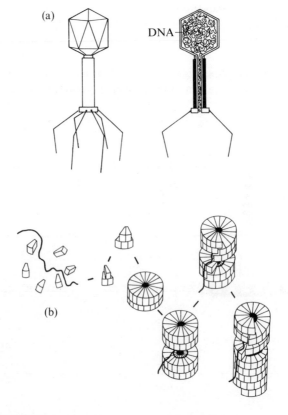

Fig. 4.4. Schematic representation of viruses. (a) Bacteriophage, type T4. *Left*: external view; *right*: internal view.

(b) Tobacco mosaic virus (TMV). The protein blocks are assembled around the RNA thread (modified from Klug 1972).

viruses, like vaccinia virus, contain both DNA and RNA. There are many strains of the same virus, which will give rise to viral particles of the same strain in the infected host cells. The viral nucleic acid is coiled in a central core; the proteins form a more or less complex outer envelope, or capsid. Fig. 4.4(a) illustrates a bacteriophage, a virus which proliferates in bacteria and finally kills them. Fig. 4.4(b) depicts the very simple tobacco mosaic virus, which infects tobacco leaves; the nucleic acid thread is surrounded by repeated units of the same protein.

4.2 The early history of molecular biology

Nucleic acids and proteins, which are at the root of modern molecular biology, were discovered in eukaryotes long before the term 'molecular biology' was coined.

Miescher was the first to isolate, from salmon sperm and from pus, a complex of nucleic acid and protein that he called nuclein [9]. Altmann, a cytologist, succeeded in separating the nucleic acid from the protein in 1887 [10]. Thymus, which is rich in nuclei, proved to be a good source of nucleic acid, which was called, for this reason, thymonucleic acid. It was not until the late 1920s that Levene demonstrated that the sugar present in thymonucleic acid is deoxyribose [11]. Today, thymonucleic acid is just one of the many deoxyribonucleic acids, the DNAs. Parallel work done on plant cells, and in particular on yeast, led to the isolation of another type of nucleic acid, pentose nucleic acid, which possesses another sugar, ribose (it contains one more oxygen atom than deoxyribose). The pentose nucleic acids are thus *ribonucleic acids* (RNAs). For a long time, it was believed that thymonucleic acids (DNAs) were typical of animal cells, while pentose nucleic acids (RNAs) were specific to plant cells.

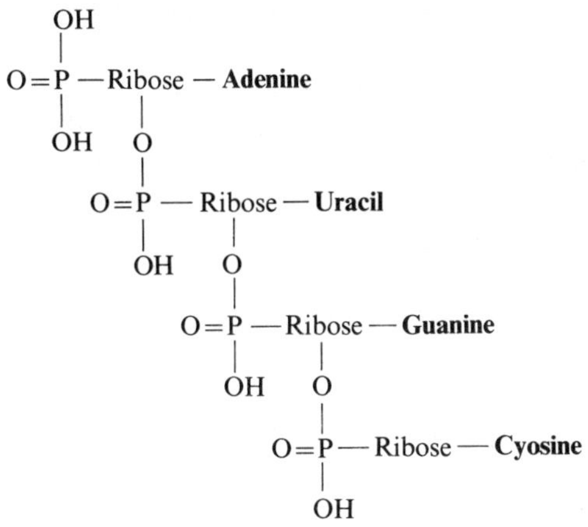

Fig. 4.5. Levene's tetranucleotide formula.

The structure of the nucleic acids was elucidated by Levene (1930) who proposed the 'tetranucleotide' hypothesis: a nucleotide is made of phosphoric acid, a pentose, and a purine or pyrimidine base. All nucleic acids contain two purines, guanine (G) and adenine (A); one of the pyrimidines, cytosine (C) is also common to both DNA and RNA. However, the fourth base is different; while RNA contains uracil (U), DNA contains its methylated derivative, thymine (T). Fig. 4.5 shows Levene's general formula describing the structure of the nucleic acids; it was based on the assumption that nucleic acids contain equal amounts of G, A, C. and T (or U). This assumption, as we shall see, was not correct.

Improvement of the methods used for nucleic acid isolation has led, in the post-war years, to a steady increase in the size of the nucleic acid molecules; the molecular weight of Levene's tetranucleotides was only about 1200 daltons. We know now that the relatively small chromosomes of the bacterium *Escherichia coli (E. coli)* is a single DNA molecule; its molecular weight is 2.5×10^9 daltons and its length, if the molecules were completely unwound, would be 1 mm. Thus nucleic acids are polynucleotides of a huge size and belong to the category of macromolecules.

Proteins were also first isolated from eukaryotic tissues. They are made of amino acids

$$R-\overset{\displaystyle H}{\underset{\displaystyle NH_2}{C}}-COOH$$

linked together by the peptide linkage (-CO-NH-) and are thus polypeptides. About twenty different amino acids can be found in proteins, and an 'average' protein is made of 300 to 500 amino acid residues. Specificity of the proteins (there are probably about 10 000 different proteins in an egg) resides in their amino acid sequence. This characteristic and specific sequence is called the primary structure of the protein. Proteins can display a great degree of complexity; for instance haemoglobin is made of four polypeptide chains: two a chains and two β chains. a and β chains have a different amino acid sequence and thus a different primary structure, although they have about the same molecular weight (about 16 000 daltons). It follows that the molecular weight of the whole haemoglobin molecule is around 64 000. Proteins undergo changes in their conformation when they are put into solution; this is discussed below.

At the time when biochemists were isolating nucleic acids and proteins, Mendel discovered the basic laws of heredity from his famous work on peas [12]. The demonstration that the hereditary characters have a material support, the *gene*, and that genes are localized in a linear fashion in the chromosomes, came mainly from the work in 1910 of T. H. Morgan and his co-workers on the fruit fly *Drosophila* [13].

The Morgan school studied in great detail *genetic recombination* i.e. the exchange of genetic material which takes place when precursor diploid cells give rise to haploid gametes (eggs or sperm). The process which leads to reduction of the chromosome number during gametogenesis is called *meiosis*. All we can observe (or rather could observe until a few years ago) when we study heredity, is a visible character (for instance, eye colour) which is transmitted according to Mendel's laws. This visible character is called the *phenotype*; it is the result of the 'expression' (or non-expression) of a gene present in our chromosomes. But how can we test the *genotype*, i.e. the load of genes present in our chromosomes? Why are different individuals (except identical twins) from the same species recognizable from each other? The reason is that genes can undergo sudden changes which will be transmitted to the offspring; these sudden changes are called *mutations*. Appropriate crosses between individuals bearing mutations, recognizable by phenotypic differences, allow the geneticists to probe the genotype. The analysis of mutants is the very basis of genetics, and it was thus an important finding when Muller discovered, in 1972, that X-ray irradiation considerably increases the number of mutants in a *Drosophila* population [14]. It was quickly suspected that genes must somehow affect proteins. This was discovered by Garrod [15], who showed that certain hereditary diseases in man result from the failure to utilize a given amino acid for protein synthesis. This failure is due to gene mutations which result in the synthesis of abnormal proteins and pathological clinical symptoms.

During the 1930s and 1940s, methods were worked out for the detection of RNA and DNA in eukaryotic cells. The most important conclusions which can be drawn from these studies by Feulgen, Caspersson, and the author of this review are the following: all cells (animal, vegetal, and bacteria) contain both DNA and RNA; DNA is localized in chromatin, while RNA is most abundant in the nucleoli and the cytoplasm; there is a correlation between the RNA (but not the DNA) content of a given cell and its capacity to synthesize proteins [16, 17, 18].

The end of the Second World War was not only a happy time for millions of human beings, it also marked the beginning of modern molecular biology and of its amazingly fast development. As we shall see, it was soon discovered that DNA is the genetic material; that each gene directs the synthesis of a specific protein; that DNA cannot directly produce this protein; and that RNA molecules are involved in the transfer of information from DNA to protein. The mechanisms of DNA replication and of mutation (the top secrets of heredity) have therefore been discovered.

If heredity is no longer the mystery it was 30 years ago, it is because the molecular biologists of the post-war period decided to leave the complex eukaryotes for the much simpler prokaryotes. Today there is a reversal in this trend; many biologists try to explain the characteristics of eukaryotes by applying to them the gigantic amount of knowledge gained by the experiments done on bacteria (in particular *E. coli*).

4.3 DNA is the genetic material

Soon after the war, Avery and his co-workers made the astounding discovery that it is possible to 'transform' a strain of the bacteria *Diplococcus pneumonia*, which is genetically unable to build up a protective capsule, into a strain which possesses this capsule by the simple addition of pure DNA isolated from a strain which is genetically capable of synthesizing the capsule [19]. This *transformation* is transmitted to the progeny, and is thus a stable hereditary character. In other words, the genes are made of DNA molecules. These DNA molecules can penetrate into a recipient bacterial strain and if it is integrated into the host-cell's DNA, it is replicated. The transformed host-cell has thus received fresh genetic information and can transmit it to its progeny; it can also express the new gene it has gained, i.e. synthesize a capsule.

Similar conclusions can be drawn from the so-called 'syringe experiment' of Hershey and Chase [20]. They worked on a very different prokaryotic system, the bacterium *E. coli* infected by a (bacterio)phage. This particular phage is made of DNA and proteins. Hershey and Chase labelled the phage DNA and proteins with two different radioisotopes and found that the phage DNA is injected into the bacterium while the protein does not penetrate into it. The result of this experiment is that complete phage particles, belonging to the same genetical strain as the infecting phage are formed in large amounts. It follows therefore that phage DNA (but not phage protein) possesses all the information needed for the formation and multiplication of complete phage particles.

Similar experiments have been tried on eukaryotic cells, but with less convincing results. There seems, however, no doubt that hereditary characters (for instance, eye colour) have been successfully transmitted in the fly *Drosophila* by addition of appropriate DNA preparations. There are many difficulties, inherent to the nature of the biological material, which explain why 'transformation' experiments are less successful in eukaryotes than in prokaryotes. DNA should penetrate into the cell, it should not be broken down by the digestive enzymes present in the lysosomes, it should be integrated in the chromosomes and finally replicated.

However, many indirect experiments demonstrate that, in eukaryotes as well as in prokaryotes, DNA is the genetic material. The DNA content of a normal diploid cell nucleus is constant for a given species and it is the same in cells of all tissues, whether they are of small or large size. The DNA content of diploid cells is exactly twice that of the haploid gametes (egg or sperm). The DNA content of the cell nucleus exactly doubles prior to cell division, so that it reaches a tetraploid value when the cell is preparing for cell division. Subsequently, the DNA is evenly distributed between the two daughter cells. Finally, in contrast to RNAs and proteins, DNA is an exceedingly stable molecule. This can easily be shown by experiments with radioactive precursors of nucleic acids and proteins. If cells are incubated with radioactive thymidine, uridine, or amino acid (the nucleosides thymidine and uridine are specific precursors for respectively DNA and RNA, while amino acids are incorporated into proteins during their synthesis) the following results are obtained: thymidine is not incorporated into DNA unless the cell is preparing for cell division and is replicating its NDA; once it has been incorporated into freshly synthesized DNA, it stays there. On the other hand, uridine is incorporated into RNA at any time during the life of the cell, but, after a few hours, the radioactive RNA is degraded by the cell enzyme (ribonucleases); the same is true for proteins, which are continuously being synthesized and degraded by proteases. Thus RNAs and proteins undergo a continuous replacement (turnover) in the cell; on the contrary, DNA remains perfectly stable (unless it undergoes one of the mutations discussed below).

This is exactly what one should expect from the genetic material.

4.4 The one-gene–one-enzyme theory

As already pointed out, we can easily see the effect of a gene in the way it is expressed. An inactive, silent gene cannot be detected by the classical methods of genetics. As we have seen, the medical observations of Garrod suggested that genetic diseases result from alternations in protein metabolism [21]. Experiments by Ephrussi and by Beadle on eye colour in *Drosophila* [22, 23] and by Beadle and Tatum on the mould *Neurospora* [24] have shown clearly that genes control the synthesis of specific proteins. The very accurate work of Beadle and Tatum on *Neurospora* dealt with 'nutritional mutants', i.e. strains which are unable to grow in a simple synthetic medium unless it is supplemented with a single chemical, an amino acid or a vitamin, for instance. It is needed for growth because the strain cannot synthesize it; it lacks an enzyme required for its synthesis. These findings led Beadle and Tatum to the formulation of the 'one-gene–one-enzyme' theory. Since this theory is equally valid for proteins which have no enzymatic activity, it has been reformulated in the following way: 'one gene–one polypeptide chain'. This means that, for example, in the case of haemoglobin (which is made of α and β chains), a gene will direct the synthesis of the α chain and another one that of the β chain. Since proteins vary considerably in size, it follows that genes (i.e. corresponding segments of the DNA molecule) also vary in size. An 'average' polypeptide chain is made of about 500 amino acid residues; the corresponding 'average' gene comprises about 3000 nucleotides.

4.5 Why work on bacteria and viruses?

We have already seen that prokaryotes are morphologically very much simpler than eukaryotes. As shown by J. Monod in particular, they can be handled like chemicals in test-tubes [25]. Their growth is very fast, and can be followed by simple physical or chemical methods. This fast growth allows a huge population of bacteria to be handled in a short time. Another advantage of bacteria is that they are haploid; their genes exist in a single copy, whereas our genes exist in two copies; so that one of them can completely mask a change undergone by its 'colleague' (called *allele*). Since bacteria are haploid, any change (mutation) which takes place in a gene will be expressed. Thus mutations, which occur at a low rate (1 in every 10^6–10^7 individuals) are easy to detect, to score, and to select in a very large bacterial population.

The main drawback of bacteria has been that they have no sexuality and thus cannot be crossed; but strains of *E. coli* have been isolated where mating between + and - individuals (comparable to males and females) occur with a high frequency. These strains have been extremely important tools for the genetic analysis of bacteria.

The simpler, non-living viruses also display heredity; innumerable strains of viruses exist, resulting from abrupt changes (mutations) in their nucleic acid moiety. For instance, we all know that there are epidemics of influenza every year and that they are due to viruses which have changed after a few years (Hong-Kong, Spanish, etc. influenza epidemics). Vaccines are often ineffective against a new strain because their specificity resides in the influenza virus proteins. If the viral DNA has undergone a mutation, the viral proteins will change, as predicted by the 'one-gene-one-polypeptide-chain' theory, and the vaccine will be less efficient or not efficient at all for the protection of human beings.

Viruses, and in particular phages, have played a major role in the development of modern molecular biology, thanks to the initial impulse of Delbrück and Luria [26]. Very extensive biochemical and genetic studies have been made on phage-infected bacteria (*E. coli*) by many people [27, 28, 29, 30, 31, 32, 33, 34]. At the same time, work of fundamental importance was done on the structure of DNA [35] and of the proteins [36, 37]; simultaneously, protein synthesis in extracts of crushed bacteria has been extensively studied [38, 39].

It is this work that has led to the formulation of Crick's 'fundamental dogma' of molecular biology, which says:

$$\text{DNA} \xleftarrow{\text{Replication}} \text{DNA} \xrightarrow{\text{Transcription}} \text{RNA} \xrightarrow{\text{Translation}} \text{Protein}$$

It means that DNA can give rise to identical DNA molecules by replication; that DNA directs protein synthesis, but in two steps; it must first be transcribed into RNA molecules which are then translated into proteins. Genetic information thus flows from DNA to RNA and finally to proteins, but not in the reverse direction, with the exception of RNA viruses which contain an enzyme (reverse transcriptase) which copies the viral RNA into DNA. We shall now examine in more detail the evidence upon which this 'fundamental dogma' rests.

4.6 The double-helix structure of DNA

In 1953, a short note in *Nature* by J. D. Watson and F. H. C. Crick [35] proposed a structure for DNA which had the simplicity and beauty of genius (Fig. 4.6). In essence, this stated that DNA is made of two complementary 'antiparallel' helices where the bases A and T on the one hand and G and C on the other are held together by hydrogen bonding. The beauty of this apparently very simple model resides in the fact that it provides easy explanations for DNA *replication, transcription*, and *mutation*.

DNA replication: how can DNA make a perfect copy of itself? This can be easily understood if one assumes an unwinding of the double helix. Thus during replication the two strands separate and enzymes present in the cell (DNA polymerases) synthesize complementary strands. In view of the necessity of base pairing (A–T; G–C) the result will be a perfect copy of the initial DNA molecule, which therefore plays the role of a 'template' for its own replication. Thanks to this mechanism, two identical DNA molecules, which possess exactly the same genetic information, will be equally distributed in the two daughter-cells at mitosis (Fig. 4.6).

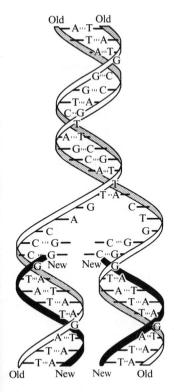

Fig. 4.6. Scheme of the semi-conservative replication of DNA.

DNA transcription, which was not extensively discussed by Watson and Crick in 1953, can also be easily explained by the double-helix model; all we have to assume is that an RNA-synthesizing enzyme (an RNA polymerase) travels along one of the two strands of the DNA molecule. This strand will again serve as a template and will be copied by the enzyme, but this time the copy will be an RNA not a DNA molecule; this RNA contains the same genetic information as the DNA strand which has been copied.

Finally, if one accepts the Watson–Crick model, *mutation* is also easy to understand: all one has to assume is that an abrupt change can take place in one of the bases: i.e. replacement of one of the four normal bases by an abnormal one. For example, nitrous acid (HNO_2) is a well-known mutagenic agent because it oxidizes and deaminates the aminopurines A and G into the corresponding oxypurines. Such a modification of the purines will distort the regular arrangements of the DNA double-helix; the result will be a genetic mutation.

The Watson–Crick model is as important as the discovery of Mendel's laws for the understanding of heredity; it gives a material, easily understandable basis to Mendel's mysterious hereditary characters.

The double-helix structure of DNA is now universally accepted; but when it was proposed it rested on certain chemical and physical evidence, the significance of which was not understood. Chargaff had discovered that DNAs from various species differ in their base content [40]; he also found that the amounts of A and T and those of G and C are always equal (A = T; G = C). However, what does vary between one species and another is the (A+T)/(G+C) ratio, i.e. the relative amount of A-T and G-C base pairs inthe giant DNA molecule. The Watson–Crick model was also in agreement with X-ray crystallographic data obtained by the patient work of Wilkins and R. Franklin on a very difficult material: bundles of orientated fibres consisting of stretched DNA molecules [41, 42].

We now know that a cell often contains, in addition to the main chromosomal DNA, 'satellite' DNAs. These are more repetitive than bulk DNA and are characterized by a high A-T or G-C content. In some satellite DNAs, A-T base pairs are repeated thousands of times; in others, G-C sequences are far more numerous than A-T-containing sequences. However, the molecular structure of these satellite DNAs (which probably play an important role in the organization of the chromosomes, since the A-T-rich satellites of mammals are accumulated in the centromeres) conforms with the universal Watson–Crick structure. This implies that DNA is a very long, but very thin, molecule; a metaphase chromosome, which is a few micrometres long, almost certainly contains a single DNA molecule but, if stretched out, this molecule would be several centimetres long. This means that, in the chromosome, DNA must undergo extensive packaging (supercoiling) into a much more compact form than the extended double helix. How the giant DNA molecule can be accommodated in a chromosome is discussed when we come to the structure of chromatin (§ 4.9).

4.7 The structure of proteins

Proteins are polypeptides, i.e. chains of a amino acids in which the carboxyl group of each amino acid is condensed with the amino group of the next one in a 'peptide bond':

$$
\begin{array}{c}
O \\
\parallel \\
-C-N- \\
H
\end{array}.
$$

The chains are always linear, never branched. The general formula for an a amino acid is:

$$
\begin{array}{c}
H \\
\vert \\
H_2N-C-COOH \\
\vert \\
R
\end{array}.
$$

They differ from each other only by the nature of the R residue. Twenty different R residues are found in proteins, defining the twenty amino acids commonly found in these macromolecules.

The carbon located between the NH_2 and the COOH group carries four different substituents, it is therefore 'asymmetric' and a amino acids can in principle exist in two configurations, L and D, the one being the mirror image of the other. In proteins, only the L form is found.

Individual proteins differ by the number, the assortment and the arrangement of the amino acids in the polypeptides. Some proteins are made of short chains, e.g. 20 amino acid residues; some have as many as one thousand in one single chain.

As polypeptides contain many single bonds, they can be folded in many different manners. X-ray diffraction data, however, showed that each polypeptide folds in a unique manner, resulting from numerous interactions of the R radicals of the amino acids of which they are made [36, 37]. Each protein has a unique three-dimensional structure, stabilized by many weak interactions (secondary bonds). The physiological properties of proteins depend on fine details of their three-dimensional structure, which results ultimately from the arrangement of the amino acids in the polypeptide chains (Fig. 4.7).

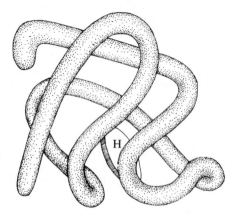

Fig. 4.7. Scheme of the three-dimensional structure of myoglobin. The haem prosthetic group is represented at H.

When a protein interacts with another protein or with a small molecule, it usually undergoes local changes of its three-dimensional structure and consequently of some of its properties. Such 'transformations' play an essential part in most physiological regulatory processes.

4.8 A brief summary of the main results obtained with bacteria and phages

As has already been pointed out, the study of *mutations* is the fundamental tool for geneticists. The average rate of mutation in bacteria, is 10^{-6} at each replication; this means that one out of every million bacteria undergoes a sudden, heritable change when the whole population doubles. Some regions of the *genome* (i.e. the totality of the genes present in a given organism) are more susceptible to mutations than others; they are the so-called 'hot spots'. The mutation rate, as we already know, can be greatly increased by the treatment of eukaryotic cells, bacteria, or viruses with a large variety of either physical agents (X-rays or ultraviolet irradiation) or chemical agents. A very large number of chemical mutagenic agents are now known (for instance, hydrocarbons present in tobacco smoke), all of which somehow affect the integrity of the DNA molecule. Mutations commonly result from either the replacement of one DNA base by another, or the incorrect association between two bases. Mutagenic agents can also break the DNA double helix, but if the break affects only one strand, enzymes present in the cell can repair the damaged DNA molecule. The injured part of the affected strand is first excised; the gap is then filled by a DNA synthesizing enzyme, DNA polymerase, and finally the repaired fragment of the DNA molecule is sealed to the uninjured DNA macromolecule by an enzyme called ligase. However, this repair process is not always perfect, and errors can be made by the repair enzymes; such errors will lead to mutations. The analysis of mutations has allowed the *mapping* of the genes on the bacterial or viral chromosome [28, 31, 43] and we now possess accurate maps showing the precise localization of many genes on the bacterial or phage chromosomes. It is interesting to note that genetical mapping analysis showed that the *E. coli* chromosome is circular before electron microscopy demonstrated that this is indeed the case [44]. The study of recombination has shown that, after breakage and reassociation of DNA segments, exchanges of parts of DNA molecules can occur inside the gene itself (intragenic crossing over).

A gene coding for the synthesis of a specific polypeptide (according to the 'one-gene–one-polypeptide-chain' theory is called a *structural* gene or *cistron*. Studies on the regulation of the expression of genes in *E. coli* and on genetic maps, led to the discovery of *operons*: genes which are closely-linked and control the synthesis of a number of enzymes usually involved in the same process and which are regulated simultaneously. For instance, 10 enzymes are needed for the synthesis of the amino acid histidine; the 10 structural genes which direct the synthesis of each of these 10 enzymes form a cluster (the histidine operon). How the genes forming such an operon are switched on or off is examined later. There are a few well-demonstrated operons in eukaryotes, although genes involved in the same function are often located on different chromosomes. For instance, a large number of haemoglobin mutations, which occur in either the a or β globin chains are known, but the a globin and the β globin genes are located in different chromosomes in man; they are thus not linked together. In order to produce equal amounts of a and β globins, regulatory mechanisms of a much greater complexity than switching on and off a whole operon must take place.

The result of a mutation in a structural gene is the production of an abnormal protein. For instance, a mutation in the a or β haemoglobin chain will result in the substitution at a given site of the corresponding globin chain of one amino acid by another. The first case where the molecular basis of a human hereditary disease has been understood is that of sickle-cell anaemia [45]; the red blood cells are abnormal in this disease because the amino acid valine has been replaced in the β chain of haemoglobin by another amino acid, glutamic acid. This change is the consequence of a mutation which took place in the gametes (egg or sperm) of the parents of the diseased person.

Highly refined genetical analysis of the *E. coli* phage system has shown that there are three mutational sites for one amino acid, and that a mutational site can exist in four alternative forms. The biochemical implications of these genetical findings are considered below. Careful study of deletions (mutations due to the loss of a single base in this molecule) has led to the important concept of *triplets*: three bases, in the DNA molecule, correspond to one amino acid. These sequences of three bases, which specify for a given amino acid, are called *codons* (see below). The information required for the synthesis of a given specific protein is 'encoded' in the DNA of the corresponding structural gene. This information is read (decoded) linearly by groups of three bases (the codons) starting from the 5' end of the DNA molecule. Insertion or deletion of a single base will change the right codon into an incorrect one and this will result in errors in coding and, as a consequence, in reading. Mutations of these types have been called 'frameshift mutations'; they result in the synthesis of an abnormal protein. Finally, genetical analysis of the *E. coli* phage system has shown that there is 'colinearity' between the gene and the corresponding polypeptide. The genetic code is deciphered starting at one end (the 5' extremity) of the DNA molecule; the primary structure of the corresponding polypeptide chain (protein) is also linear, each amino acid corresponding to the codon which has been read.

4.9 DNA replication

As has been previously pointed out, the Watson–Crick model provides a remarkably simple explanation for DNA replication: unwinding of the double helix would

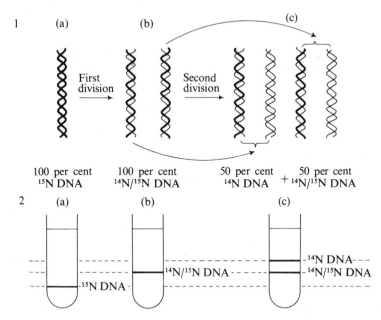

Fig. 4.8. Diagrammatic representation of the Meselson and Stahl experiment. In (a), bacteria are grown in a ^{15}N medium. In (b) and (c), the ^{15}N bacteria so obtained are in a ^{14}N medium.
   1. Theoretical prediction of DNA structures if the semi-conservative replication model is true.
   2. Sedimentation bands obtained after gradient-centrifugation of DNA extracted from the (a), (b), and (c) bacteria respectively.

be followed by the copy of each of the separated strands. If this is so, when the DNA molecule has been completely replicated, one of the strands should be of parental origin, while the other has been synthesized *de novo*, thus, each parental strand, after unwinding of the double helix, would serve as a 'template' for the synthesis of the complementary strand (Fig. 4.6, p.99). Such a type of replication has been called *semi-conservative*. Brilliant experiments by Meselson and Stahl [46], on the bacterium *E. coli*, have demonstrated that DNA replication is indeed of the semi-conservative type (Fig. 4.8). They grew *E. coli* in a medium containing heavy nitrogen (^{15}N) until the DNA bases contained ^{15}N instead of ^{14}N, and then tranferred the bacteria to normal ^{14}N-containing medium. After one division (one replication cycle), of the bacteria they isolated the DNA, analysis of which showed that one of the strands was 'heavy' (containing ^{15}N) and the other 'light' (containing ^{14}N). The heavy strand was the one which had served as a template for the replication of the light, newly synthesized strand. Similar experiments on eukaryotes leave no doubt that semi-conservative DNA replication is an universal phenomenon.

However, DNA replication is not so simple as it seems; synthesis of the new strand requires the intervention of an enzyme, DNA polymerase [47]. In eukaryotes, there are three different DNA polymerases (α, β, and γ), while an additional DNA polymerase is located in the mitochondria and serves for mitochondrial DNA replication. It is probable that only one of the three DNA polymerases is involved in chromosomal DNA replication, the others are thought to be responsible for the aforementioned DNA repair processes. To be active, DNA polymerases require the presence of DNA, which serves as the template which will be copied. The building blocks for DNA synthesis are the four deoxyribonucleotide triphosphates (dATP, dGTP, dCTP and dTTP). The reaction catalysed by the DNA polymerases can thus be written:

$$\text{DNA} + n \begin{cases} \text{dATP*} \\ \text{dGTP*} \\ \text{dCTP*} \\ \text{dTTP*} \end{cases} \xrightarrow{\text{DNA Polymerase}} \text{DNA} + \text{*DNA} + n\text{PP (pyrophosphate)},$$

where * indicates that the precursors were labelled and that the newly synthesized DNA molecule is also labelled.

Recent work has shown considerable complexity in DNA replication, both in bacteria and in eukaryotes. As expected, uncoiling proteins (which open the double helix) have been discovered; but an unexpected finding was that the DNA which is synthesized in a short time (1–2 minutes) after the addition of labelled precursors is made of small fragments, the so-called Okazaki fragments [48]. DNA replication is thus a *discontinuous* process; it starts at many *initiation points*, where DNA polymerase synthesizes only short stretches of complementary DNA (the Okazaki fragments). These short stretches are then linked together by ligases, and are finally integrated into macromolecular DNA. At the initiation points, DNA synthesis requires the unexpected presence of an RNA molecule linked to DNA. This acts as a 'primer', and might represent a signal that will be destroyed when it is no longer needed. That DNA replication is really discontinuous has been shown by electron microscopy, which has demonstrated the presence of 'forks' in replicating the DNA molecules [44]. These correspond to newly synthesized polynucleotide chains, and start from the initiation points (also called *replicons*).

In contrast to bacteria, DNA replication takes place, in eukaryotic cells, during a limited part of the *cell cycle* (the time-interval between two mitoses). When a daughter cell arises from mitotic division, its nucleus contains the diploid number of chromosomes (called the $2C$ value). For several hours the DNA content remains constant at this $2C$ value; this is the so called G_1 phase (G = gap) the cell then synthesizes DNA, according to the semi-conservative type of synthesis, and exactly doubles its DNA content; this synthetic phase (S phase) thus brings the DNA content from $2C$ to $4C$. There is another gap (G_2) before the cell actually enters mitosis (M) which will then restore the DNA content of the nucleus to the $2C$ value. It is believed that G_1 is necessary for the synthesis of the enzymes and precursors needed for DNA replication, while G_2 is the time required to prepare the cell for mitosis. This interpretation is made likely by the fact that newly fertilized eggs, which divide very quickly, have no G_1 phase and their G_2 phase is not distinct from prophase. The absence of G_1 in cleaving eggs is easily explained by the fact that the unfertilized egg already contains a large store of all the enzymes and precursors required for DNA synthesis. When cells no longer divide, like our neurons, they are said to be arrested in G_0.

The length of the three phases is of course related to the speed of DNA synthesis, which itself depends more on the frequency of the initiation points than on the speed of polynucleotide chain elongation, which is catalysed by DNA polymerase moving along the DNA template. For instance, in *Drosophila*, the S phase lasts less than 10 minutes in eggs, and about 10 hours in the adult. However, the speed of DNA elongation is the same: about 2600 nucleotide pairs per minute. In contrast, the 'interfork' distance is much shorter in the egg than in the adult, showing that the former has many more initiation sites than the latter. It has recently been shown that cells which divide actively (i.e. eggs, yeast cells, cancer cells, etc.) contain *initiation factors*; these are proteins which, in a still poorly understood way, increase the number of initiation points where DNA polymerase can bind to DNA.

That the composition of chromatin is much more complex in eukaryotes than in prokaryotes has already been mentioned. In the former, DNA is bound to basic proteins, the five *histones* H_1, H_2a, H_2b, H_3, and H_4, but there are in addition many non-histone proteins bound to the DNA. Recent work has demonstrated that chromatin has the same organization in all eukaryotes; both electron microscopy and biochemical methods have shown that it is in the form of a string of beads. The units of this very regular pattern are called *nucleosomes* (or bodies) and consist of 'core' particles united by DNA threads called 'linkers'. The cores are aggregates of dimers of the H_2a, H_2b, H_3, and H_4 histones, wrapped by a DNA spiral of 140 base pairs. The linkers, which are also made of DNA and are probably associated to histone H_1, vary somewhat in length from one species to another (between 15 and 100 base pairs). It is believed that phosphorylation (addition of phosphate) of histone H_1 plays an important role in the further packaging of chromatin when it condenses for the formation of chromosomes in prophase or in sperm heads (sperm chromatin is so condensed that it is almost crystalline). There is also some evidence for the view that phosphorylation of histone H_1, by inducing chromosome condensation, might be a signal for the transition from G_2 to M and thus for entry into mitosis.

We possess many inhibitors of DNA synthesis, some of which are widely used in cancer chemotherapy. All of them are mutagenic agents and may turn out to be carcinogenic in the long run. As the result of chromosomal breakages, irradiation also induces mutations, but, as we have seen, they will be more or less successfully impaired by the cell repair enzymes. Prolonged irradiation can also induce cancers; cancer chemotherapy and radiotherapy are thus faced with an awkward dilemma.

4.10 DNA transcription (RNA synthesis)

There are many kinds of RNAs in the cell (Fig. 4.9) all of which are synthesized on a DNA template by RNA polymerases. In most animal cells, except for the eggs of several species during the very early stages of development, almost all the RNA is synthesized on chromatin and transcription of mitochondrial DNA plays only a minor role in RNA synthesis. However, in green plant cells, the contribution of chloroplastic RNA synthesis, as directed by the chloroplast genome, is far from negligible and can even be preponderant.

The cell nucleus contains a very heterogeneous population of RNA molecules called *hnRNA* (heterogeneous nuclear RNA): a mixture of RNA molecules with molecular weights ranging from several million to a few thousand daltons. These molecules are short-lived and are degraded, partially or completely, within a few minutes after their synthesis. However, three different kinds of RNAs have greater stability; they move out of the nucleus and accumulate in the cytoplasm, where they play a fundamental role in the translation of the genetic code. These three major RNA species are the *ribosomal* RNAs (rRNAs), the *messenger* RNAs (mRNAs) and the *transfer* RNAs (tRNAs). They arise in the nucleus, in the form of larger precursor molecules, which are cut down to their final size by nuclear enzymes (ribonucleases); this 'processing' of the large precursors follows their transcription and is thus a post-transcriptional event.

mRNAs are a faithful copy of the structural genes (cistrons) coding for individual proteins [49]; they possess therefore the same information as the DNA gene itself. Their role is to carry the genetic messenger from the nucleus to the cytoplasm, where it will be deciphered. The first eukaryotic mRNA to be isolated was that of haemoglobin [50]; prokaryotic mRNAs had already been isolated by F. Gros and S. Brenner [51, 52]. Isolation of mRNAs, which only exist in small amounts in most cells, is facilitated by the fact that many of them possess, at their $3'$ end, a sequence made of repetitive adenylic acid sequences (poly-A). This poly-A tail allows the experimenter to isolate and purify the polyadenylated mRNAs. Polyadenylation (i.e. the addition of a poly-A sequence) is a post-transcriptional phenomenon and can take place in the cytoplasm of sea urchin eggs. It seems that

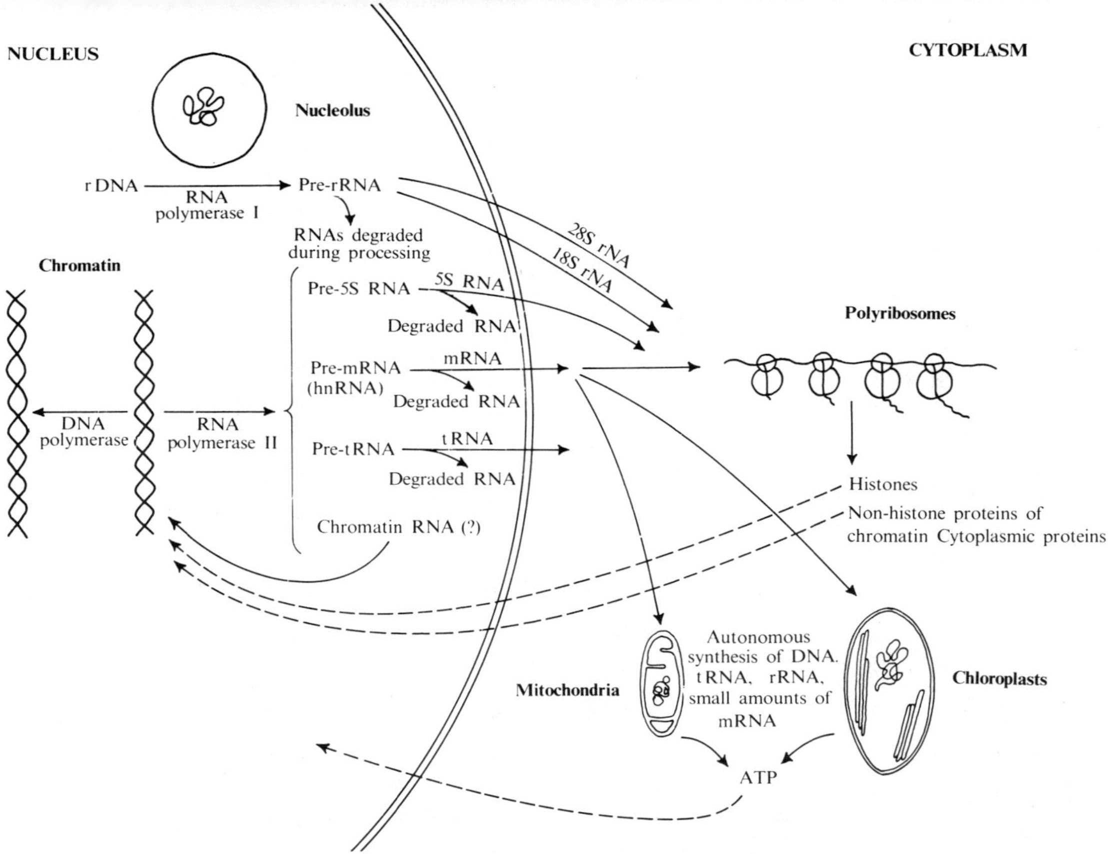

Fig. 4.9. Diagrammatic representation of RNA biosyntheses in a cell.

the role of the poly-A sequence is to protect the mRNAs against degradation in the cytoplasm by ribonucleases. Histone mRNAs differ from the others by the absence of the poly-A terminal sequence; their stability is lower than that of the poly-A-containing mRNAs.

Until recently, it was thought that mRNAs result from the processing of giant hnRNA molecules; but it now seems clear that in the case of haemoglobin mRNA at least, the precursor is not very much larger than the final mRNA molecule. This finding casts some doubts about the biological role of hnRNA. If it is not a precursor of the mRNAs, it might play a role in the regulation of gene expression in eukaryotes. In general, copies of giant hnRNA molecules are not found in the cytoplasm; however, the presence in the cytoplasm of very large RNA molecules has been reported in both sea urchin eggs and in insect salivary glands.

The cytoplasmic mRNAs form a very heterogeneous population. In sea urchin eggs, where all or almost all of the genes are transcribed, it has been estimated by E. Davidson [53] that the cytoplasm contains between 10 000 and 15 000 different mRNA species, each coding for a specific protein. In this large population, some mRNAs are present in few copies only, while others are abundant. As explained in the next section, only a part of the mRNA molecule contains genetic information; the remaining part of the molecule is made of non-coding sequences.

The *rRNAs* represent the bulk (80–90 per cent) of total cytoplasmic RNA. They are found in the *ribosomes* (see § 4.11), and they belong to three different classes, which have been called 28S, 18S, and 5S rRNAs according to their sedimentation constant (S) in the ultracentrifuge. Their respective molecular weights are: 1.3×10^6, 0.7×10^6, and 40 000 daltons. The 28S and 18S rRNA are synthesized and processed in the nucleoli. Their synthesis takes place on the *nucleolar organizers*, which are made of a particular kind of DNA called *ribosomal* DNA (rDNA); this rDNA is often richer in G–C than the bulk of chromosomal DNA, and this characteristic facilitates its isolation and purification. The genes which are transcribed into 5S RNA (5S genes) have a different localization than the 28S and 18S ribosomal genes and they can be present in several of the chromosomes. All these genes are highly *reiterated*; this means that there are many copies of the ribosomal and the 5S genes in the cells. In contrast, the structural genes (coding for mRNAs and proteins) exist in only one or very few copies in chromatin. In the *Xenopus* toad, somatic cells (red blood cells, for instance) contain 1000 copies of the 28S and 18S genes and 24 000 copies of the 5S genes. Redundancy is of necessity, because every cell must produce many ribosomes; large amounts of 28S, 18S, and 5S rRNA are thus required. The need for synthesizing a huge number of ribosomes is still greater in ovarian eggs (oocytes); in *Xenopus*, they measure 1.2 mm in diameter and are full of ribosomes. This requirement for increased rRNA synthesis during growth of the oocyte is met by a process called *amplification. Xenopus* oocytes possess about 1000 nucleoli and the nucleolar organizers of each of these nucleoli contains, like those of the somatic cells, about 1000 copies of the 29S and 18S rRNA genes. The final result is that the oocyte possesses about 10^6 copies of these genes.

Amplification results from a synthesis of rDNA, which takes place independently from chromosomal DNA replication. Thanks to their amplification in oocytes and their redundancy in all other cells, D. Brown has been able to isolate the ribosomal and 5S genes in a pure form and to analyse their structure [54]. In both cases, the ribosomal cistrons are separated by 'spacers' which have a different base composition from that of the genes themselves: they display a much greater variability in their length and chemical composition. The 28S and 18S rRNAs arise from the processing of a larger 45S (40S in *Xenopus*) precursor synthesized by the nucleolar organizer's ribosomal genes. The various stages of this processing are very well known and are shown schematically in Fig. 4.10 together with the structures of the ribosomal and 5S genes.

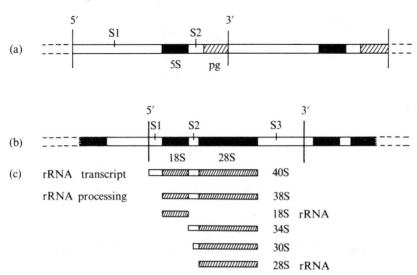

Fig. 4.10. Ribosomal genes (in *Xenopus laevis* oocytes). (a) Tandem repeating unit of 5S DNA. (Modified from the Annual Report of the Carnegie Institution, 1975–6.) S_1, A + T - rich spacer region; 5S, 5S gene; S_2, short spacer region; pg, 'pseudogene' with sequences similar to those of the 5S gene (function unknown).

(b) Repeat unit of rDNA. (Modified from P. K. Wellauer *et al.* 1976.) S_1, initial spacer (transcribed); 18S, 18S gene; S_2, intergene spacer (transcribed); 28S, 28S gene; S_3, terminal spacer (not transcribed).

(c) Processing of the 40S rRNA transcript. (Modified from P. K. Wellauer and I. David 1974.)

The precursors of the tRNAs are, like those of 5S rRNA, only slightly longer than the final tRNAs which are small, 4S molecules (molecular weight 25 000 daltons). We know the complete base sequence of many tRNAs; they are made of about 80 nucleotides with short sequences complementary to one another. It was therefore assumed that these complementary sequences were associated in short double helical segments as shown in the 'clover-leaf' scheme (Fig. 4.11) which applies to all known tRNAs. This model emphasizes common features of all tRNAs but does not reflect the tertiary structure of the molecule. The actual three-dimensional structure of tRNA was deduced from X-ray diffraction data on tRNA crystals. In it, the clover leaf is foled upon itself so as to form two helices at right-angles. In addition to the usual A, G, C, and U bases, they have a rather high

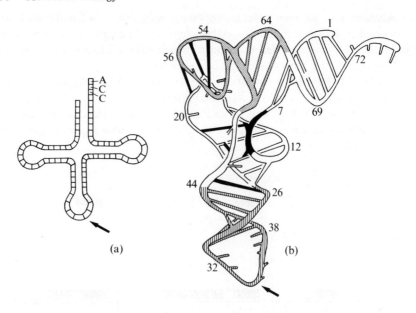

Fig. 4.11. Transfer RNA. (a) Scheme of a clover-shaped tRNA, with its anticodon (arrow) and its terminal CCA sequence.

(b) Three-dimensional structure of the yeast phenylalanine-tRNA; the anticodon is shown by the arrow. Tertiary structure interactions are illustrated by black rods. (From S. H. Kim *et al.* 1974).

content (about 10 per cent) of 'rare' bases which are slightly different from the classical ones. The function of the tRNAs is to bind a given amino acid at one of its ends and a mRNA triplet (codon) at its other end. All tRNA end with a –C–C–A sequence, which is required for amino acid binding; in the middle loop of the molecule, there is a specific sequence of three bases, the *anticodon*. This anticodon binds to a specific triplet (codon) of the mRNA during translation (protein synthesis). In the three-dimensional structure, the anticodon is located at the tip of one of the helices and the amino acid is bound at the other end of the other helix.

Charging a tRNA with an amino acid requires the intervention of enzymes specific for each amino acid. These are the tRNA-amino acid *synthetases*, which catalyse the following reactions (where AA denotes amino acids):

$$ATP + AA \rightarrow AA\!-\!AMP + P$$
$$AA\!-\!AMP + tRNA \rightarrow tRNA\!-\!AA + AMP.$$
(charged tRNA)

For reasons that we shall see later, there can be more than one tRNA for the same amino acid; thus tRNAs, which have a different base sequence but can nevertheless be charged with the same amino acid, are called *iso-accepting* tRNAs. Finally, like the ribosomal and 5S genes, the tRNA genes are present in numerous copies and are thus reiterated (about 20 000 copies in a diploid cell, each tRNA species being reiterated about 200 times).

Transcription is an enzymatic process, catalysed by the *RNA polymerases*. Like the DNA polymerases, they require a DNA template, but, in their case, the building blocks are the ribonucleotide triphosphates (ATP, GTP, CTP, and UTP). The reaction can be written:

$$\text{DNA} + n \begin{cases} \text{ATP*} \\ \text{GTP*} \\ \text{CTP*} \\ \text{UTP*} \end{cases} \to \text{DNA} + \text{*RNA} + n\text{PP}.$$

Bacteria possess a single RNA polymerase, whose molecular structure is well known. In contrast, eukaryotes have three different polymerases: I, II, and III. RNA polymerase II is specialized in the synthesis of mRNAs; it is localized in chromatin and can be specifically inhibited by a substance extracted from a poisonous mushroom, a-amanitin. RNA polymerase I is the enzyme which synthesizes the 28S and 18S rRNAs and is localized in the nucleoli. Finally, RNA polymerase III is specialized in the synthesis of the small RNAs (5S rRNA and the tRNAs). All transcription can be halted by the addition of actinomycin D, a drug which intercalates between the G–C pairs of DNA and thus distorts the double helix. Low concentration of actinomycin D are often used to inhibit preferentially 28S and 18S rRNA synthesis, and it is thought that this synthesis is particularly sensitive to actinomycin D because the ribosomal genes are unusually rich in G–C, the target for actinomycin D.

As has already been mentioned, mitochondria and chloroplasts have their own DNA. Mitochondrial DNA, like bacterial DNA, is circular and is relatively small, with a molecular weight of around 10^6 daltons in vertebrate cells. Chloroplastic DNA can be much larger. Mitochondria and chloroplasts also possess their own RNA polymerase, which is very similar to the bacterial enzyme; like bacterial RNA polymerase, it is inhibited by rifampicin, which has no effect on RNA polymerases I, II, and III. The main products of mitochondrial DNA transcription are the mitochondrial rRNAs, several tRNAs different from those synthesized by the cell nucleus, and a small number of mRNAs. As can be seen, transcription in mitochondria and bacteria has many similarities, and this is why many biologists believe that mitochondria are descendants from symbiotic bacteria which invaded primitive eukaryotes.

Many questions regarding transcription remain unanswered. What is the mechanism for selective transcription of only one of the two DNA strands? How does DNA polymerase find the correct place on the DNA molecule where transcription should be initiated? Is this specificity due to non-histone proteins associated to DNA, or does the enzyme recognize specific DNA sequences which play the role of signals? How does the very precise processing of the rRNA precursor take place? Is it due to the existence of highly specific ribonucleases or to a protection afforded by proteins bound to the precursor? How can large RNA molecules (larger than 10^6 daltons in the case of 28S rRNA) migrate from the nucleus to the cytoplasm through the nuclear membrane pores? We know that the upper limit for the penetration into the nucleus of proteins injected into the cytoplasm is around

70 000 daltons; the large rRNAs must therefore undergo changes in their conformation in order to cross the nuclear membrane barrier. This is a quantitatively important process, since it has been calculated that, in *Xenopus* oocytes, about two molecules of 29S and 18S rRNA molecules move out of each nuclear pore every minute. Since there are about 25 millions pores in the nuclear membrane, it follows that about a million 28S and 18S rRNA molecules leave the nucleus for the cytoplasm every second. This very high rate of rRNA synthesis is made possible by ribosomal gene amplification as well as by accumulation of RNA polymerase I molecules in the 1000 nucleoli present in the nucleus of the oocyte.

4.11 Translation of the genetic message (protein synthesis)

While DNA replication and transcription are the two major functions of the cell nucleus, protein synthesis takes place in the *cytoplasm*. Proteins are built up by complex cell organelles called *polyribosomes* or polysomes which result from an assembly of ribosomes, mRNA and a variety of tRNAs charged with their appropriate amino acids.

Ribosomes are small particles made of approximately equal amounts of rRNAs (28S, 18S, and 5S) and proteins. They are smaller in prokaryotes than in eukaryotes, their sedimentation constants being 70S and 80S respectively. However, in both cases, the ribosomes are made of two *subunits*, which can easily be dissociated from each other (for instance in the absence of magnesium ions). These subunits are of unequal sizes; the large subunit of eukaryotic ribosomes has a sedimentation constant of 60S and contains 28S rRNA; the small subunit has a 40S value and possesses 18S rRNA. Many proteins are associated with the rRNAs in the ribosomes, and these are different in the two subunits. Most of the proteins of eukaryotic ribosomes are very similar to, or even identical with, those present in the nucleoli; and, as we have already seen, the nucleolus is the site for the synthesis of rRNA precursor and for its processing into the ribosomal 28S and 18S RNAs; thus nucleoli and ribosomes have much in common. In *E. coli*, thanks particularly to the work of Nomura [55], we know much more about the nature and arrangement of the ribosomal proteins. Isolated and purified ribosomal proteins can, in the presence of the rRNAs, be assembled *in vitro* into complete, functional ribosomes.

The integrity of the rRNAs is required for ribosomal activity, but we know very little about the precise role of these nucleic acids, despite the fact that they represent 80–90 per cent of total cellular RNA. That ribosomal proteins are also important for proper functioning of the ribosomes has been elegantly demonstrated, for bacterial ribosomes, by experiments where a single ribosomal protein has been modified by treatment with the antibiotic streptomycin. As a result, all protein synthesis stops in bacteria (but not in eukaryotes, where the ribosomal proteins are different from those of the prokaryotes). Free ribosomes, which are not engaged in protein synthesis, and free ribosomal subunits are found in the cytoplasm. In eukaryotes, polyribosomes can be either free in the cytoplasm or attached to the endoplasmic reticulum membranes. Such membrane-bound polyribosomes are specialized in the synthesis of proteins which will be exported out of the cell, e.g. pancreatic enzymes.

Fig. 4.12 depicts schematically the structure of a polyribosome. A number of ribosomes are linked together by a thread of mRNA which is bound to the smaller

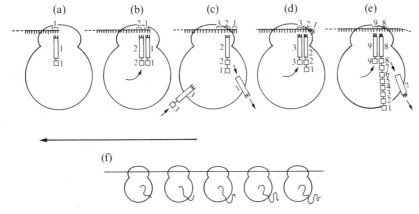

Fig. 4.12. Diagrammatic representation of the main steps of protein synthesis on a ribosome. (a) The first mRNA codon is recognized by the first tRNA (rectangle 1) charged with the first aminoacid, *N* formylmethionine (square 1).

ribosomal subunit; the number of ribosomes associated in a polyribosome and the size of the protein which will be synthesized depend upon the size of this mRNA thread. A charged tRNA molecule attaches to the ribosome, and with its anticodon loop, binds with the complementary sequence (codon) of the mRNA. As in DNA, this binding is due to base pairing. For instance, if the mRNA contains a AAA triplet, only the TRNA possessing the complementary UUU anticodon will bind to it. The underline to Fig. 4.12 describes the main steps of the very complicated process of protein synthesis in a polyribosome. The polypeptide elongates stepwide by successive additions of individual amino acids in the order specified by the message; the growing polypeptide is never released before it is completed; at each elongation step, it remains bound by its carboxyl end to the tRNA of the last amino acid introduced. There are two sites occupied by tRNAs on the ribosomes: site A and site P.

The beginning of a polypeptide chain is coded by 'initiator codons' AUG or GUG which specify formylmethionine. Initiator tRNA carrying formylmethionine binds to the RNA-ribosome complex at site P. The tRNA carrying the next amino acid binds at site A. The peptide bond then forms, and the resulting dipeptide is thus bound to the tRNA which is still at site A. 'Translocation' now takes place; it involves rejection of the unloaded initiator RNA from site P, the transfer of the tRNA carrying the dipeptide from site A to site P while the messenger moves by three nucleotides, bringing the codon of the next amino acid at site A; the corresponding charged tRNA binds, the peptide bond forms, translocation occurs, etc. This process continues until the codon brought to site A is either UAA, UAG, or UGA, which means that the polypeptide is completed. The bond between the polypeptide and the tRNA of the last amino acid breaks, and the finished poly-peptide chain is released (termination).

By what mechanism is the right positioning of the amino acids in the growing polypeptide chain ensured? No errors can be allowed in the amino acid sequence of a protein; a protein with an ACB sequence instead of ABC would be worthless. In order to understand how polyribosomes are capable of synthesizing *specific* proteins, which have a unique and correct amino acid sequence, more must be said about the mRNAs. This amazing specificity in the positioning of the amino acids (remember that there are between 10 000 and 15 000 different mRNAs and proteins in an egg cell) is due to the existence of the *genetic code*. We have already seen that genetic analysis has demonstrated the existence of triplets (codons) in the DNA molecules. The study of frame-shift mutations due to intercalation or deletion of a single base has shown us that a given amino acid corresponds to a sequence of three nucleotides (triplet). These genetic studies have led to the idea that the information present in a structural gene, and needed to specify the synthesis of the corresponding protein, is 'encoded' in the DNA molecules in the form of a succession of triplets. During transcription, the successive triplets which form the genetic code are copied into mRNA. Thus a given mRNA which is specific for a certain protein has the complementary base sequence of the DNA on which it has been copied, except of course that the base thymine (T) of DNA is replaced by uracil (U), which differs only by one methyl group, in RNA.

One of the most outstanding successes of molecular biology is to have completely deciphered the genetic code. This has been achieved thanks to biochemical work done on artificial polynucleotides by Nirenberg, Ochoa, and others [38, 39]. The first experiments dealt with a synthetic, completely repetitive polynucleotide made only of uridylic acid (poly-U). If this poly-U was added to ribosomes, with ATP as an energy source and a mixture of all natural amino acids, a single 'protein' was synthesized. This was a repetitive polymer of the same amino acid, phenylalanine (polyphenylalanine). Thus poly-U codes for polyphenylalanine; this means that the triplet UUU is the codon for phenylalanine. Similar experiments with polyadenylic acid have led to the synthesis *in vitro* of another monotonous polypeptide, poly-lysine; AAA is thus the codeword for lysine. Since codons are made of three nucleotides, it follows that there are 4^3 (64) possible different codons. Utilization of artificial polynucleotides containing two or three bases in a defined order has allowed the complete elucidation of the code (Table 4.1). Later work has shown that the genetic code is the same for all living organisms and is thus universal.

Examination of the code letters shows that the genetic code is *degenerate*; for instance, UUA, UUG, CUU, and CUC are different codons for the same amino acid, leucine. This explains the existence of the iso-accepting tRNAs previously mentioned. Four codons out of the 64 do not code for any of the amino acids present in proteins.

Genetic analysis of certain phage mutants has demonstrated the existence of so-called 'nonsense' mutations, and has greatly helped in the identification of the four mysterious codons; one of them, AUG is the *initiation codon*. The ribosomes will not 'read' the nucleotide sequence of the mRNA until they meet the first AUG

Table 4.1 *The genetic code*

5'-OH terminal base	Middle base				3'-OH terminal base
	U	C	A	G	
U	Phe	Ser	Tyr	Cys	U
	Phe	Ser	Tyr	Cys	C
	Leu	Ser	CTS	CTS	A
	Leu	Ser	CTS	Trp	G
C	Leu	Pro	His	Arg	U
	Leu	Pro	His	Arg	C
	Leu	Pro	Gln	Arg	A
	Leu	Pro	Gln	Arg	G
A	Ile	Thr	Asn	Ser	U
	Ile	Thr	Asn	Ser	C
	Ile	Thr	Lys	Arg	A
	Met†	Thr	Lys	Arg	G
G	Val	Ala	Asp	Gly	U
	Val	Ala	Asp	Gly	C
	Val	Ala	Glu	Gly	A
	Val¹	Ala	Glu	Gly	G

¹ Chain initiation (see text)

CTS, chain termination signals

Ala, alanine	Glu, glutamic acid	Lys, lysine	Thr, threonine
Arg, argine	Gly, glycine	Met, methione	Trp, tryptamine
Asp, asparagine	His, histidine	Phe, phenylalanine	Tyr, tyrosine
Cys, cysteine	Ile, isoleucine	Pro, proline	Val, valine
Glu, glutamine	Leu, leucine	Ser, serine	

codon. The initiation codon AUG corresponds to a derivative of the aminoacid methionine, *N*-formylmethionine. Thus the first tRNA (a in Fig. 4.12) is formyl-methionine–tRNA. The three other nonsense codons are *termination codons*. Elongation of the polypeptide chain stops when the ribosome meets one of these codons; then the polyribosome falls to pieces because the ribosomes 'run off'. Initiation and termination codons allow the synthesis of proteins which have the right size; a protein with one extra amino acid or with one too few would be abnormal and physiologically useless.

In summary, protein synthesis is the reading of the message encoded in the base sequence of a mRNA; it can be compared to listening to a message (music, sentences) recorded on a magnetic tape.

mRNA plays a key role in the specificity of protein synthesis and is at present the subject of extensive studies. We know that the mRNAs are associated with a small number of proteins in the polyribosomes; the role of these proteins remains obscure, but it is probable that they protect the mRNA against fast degradation by the cytoplasmic ribonucleases. We have already seen that almost all mRNAs end with a poly-A sequence at their 3' end (with the important exception of the histone messengers). At the other end of the coding sequence (the 5' end), mRNAs have a structure called the 'cap'; this is a particular base sequence characterized by the presence of a methylated purine base (generally a methylguanine). Removal of this cap prevents the binding of the mRNA to the ribosomes, and the possibility of polysome formation. The nucleotide sequence of an increasing number of viral and eukaryotic mRNAs is now being analysed in various laboratories [56, 57] and it seems to be a general fact that mRNAs possess, ahead of their first AUG initiation site, a nucleotide sequence of variable length which has no coding significance. However, it could be that part of this non-coding sequence adjacent to the cap might be required for binding the mRNA to the 40S subunit of a first ribosome.

This binding leads to the formation of an *initiation complex* consisting of formylmethionine–tRNA, the AUG codon of mRNA, and the 40S subunit of a ribosome. The formation of this ternary complex is far from simple. It requires, among other things, the presence of GTP (which is hydrolysed to GMP: guanosine monophosphate) and of several proteins called *initiation factors*, loosely associated with the ribosomes.

Other proteins, which are also loosely attached to the ribosomes, are required for the elongation of the protein chain and for its termination (*elongation and termination factors*). A discussion of the role played by these factors in protein synthesis would draw us too far. Finally, the linking together of the amino acids by peptide bonds requires energy, which is provided by ATP generated in the mitochondria.

When protein synthesis comes to an end, the finished polypeptide chain is released into the cytoplasm, where it looses its initial formylmethionine and undergoes the conformational changes which have been discussed in § 4.7. Many proteins undergo further 'post-translational' changes. Addition of sugars, in the Golgi bodies, will transform some of them into glycoproteins, which as we have seen are important constituents of the cell membrane and the intercellular matrix; other proteins receive phosphate groups from ATP and are converted into phosphoproteins by enzymes called proteinkinases.

We have at our disposal many *inhibitors* of protein synthesis, which either prevent the initiation or the elongation of the polypeptide chain. Those particularly used for eukaryotes are *puromycin* and *cycloheximide*; *chloramphenicol* arrests protein synthesis in bacteria.

As we have seen, mitochondria and chloroplasts synthesize their own specific rRNAs. They have their own ribosomes, which are smaller than 80S ribosomes present in eukaryotic cytoplasm (70S for chloroplastic and 55S for mitochondrial ribosomes); and since isolated mitochondria or chloroplasts can synthesize proteins, these ribosomes are capable of protein synthesis. However, the major part of the proteins present in these organelles is synthesized by cytoplasmic polyribosomes and then migrates into them. It has been shown that chloramphenicol inhibits both mitochondrial and chloroplastic protein synthesis, and this is another argument in favour of a symbiotic origin of these two organelles. The utilization of specific inhibitors like cycloheximide and chloramphenicol, as well as the analysis of mutations which affect mitochondrial or chloroplastic DNA, has proved the basis for the conclusion that mitochondria and chloroplasts have only a restricted degree of autonomy; it is certain that they largely depend on cytoplasmic protein synthesis directed by the nuclear genes for the maintenance of their own proteins and their functional activity.

4.12 Gene regulation in prokaryotes and viruses

All proteins are not synthesized at a constant rate. This rate is, in general, dependent on the rate of mRNA production by the corresponding gene and is thus controlled at the transcriptional level. Extensive work was done on the *induction* of enzymes in *E. coli* by J. Monod [25]. He discovered that the addition to bacteria of a substrate that they can utilize leads to the synthesis of a battery of enzymes involved in the utilization of this substrate. These findings have led to the already mentioned concept of the operon as a battery of closely linked structural genes which are simultaneously switched on or off. Enzyme synthesis can also be repressed, instead of induced, in bacteria. Repression takes place when the end product of the enzymatic reaction accumulates in the medium. For instance, the synthesis (and not only the activity) of bacterial phosphatase is inhibited when inorganic phosphate, the end-product of the enzymatic reaction, is added to the culture medium.

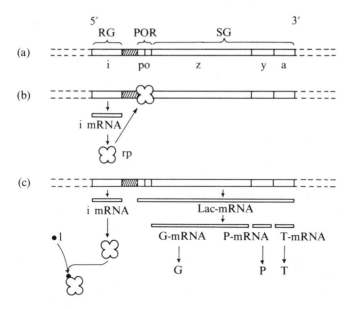

Fig. 4.13. The Lac operon of *Escherichia coli*. (After F. Jacob and J. Monod.) (1) Schematic representation of the Lac region: the i regulatory gene (RG); the promoter–operator region (POR); and the structural genes (SG) z, y, and a.

(b) In the absence of lactose, the repressor protein (rp) is fixed on the PO region: the structural genes are not transcribed.

(c) In the presence of lactose (l), the repressor protein combines with it and undergoes allosteric modification. The structural genes are then transcribed into a large Lac-messenger RNA which is processed into galactosidase- (G-mRNA), permease- (P-mRNA), and transacetylase-messenger (T-mRNA) RNAs. These messengers are then translated into the enzyme (G, P, T).

How can a whole operon be switched on and off when the substrate or the end product of the reaction is added to the bacterial culture? This has been explained by Jacob and Monod. Their model is shown in Fig. 4.13, again has the simplicity of genius.

In the Jacob–Monod model, the whole operon is controlled by an *operator* gene which is adjacent to the structural genes forming the operon; this operator is itself controlled by a *regulatory* gene, which lies at some distance from the operon. The regulatory gene produces a protein, the *repressor*, which can undergo conformational changes (by allostery) in the presence of either the substrate or the end-product of the enzymatic reaction controlled by the operon. If the repressor binds to the operator, the whole operon will stop working and the enzymes will no longer be synthesized by the structural genes. If the repressor is modified in such a way that it cannot bind to the operator gene, the operon will direct the synthesis of the whole array of enzymes. The reality of this *negative control* has been completely demonstrated; repressors for various operons have been isolated and it has been demonstrated that they are indeed proteins which have a very great affinity for the DNA of the operator.

Further analysis of gene control in phage-infected bacteria has shown that, besides this negative control, there are also *positive* controls [58]: the products of certain genes exert an activating effect on other genes. Gene regulation, even in the case of a simple virus like the phage of *E. coli* (its DNA has a molecular weight of only 3×10^7) is a very complicated circuit of positive and negative interactions between various genes; in fact, this control is so intricate that the only way to express it correctly is by a mathematical model which requires the assistance of computers.

4.13 Present work on the molecular biology of eukaryotes

It would be a great mistake to believe that molecular biologists are no longer studying prokaryotes and viruses. The *Journal of Molecular Biology*, which has specialized in this field, is as active as ever and our knowledge is still steadily increasing. However, having always worked with eukaryotes, I hope that I shall be forgiven if the last part of this review is devoted to the fields of research which are closer to my own interests.

A major problem is the understanding of gene regulation in eukaryotes. This must be of tremendous complexity because, as we have just seen, gene regulation in the very simple and well-studied phage is so difficult to grasp that the use of mathematical models is becoming a necessity. The DNA content of this phage is almost negligible when compared to that of eukaryotic cells, which, as we know, possesses enough information for the production of 10 000–15 000 different proteins. There are too many differences between eukaryotes and prokaryotes to use a relatively simple model, such as the Jacob–Monod theory, without increasing its complexity. In particular, the proteins which are associated with DNA (histones and non-histone proteins) in eukaryotic chromatin must certainly play an important role in gene regulation; such proteins, if present at all, are of minor importance for the functioning of the *E. coli* chromosome. As already pointed out, another important difference is that very few cases of authentic operons are known in eukaryotes. The most convincing case is that of the histone operon, and it seems probable that the synthesis of all the five histones lies under the control of the same regulatory genes (Fig. 4.14, p. 121).

Finally, enzyme synthesis is not induced, in eukaryotic cells, by the simple addition of the substrate for this enzyme, although it can be induced, in appropriate target cells, by the addition of hormones. We consider below what is known today about the mechanisms of hormonal stimulation.

A number of models have been proposed to explain gene regulation in eukaryotes [59, 60, 61]. The most elaborate is that of E. Davidson and R. Britten [62], who have attempted to explain cell differentiation (which results from selective gene activation) by a model derived from the Jacob and Monod theory. In this hypothesis, in addition to the structural genes, the existence of three different kinds of regulatory genes is postulated. The complexity of this interesting model precludes its discussion here. Its main weakness lies in the fact that because of its refined complexity it cannot yet be tested experimentally.

Only some of the prevalent lines of approach towards the understanding of the molecular bases of eukaryotic cell functioning will be presented here.

(a) *Organization of the eukaryotic genome*

All the models so far proposed for explaining gene regulation in eukaryotes are derived from the Jacob and Monod model for bacteria and phages; they assume, in general, that the coding sequences are adjacent to regulatory sequences in the DNA molecule. This assumption has been experimentally tested by Davidson and Britten in the following way. First, the huge DNA molecules were broken down into small pieces (of molecular weight around 300 000); DNA was then 'denatured'. This means that the two helices were separated by heating and the separated double helices were finally allowed to reassociate during slow cooling (annealing). The kinetics, as well as the theory of DNA denaturation and reassociation, have been carefully worked out. The analysis of DNA prepared from a number of eukaryotes has been performed with this method and, it has been found that DNA is not a homogeneous molecule. It contains three main types of base sequences: *highly repetitive* sequences which reassociate almost immediately after separation of the two strands; *middle-repetitive* sequences which reassociate less quickly, and finally, *unique* or 'single-copy' sequences which reassociate very slowly or not at all.

The highly repetitive sequences, which are very monotonous segments of the DNA molecules, correspond to the already mentioned satellite DNAs. They are found in the centromeres (kinetochores) and in some parts of the chromosomes; they correspond to the heterochromatin of the cytologists, which contains very few or no structural genes. It is believed that the main function of the highly repetitive DNA sequences is to maintain chromatin in the condensed state characteristic of heterochromatin. However, DNA also contains highly repetitive sequences which are not clustered in heterochromatin, but distributed all over the genome. Since they renature almost immediately after denaturation, they have been called 'snap-back' or 'hairpin' sequences; their role is not known for certain, but it is often assumed that they might be signals in the DNA molecule indicating where transcription should begin.

As shown by Davidson and Britten, middle-repetitive and single-copy sequences are regularly interspersed in the genomes of all eukaryotes. In the great majority of the eukaryotes, the short repetitive sequences have about 300 base pairs, while the single-copy sequences are much longer (800 to several thousands of base pairs). DNA is organized in a different way in *Drosophila* and a few other insects; the two kinds of sequences are also interspersed but the length of the middle-repetitive sequences (3000 base pairs) is much longer. The reason for this difference between *Drosophila* and the vast majority of animal species remains completely unknown.

It is tempting to speculate that single-copy DNA corresponds to structural genes, while the adjacent middle-repetitive sequences would play a regulatory role. However, such a simple explanation does not fit with the facts.

It is possible to make hybrids between DNA and RNA provided that their base sequences are complementary (as is the case for the DNA of a structural gene and the mRNA which has been transcribed on this gene). Such hybridization experiments have shown that only 1 to 5 per cent of the single-copy DNA hybdridizes with cytoplasmic RNA. Thus, 95 per cent or more of single-copy DNA does not correspond to structural genes, since it is never transcribed into RNA. It might be that the single-copy DNA which is not transcribed plays a role in the regulation of the activity of the structural genes; the latter thus represent less than 5 per cent of total single-copy DNA. The middle-repetitive sequences, like the highly repetitive ones, might be mainly concerned with the maintenance of chromatin structure. We do not know the answer to this problem yet, but it can be expected that further work will solve this paradox.

There is another paradox, which points towards the general conclusion that the great majority of DNA is genetically inactive; this is the so-called *C-paradox* (*C*, as we have seen, corresponds to the chromosome complement, or DNA content of a haploid cell). It is well established that cells (including sperm or eggs) of some salamanders contain 10 times more DNA than those of frogs. Since it is absurd that a salamander should contain 10 times more genes than a frog, from the viewpoint of a geneticist, the vast majority of eukaryotic DNA is thus just 'junk'. However, it would be premature to despair. It has recently been shown that the haemoglobin genes (which are bona fide structural genes) are flanked by middle repetitive DNA sequences; whether the latter play a regulatory role in the expression of the haemoglobin genes remains to be seen.

There are other strange facts not yet understood; for instance, E. Davidson has shown that some of the middle-repetitive sequences of DNA are transcribed into RNA [53]. The products of this transcription (the 'transcripts') can be found in the cytoplasm of unfertilized eggs, but what is more unexpected is the fact that they are still detectable in the embryo. What could be the biological significance of cytoplasmic RNAs transcribed on repetitive genes? Since we do not know and because this is the fashionable hypothesis today, a role in gene regulation has been suggested.

Finally, it is worth repeating that the complexity of the mRNAs present in the polyribosomes of sea urchin eggs is sufficient to code for 10 000 to 15 000 proteins. The complexity of the hnRNA is still 10 times greater and most of it is quickly degraded in the nucleus. Why does the cell use so much energy for the synthesis of a large amount of RNA which will immediately be destroyed?

This appears to be a tremendous waste of energy. It is analogous to our modern consumption-orientated society. Evolution should have taken care of this problem long ago and it still awaits a satisfactory solution.

(b) *Isolation of pure genes*

We have already mentioned the fact that ribosomal and 5S genes have been isolated by D. Brown [54] and the structure of these genes has been shown in § 4.9. Similar work has been done on the histone genes by Birnstiel, Kedes, and others [64, 65, 66], who have shown that the respective position of the H_1, H_2a, H_2b, H_3, and H_4 genes in the histone operon is as in Fig. 4.14.

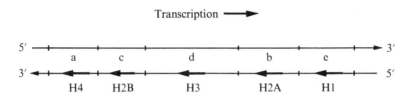

Fig. 4.14. Representation of arrangement and polarity of transcription of the histone gene cluster. (From Gross *et al.* 1976.)

How has it been possible to isolate and dissect these genes in order to establish their molecular anatomy? Isolation has been facilitated by the fact that they are reiterated genes and thus exist in many copies. Today, more and more work is being devoted to the isolation of single-copy genes (coding for haemoglobin or ovalbumin, for instance) and very encouraging progress is being made, despite the difficulty of finding a single structural gene in the DNA 'junk' – the proverbial pin in a haystack! Analysis of the fine structure of the gene is a product of 'genetic engineering' and this will be discussed in more detail at the end of this chapter. Briefly, the isolated gene is introduced (inserted) into the DNA of a phage or a virus-like particle called a *plasmid* which can integrate its DNA into the chromosome of a bacterium such as *E. coli*. The inserted foreign gene will then be replicated when the bacterial culture grows and can be isolated in large amounts as a result of its amplification during bacterial growth. The isolated purified genes are finally submitted to the action of highly specific nucleases (restriction enzymes) which cut DNA at precise and known positions in the polynucleotide chain, and the specific fragments thus obtained can be analysed by standard biochemical methods.

A promising system for the accurate study of the *transcription* of a purified gene has been recently discovered by D. Brown and J. B. Gurdon and has been applied to the study of 5S and histone genes [67]; it consists of injecting pure DNA into the nucleus of a large oocyte (ovarian egg) of the frog *Xenopus*. This nucleus contains very large amounts of RNA polymerases (2×10^8 molecules of RNA polymerase II, which transcribes 15 nucleotides per second) and it has been possible to show that transcription of the injected pure 5S gene is absolutely correct.

These two approaches (analysis of genome organization, isolation of pure genes), together with the comparison of the gene protein (for instance, haemoglobin) in a variety of animal species, are very important tools for our understanding of the molecular mechanisms of evolution. Evolution of the proteins almost certainly results from the duplication of an ancestral gene, followed by mutations. The study of the ribosomal and 5S genes has shown that these genes have been fairly well conserved during evolution. In contrast, the spacers which link these genes together are extremely variable; their length and base sequences can differ from one individual to another and even, in the case of the ribosomal genes, from one nucleolar organizer to another. Thus in contrast to the genes, the spacers have widely diverged during evolution. This is also true for mitochondrial DNA which also contains spacers. It is possible to study, by hybridization (crosses) between two related species, the hereditary transmission of the ribosomal and mitochondrial DNAs. These experiments [68, 69] have demonstrated that the ribosomal genes are transmitted by the chromosomes according to Mendelian heredity, whereas mitochondrial DNA is transmitted by the cytoplasms since there are no copies of the mitochondrial genes in chromatin.

For the cell biologist, it is important to know the localization of the genes on the chromosomes; this can be established, in the case of reiterated genes (28S, 18S, 5S, histones), by hybridization *in situ*. The principle of this technique is as follows: first DNA is denatured (by heating or alkali treatment) in the cell after approriate fixation in order to maintain the structure of chromatin; secondly, the cell is treated with purified radioactive RNA (for instance, radioactive rRNA), which will hybridize specifically with the corresponding genes (in the case of rRNA, the nucleolar organizers); and finally the radioactive hybrid is detected by a photographic process (autoradiography). This procedure becomes still more powerful when it is applied at the electron microscope level.

Finally, for the molecular biologist, there is an intriguing possibility for the future: the complete synthesis of a gene, using the methodology of organic chemistry. The way towards this ambitious goal has already been paved by G. Khorana and a large team of co-workers, who have recently succeeded in synthesizing a tRNA gene [70]. Since tRNA is a small molecule, the corresponding tRNA gene is also small and this is why this *tour de force* has been possible; however, there is no doubt that other, larger genes, will be synthesized *de novo* in the future.

(c) *Control of transcription. The mechanisms of hormonal stimulation*
The control of transcription, and in particular that of mRNA synthesis, is of paramount importance to our understanding of cell differentiation. For example, all of our cells contain the haemoglobin genes; but these genes are expressed only in the red blood cell line. In all our other cells, the gene is repressed and there is no (or perhaps exceedingly little) haemoglobin synthesis. We still know very little about the control of gene expression, but it seems certain that it has nothing to do with

the organization of chromatin in the nucleosomes, since both the genetically active euchromatin and the inactive heterochromatin are made of nucleosomes. This suggests that the histones associated in nucleosomes do not play an important role in the specific repression or derepression (activation) of the structural genes.

Molecular hybridization is a very powerful tool if one wishes to know how many copies of a given structural gene and its specific mRNA are present in a cell; this is easy to discover provided that one has first succeeded in isolating the mRNA in a purified form. Since it is possible to isolate haemoglobin mRNA and to make it radioactive, it is an easy matter to calculate the number of haemoglobin genes present in DNA: all one has to do is to hybridize this mRNA to total DNA and to measure by appropriate techniques the radioactivity of the hybrid. One can also transform haemoglobin mRNA into its complementary radioactive DNA copy (cDNA) with an enzyme present in RNA-containing viruses, reverse transcriptase, which synthesizes DNA on a RNA template. Radioactive haemoglobin cDNA will hybridize specifically with haemoglobin mRNA; measurement of the radioactivity will allow the experimenter to titrate the number of haemoglobin mRNA molecules present in any kind of cell.

These techniques have disclosed some important facts: structural genes (the haemoglobin genes, for instance) exist in only a small number of copies (one or two) per haploid genome, and differentiation of haemoglobin-producing red blood cells is not due to a selective amplification of the haemoglobin genes; these cells have the same low number of haemoglobin genes as all the other cells. Thus red blood cell differentiation and haemoglobin synthesis are due to the increased *transcription* of the haemoglobin genes which produce many more molecules of haemoglobin mRNA in the future red blood cells than in any other cell. Increased transcription of the haemoglobin genes during red blood cell differentiation has been firmly established by the cDNA-haemoglobin-mRNA hybridization.

Another very interesting biological system, the synthesis of the egg white proteins (in particular ovalbumin) in the hen's oviduct, is being studied in detail by O'Malley [71]. The synthesis of these proteins can be induced by the injection of oestrogens into a non-egg-laying hen. These steroid hormones act in the following way: initially they cross the plasma membrane and penetrate into the cytoplasm, where they bind to a specific protein receptor which has a very high affinity for the hormone. The oestrogen–receptor complex undergoes certain modifications in the cytoplasm and then moves into the nucleus, where it binds to the chromatin, and, by a mechanism which is still unclear, specifically derepresses (activates) the ovalbumin genes. The result is the synthesis of ovalbumin mRNA in the nucleus. This mRNA moves out of the nucleus and combines with cytoplasmic ribosomes to form polyribosomes; the latter finally produce large amounts of ovalbumin. Thus ovalbumin synthesis is controlled at the transcriptional level; this is proven by the fact that actinomycin D which inhibits transcription, completely suppresses ovalbumin synthesis in response to oestrogen stimulation. Utilization of the DNA-RNA hybridization techniques has given us quantitative data on this process. They have shown that the hen possesses only one copy of the ovalbumin gene per haploid genome and that the number of these genes (two copies per nucleus) is the same in the oviduct and in all the other cells of the hen. The other cells do not respond to the hormone because they lack the structural gene for ovalbumin. It has been calculated that in the hen oviduct the steroid hormone derepresses, in 30 minutes, the two copies of the ovalbumin genes. They synthesize 12 molecules of ovalbumin mRNA every minute and each of these mRNA molecules is translated 50 000 times before it is degraded. The final result is that, in the fully stimulated oviduct, ovalbumin represents 70 per cent of the total proteins synthesized. It has also been shown that the hen oviduct cells contain a population of about 13 000 different species of mRNA molecules and that ovalbumin mRNA represents only 0.02 per cent of this total mRNA population in the unstimulated oviduct, but as much as 69 per cent of the total mRNA after oestrogen treatment.

The steroid hormone has thus induced a specific depression of a single structural gene. The mechanism of this specific gene activation, which, as we shall see, is of fundamental importance for the understanding of cell differentiation, is presently the subject of many investigations. Particularly promising, but not yet entirely convincing, is the work done on *reconstituted chromatin* by Paul and Gilmour in the haemoglobin synthesis system [72], and by O'Malley on ovalbumin synthesis [71]. It is possible to separate from each other the various constituents of chromatin (DNA, histones, non-histone proteins) and to mix them together at will. It is also possible to transcribe DNA or chromatin (complete or partially reconstituted) *in vitro* by adding RNA polymerase and the required precursors (nucleotide triphosphates) for RNA synthesis; the presence in this system *in vitro* of specific mRNAs for haemoglobin or for ovalbumin can be detected by molecular hybridization with the appropriate radioactive cDNAs. Such experiments have shown that the addition of non-histone nuclear proteins from haemoglobin producing cells to a DNA-histone mixture is followed by the synthesis of haemoglobin *in vitro*, but that the addition of non-histone proteins from brain or liver is ineffective in making the haemoglobin gene available for transcription. Similar results have been obtained for the ovalbumin gene transcription. Thus it seems that the factor responsible for specific gene activation is a non-histone protein; however, the experiments on reconstituted chromatin are still open to criticism for technical reasons. It would thus be hazardous to dismiss lightly the possibility that nuclear RNAs (hnRNA, nuclear RNAs of low molecular weight) are also involved in the specific control of gene activity in eukaryotes.

Certain cells, such as the salivary glands of certain insects or the amphibian oocyte, constitute a favourable material for the study of transcription because of the large size of their chromosomes. Salivary glands have giant chromosomes, made of many DNA strands (polyteny) which have been formed by repeated DNA replication without cell division (endomitosis). These giant chromosomes have a banded structure; the bands or, more exactly, a small part of each band, correspond to a gene. The treatment of the larvae with a hormone called ecdysone induces the uncoiling and swelling of a small number of the bands; these hormone-induced *puffs* are the site of extensive RNA and protein synthesis and have been studied by Edström using micromethods [73]. Puff formation is inhibited by actinomycin D and is thus the morphologically visible result of selective gene activation.

Growing oocytes (ovarian eggs) are the site of considerable RNA and protein synthesis; their large chromosomes have a peculiar structure and are made of a string of DNA-containing beads (the *chromomeres*) from which loops extend into the nuclear sap. These loops possess a DNA axis and are exceedingly active in transcription, owing to the very high content of RNA polymerase II in the nucleus of the oocyte. The chromosomes of the oocyte, which have the 4*C* value, are called *lampbrush* chromosomes because of their particular structure (an axis made of chromomeres and projecting loops). There are about 20 000 loops; however, since their initial transcripts are very large (2×10^7 to 4×10^7 nucleotides) RNA molecules, each loop is not a structural gene, but a more complex transcriptional unit. Recent analysis has shown that although only 5 per cent of the DNA present in the loops is transcribed into RNA, the synthetic activity is very high: 6×10^8 nucleotides per second, or 20 RNA molecules per minute. The giant product of DNA loop transcription is very complex and unstable; its average half-life does not exceed 30 minutes, despite the fact that the newly synthesized RNA immediately binds to proteins of the nuclear sap in order to form ribonucleoprotein (RNP) particles. It is likely that the more stable of these RNP particles move into the cytoplasm, where they will be stored as maternal mRNAs; however, some of the transcripts arise from repetitive DNA sequences present in the loop and might play a role in gene regulation. As one can again see, 95 per cent of the DNA is not transcribed into RNA and is believed to be responsible for the maintenance of the lampbrush structure of the chromosomes, but this remains hypothetical.

(d) *Protein synthesis in the absence of the nucleus. Post-transcriptional controls*

The best way to completely suppress transcription of the nuclear genes is to remove the cell nucleus; this can be done by microsurgery in the case of large unicellular organisms (amoebae, the alga *Acetabularia*) or eggs, and by chemical means for ordinary cells. The treatment of many cells with the drug cytochalasin B (which inhibits the assembly and contractility of actin filaments) leads to the extrusion of the nucleus with a small amount of cytoplasm; this small nucleated fragment is called a *karyoplast*, while the remaining cytoplasm is a *cytoplast*. Loss of the nucleus also takes place when the mother cells of our red blood cells differentiate into the so-called reticulocytes (immature red blood cells). The latter are anucleate, but contain polyribosomes capable of haemoglobin synthesis.

The survival time of enucleated cytoplasm is highly variable: cytoplasts die within 12-20 hours, anucleate fragments of the giant unicellular alga *Acetabularia* can survive for as long as 3 months. This variation in the life-span of enucleated cytoplasm depends, as we and others have shown, on the stability of the mRNAs of nuclear origin that they contained at the time of enucleation [74]. Survival is long when, as in eggs or *Acetabularia*, the pre-existing mRNAs are protected against cytoplasmic degrading enzymes by combination with proteins in RNP particles.

Enucleation, of course, suppresses the outflow of all kinds of RNAs from the nucleus and the RNA content, except in the case of the alga *Acetabularia*, which has many chloroplasts and is discussed below, decreases as the result of enucleation. In contrast, protein synthesis continues in all cases; in fact, reticulocytes constitute the ideal material for the study of haemoglobin synthesis directed by haemoglobin mRNA which has accumulated in the polyribosomes before the loss of the nucleus.

Anucleate fragments of eggs can respond to external stimuli by increasing the rate of protein synthesis: for instance, if anucleate fragments of sea urchin eggs are treated with hypertonic sea-water, their protein synthesis increases many times within a few minutes [75]. Among the newly synthesized proteins are tubulin and histones; this means that the anucleate fragments contained in a masked, inactive form the mRNAs for these specific proteins. Treatment with hypertonic sea-water thus induces the unmasking of the inactive mRNAs. The same kind of *translational control* exists in amphibian oocytes: treatment with the hormone progesterone induces the synthesis of many different proteins, whether the nucleus is present or has been removed. The mechanism of action of progesterone, in this case, is clearly very different from the one that has been outlined for the oestrogen–hen-oviduct system in the preceding section.

The case of the giant alga *Acetabularia* deserves special mention. As first shown by Hämmerling [76], an anucleate fragment of the alga not only survives for a long time, but is even capable of regenerating a complex structure called the 'cap' or umbrella. Biochemical analysis of the anucleate fragments has disclosed extensive DNA, RNA, and protein synthesis [77]. The unexpected DNA and RNA syntheses which take place in the absence of the nucleus were understood when we found that *Acetabularia* chloroplasts contain both DNA and RNA, and that they can increase in number in anucleate fragments of the algae. Recent work by H. G. Schweiger has shown that chloroplasts autonomy is far from complete [78], since the synthesis of several chloroplastic proteins is under nuclear control.

As one would expect, the synthesis of cytoplasmic rRNAs and mRNAs occurs only in nucleate halves. The remarkable regenerative capacity of anucleate fragments of the alga is the result of a storage, in a very stable form, of mRNAs previously synthesized in the nucleus and transferred to the cytoplasm.

We still know very little about the biochemical mechanisms of the post-transcriptional control of protein synthesis. It is usually believed that the most important of the control mechanisms is the phosphorylation of the initiation factors required for the initial step of protein synthesis. Phosphorylation or dephosphorylation of such initiation factors by protein kinases and phosphatases respectively might inactivate or activate them and thus switch on or off the whole machinery for protein synthesis.

(e) *Present trends in molecular embryology*
Embryonic development — the transformation of an egg into a hen — has always been a fascinating problem for biologists and philosophers. The fertilized egg contains both the paternal and maternal genes required for the fulfilment of the whole developmental programme. Thanks to the elegant experiments of J. B. Gurdon [79], we now know that the initial genes do not undergo any important changes during development; if one enucleates an unfertilized *Xenopus* egg and injects into it a nucleus taken from a fully differentiated adult cell, a normal tadpole can ensue. The adult nucleus thus still contains all the genetic

information needed for development and differentiation into brain, muscles, red blood cells, etc.

Many experiments have proven that the initial period of egg development (repeated cleavage into smaller cells) is a nucleus-independent, purely cytoplasmic process; an anucleate unfertilized egg if activated by physical or chemical means (pricking, treatment with hypertonic sea-water, for instance) can undergo repeated cleavages despite the fact that it has no chromosomes. DNA transcription is thus not required during this early phase of development. In contrast, later development, which involves cell movements (gastrulation, neurulation, and formation of organ rudiments) requires DNA transcription; these morphogenetic processes do not take place if RNA synthesis is inhibited by actinomycin D treatment. Chromosomal gene transcription thus acquires more and more importance as development proceeds and leads to embryonic differentiation.

These biological findings can be explained at the molecular level. As has already been mentioned, during its growth in the ovary the oocyte builds up a huge store of ribosomes and mRNAs. The latter are 'masked' in the form of RNP particles in the unfertilized egg and are 'set free' after fertilization. These mRNAs possess the information needed for repeated cleavages but not for later stages of development; fresh transcription must take place when the early cleavage period is over. It should be added that, although there is some mRNA synthesis during cleavage, the newly synthesized messengers are not needed for cleavage itself; they are required for later stages of development.

Many years ago, it was established that eggs of many species are a 'mosaic of germinal localizations'; this means that certain parts of the egg give rise to a given organ or tissue. For instance, at the eight-cell stage of an ascidian egg, only two cells contain the material necessary for the formation of the muscles of the larva. If one destroys these two cells by pricking, a larva will form, but it has no muscles and is thus unable to swim. We do not yet know whether these germinal localizations are local accumulations of preformed mRNAs, synthesized during oogenesis and needed for muscle differentiation, or whether the germinal localization for muscle formation contains substances which specifically activate the genes which control muscle formation. This unsolved dilemma is identical, in molecular terms, to the old question of the eighteenth-century philosophers: is embryonic development due to preformation or to epigenesis?

At later stages of embryogenesis, realization of the developmental program results from variable gene activity and, as in the case of the steroid-induced synthesis of ovalbumin in the oviduct, transcriptional controls (by non-histone chromosomal proteins or nuclear RNAs) become more and more important. However, these controls must be exerted in the developing embryo, in an orderly fashion; biochemical activities follow 'gradients'. For instance, the intensity of protein synthesis, in a vertebrate embryo, progressively decreases from the head to the tail and from the dorsal to the ventral parts; the embryo is thus the site of cephalo-caudal (anterio-posterior) and dorso-ventral gradients. Every cell is thus located at a given place in these gradients and is thought to receive 'positional information' from the neighbouring cells; however, much remains to be done before the nature of the signals received (probably at the cell membrane level) by a cell from its neighbours is elucidated.

Another important factor in embryogenesis is *induction*: the nervous system, for instance, does not differentiate spontaneously; its presumptive cells must receive an inducing stimulus from neighbouring cells, which will differentiate in other organs. The same is true for many organs, such as the eye-lens, pancreas, kidney, etc.; they are incapable of self-differentiation unless they receive an inducing stimulus from other cells. This stimulus is certainly chemical in nature and it seems very likely that all of the inducing agents are either proteins or glycoproteins.

Differentiation of embryonic cells can also be studied by cultivating them *in vitro*. Under such conditions, the cells are no longer under the influence of morphogenetic gradients and inducers. For this reason, conclusions drawn from experiments made on isolated cells cannot be extended to the whole developing embryo without caution. However, the fact that undifferentiated embryonic cells can differentiate *in vitro*, according to their embryonic origin, into muscle, cartilage, pigment cells, neurons, red blood cells, etc. provides a powerful tool for the experimental analysis of cell differentiation.

Of particular interest is the fact that it is possible to suppress differentiation *in vitro* by two different methods: treatment with bromodeoxyuridine (BrdUR) and somatic hybridization.

BrdUR is very similar to thymidine, the classical precursor for DNA synthesis; cells treated with BrdUR mistake this 'analogue' for thymidine, and synthesize abnormal BrdUR-containing DNA. If BrdUR has been substituted for thymidine to some extent in the DNA molecules, the cells fail to differentiate; however, the addition of thymidine restores the capacity to differentiate. These experiments clearly show that the integrity of DNA (structural or regulatory genes) and correct transcription are needed for the expression of the differentiated state (synthesis of a so-called 'luxury' protein, for instance haemoglobin during red blood cell differentiation).

Somatic hybridization is the experimentally induced fusion of two different cells. The hybrid cell has, of course, two nuclei which possess different genetic information and is called a *heterokaryon*. If, as was first performed by Ephrussi [80], one fuses together a differentiated cell (a pigmented cell, for instance) with an undifferentiated embryonic cell, the differentiated character is 'extinguished'; the pigment is no longer synthesized and quickly disappears. The same result is obtained if one fuses together a differentiated cell and a cytoplast (i.e. anucleate cytoplasm) prepared from an undifferentiated cell. It thus appears that the cytoplasm of the undifferentiated cell contains a repressor which can block the activity of the genes responsible for cell differentiation (the genes controlling pigment formation in the present example).

Thus cell differentiation, which remains a major enigma for molecular biology, apparently results from a complex interaction between nuclear genes and cytoplasmic factors.

Other factors probably act at the level of the cell membrane; this is the case for the so-called cyclic nucleotides, which are modified forms of two nucleotides present in nucleic acids (adenylic and guanylic acids). Cyclic AMP (cAMP) plays a very important role in the response of the cells to non-steroid hormones (adrenaline, insulin, thyroxine, etc.) and for this reason has been called a 'second messenger'. These hormones bind to receptors present in the cell membrane, which contains enzymes capable of increasing or decreasing the levels of cAMP and of cyclic GMP (cGMP) in the cell. The cyclic nucleotides, in particular cAMP, control the activity of proteinkinases, the enzymes which add phosphate to proteins. Phosphorylation or dephosphorylation of enzymes can strongly affect their activity. The cAMP-cGMP equilibrium thus controls many properties of the cell; in general, an increase in cAMP slows down cell proliferation and favours cell differentiation, while an increase in cGMP has the opposite effects.

(f) *Cancer*

Generally speaking, a malignant cell has lost the capacity to differentiate and has retained a full capacity for proliferation; such a cell has escaped from the control mechanisms exerted among other things by the cyclic nucleotides, which regulate growth and differentiation in normal cells.

Cancer will not be discussed at length here since it is one of the main subjects of Chapter 5. It is now well established that there are physical and chemical cancer-inducing agents. Some cancers (leukaemia in mice, for instance) are of viral origin and are due to well identified DNA- or RNA-containing viruses. Others are due to chemical carcinogens, for instance hydrocarbons present in tobacco smoke and dyes used by hairdressers or added to food, metals, etc. These carcinogens can 'awake' a latent virus which is already present in human cells, but did not proliferate. There is growing evidence that most carcinogens are mutagens; the correlation between carcinogenicity and mutagenicity is becoming more and more impressive. A fast test for mutagenicity has been developed by B. Ames using a bacterial system [81]; other tests for screening mutagens in eukaryotic cells are being devised. But final proof that a mutagen is a carcinogen requires experiments on animals, where the appearance of a tumour often takes a considerable time.

Present work, at the cellular and molecular levels, deals mainly with the bio-chemical changes which occur in 'transformed' cells; malignant transformation can be obtained by the appropriate treatment of normal cells with viruses or carcinogens *in vitro*. These studies have shown that the properties of the cell surface are changed in transformed cells. They also indicate that certain chromosomes, at least in the mouse, bear 'cancer genes'. Experiments on cell fusion by H. Harris have shown that malignancy is a *recessive* character, since if one fuses a malignant with a normal cell, the resulting hybrid is not malignant [82]. Cell hybridization is a powerful tool for the study of human genetics and thus for the search of possible cancer genes. If one fuses a mouse cell and a human cell, the two nuclei of the heterokaryon fuse together, giving rise to a hybrid *synkaryon*; such hybrid cells can proliferate under culture conditions *in vitro*. During the successive cell divisions of the syn-karyons, human chromosomes are progressively lost, so that when only one of them is left it becomes possible to analyse its genetic constitution by standard genetic methods.

Cancer cells and embryonic cells have much in common, including the presence of identical proteins (antigens) in their cell membrane; these proteins are lost when the embryonic cells differentiate. This similarity between embryonic and cancer cells is particularly obvious in malignant *teratomas*. These tumours originate from the testis or the ovary and some of their cells can differentiate to form monstrous, disorganized tissues consisting of several hearts, hairs, brains, etc. Other teratoma cells remain undifferentiated but are malignant, so that tumours develop which kill the animal in which such cells have been injected. These malignant cells have the same surface antigens as early embryonic cells and, in both cases, these antigens are lost when the cells differentiate. Recent work by B. Mintz has shown that if a malignant teratoma cell is added to normal early embryonic cells and if this mixture of cells is allowed to develop, the teratoma cell looses its malignant character; it neither kills the neighbouring cells nor the host, but develops perfectly normally [83]. This weighs against the idea that the malignancy of teratomas is due to gene mutations and shows, at least, that the malignant character can be checked by intimate contact of the tumour cells with normal early embryonic cells.

These experiments concerning cell fusion (Harris) and chimerae between malignant and embryonic cells (Mintz) might in the future provide new ways towards the cure of cancer. It would, however, be overoptimistic to believe that a cancer cell would loose its malignancy if it were forced to differentiate. It is possible to obtain differentiation *in vitro* of tumour cells from the nervous system (neuroblastomas) and of some leukaemic cells (Friend's erythroleukaemia) into neuron-like and erythrocyte-like cells respectively by treatment with chemicals such as cAMP, but these morphologically and biochemically differentiated cells remain malignant and form tumours if they are injected into an appropriate host.

(g) *Ageing*

This is clearly a 'molecular disease'. Senescence is largely due to the fact that the organism has reached the end of the developmental programme inscribed in its DNA; this programming hypothesis explains why the average life-span differs widely from one species to another (for instance, between mice and men). But ageing is also due to an increasing number of errors made in the complex machinery which goes from the DNA gene to its expression. Deleterious somatic mutations accumulate during life (X-ray irradiation for instance shortens the life-span); DNA repair mechanisms might become less efficient in old cells; the transcription mechanisms and their regulation might lose their precision during senescence; and, finally, errors can be made in the translation of the genetic message. When such errors accumulate, they end in a catastrophy, i.e. death (Orgel's error catastrophy theory [84]). Thus ageing results from the summation of many different causes.

Much work is being done now about ageing of cells cultivated *in vitro*; embryonic cells cultivated *in vitro* show several signs of senescence (i.e. decrease in the proliferation rate) and die after a finite number of generations. It is not certain that senescence in the whole organism and in isolated, cultured cells are identical phenomena, but experiments in which senescent cells have been fused with young cells, or even anucleate 'cytoplasts' from young cells, are worth mentioning, since they indicate that an old cell nucleus can be reactivated by young cytoplasm. This finding might be important for future research in gerontology; cytoplasm from young cells might contain factors capable of reprogramming the DNA of old cells.

(h) *The immunological paradox*

This question is dealt in detail in Chapter 5. The immunological response is a fascinating problem for molecular biologists because human lymphocytes can react to almost any foreign substance (*antigen*) by synthesizing highly specific protein 'antibodies' (immunoglobulins) which bind specifically to the foreign antigen. Two theories have been proposed to explain the astounding specificity of the immune response. Some immunologists believe that human germ cells (eggs and sperm) already possess the genes needed in order to produce all possible antibodies and it has been estimated that the minimal number of genes required for this purpose is of the order of 10 000. In opposition to this 'germinal' theory is the 'somatic mutation' theory. This states that immunological specificity results from mutations occurring in the mother cells of the lymphocytes or in the lymphocytes themselves, so that new genes, implicated in the immune response, would make their appearance during our life-time as the result of mutations.

It is probable that a choice between these two opposing theories will be possible before long, thanks to the work done by molecular biologists; several attempts are now being made to isolate and purify the genes which code for the immunoglobulins. When we know their exact base sequences, we shall be in a position to choose between the germinal and the somatic mutation theories.

4.13 Prospects for the future

The development of molecular biology has been so explosive during the past 25 years that it would be foolish to predict its future in the next 20 or 25 years. It is likely that many, and possibly all of the points which have been presented as 'still obscure' in § 4.12 will be solved within a few years. The methods which will be used to understand the molecular mechanisms of cell differentiation will probably remain those which have led to success in the case of prokaryotic heredity, i.e. a combination of genetic analysis and the use of more and more sophisticated biochemical and biophysical methods. There is no doubt that the isolation of an increasing number of mutant cell strains and the possibility of hybridizing them by cell fusion will immensely increase our knowledge of eukaryotic genetics. In the case of man, studies on man–mouse somatic hybrids (where, as we have seen, human chromosomes are lost when synkaryons are replicated) have already provided much more precise information about the genetic organization of our chromosomes than the collection of data about congenital diseases in townships, churches, armies, etc; the mapping of the genes located in human chromosomes is making rapid progress and will undoubtedly continue. The refinement of present methods for the analysis of chromosomal fine structure will also progress quickly; flow cytophotometric methods allow the rapid sorting out of different families of cells in populations which, a few years ago, would have appeared homogeneous. At the molecular level, improvements in X-ray and neutron crystallography will give us much more precise information about the structure of nucleic acids, proteins, ribosomes, chromatin, etc. Thus, as has been the case for embryology since the days of Leeuwenhoek, progress in molecular biology will be linked to improvements in technology.

But we also need new theoretical ideas. Collaboration between theoretical physicists and experimental biologists should lead to new models in order to explain, for instance, cell differentiation. The critical analysis of such models will require increasing use of computer simulation.

There is no doubt that an increasing number of genes, including structural genes present in only a very few copies in the eukaryotic genome, will be isolated and their complete base sequence elucidated. Regulatory genes are more elusive than structural genes coding for a well-known protein, such as haemoglobin, but I have little doubt that the quest for regulatory genes will also be successful within a few years. This should lead to a much better understanding of gene regulation, the molecular mechanisms of cell differentiation and, in consequence, the reasons why cancer cells usually fail to differentiate.

The reader has perhaps been struck by the limited number of practical applications which have been derived from the tremendous progress that has been made in our fundamental knowledge of molecular biology. The main reason for this surprising fact can probably be sought in the amazing rapidity of research progress in this field; competition between the largest laboratories has been — and often remains — so strong that the great leaders of molecular biology have had little opportunity to devote their time to possible applications and, in particular, to the medical sciences. They are, however, becoming more and more conscious of the possible importance of the practical applications of molecular biology to medicine. For instance, Max Perutz, who was awarded a Nobel Prize for elucidating the structure of haemoglobin by X-ray crystallography, has recently published a paper in *Nature* to the 'haemoglobinopathies' [85]; there are many kinds of anaemias due to genetic reasons and we are understanding their molecular causes more and more. The amino acid sequence of a large number of abnormal haemoglobins is well known. In the so-called thalassaemias, one of the two globin chains (a or β) can be completely missing for a variety of reasons; for example, deletion of the a or β globin gene in DNA; lack of transcription into globin mRNA of an existing, but inactive globin gene; synthesis of an abnormal globin chain which will be destroyed by degradation enzymes. The prevention, by abortion, of the spreading of genetical diseases will in the future be based on ever more accurate tests, i.e. precise analysis of chromosomal abnormalities by automated cytochemical methods in foetal cells, collected by amniotic puncture and the detection of traces of abnormal enzymes or other proteins. Molecular hybridization *in situ* might also become an important tool for the early detection of genetic diseases.

Molecular biologists dream of introducing normal genes into genetically abnormal cells in order to cure genetic diseases. This is a difficult task, not only because the penetration of DNA into cells is poor, but also because most of the DNA that succeeds in penetrating into the cells is degraded by enzymes present in the lysosomes and does not enter the nucleus. Furthermore, in order to achieve success, the exogenous DNA should be integrated in the chromosomes of the host-cell in such a way that it can be replicated and transmitted. Purified genes might be introduced into cells by adding to the recipient cell DNA entrapped in *liposomes*: these are lipid bilayers which can be artificially made and which can fuse with the lipid bilayer of the cell membrane. There is no doubt that many efforts will be devoted to the transformation of cells by the addition of DNA, but it is impossible to foresee how successful they will be.

An easier way to approach the same problem is the injection of DNA into eggs. Many years ago, experiments in the author's laboratory showed that *E. coli*-labelled DNA is incorporated into the nuclear DNA of amphibian eggs. Similar conclusions have been drawn recently by Gurdon and Brown [67], who injected purified 5S RNA genes into frog eggs. When the number of purified genes at our disposal increases, more and more experiments of this type will become possible. Injection of pure DNA preparations into mammalian eggs of known genetical constitution presents no major technical difficulties and might possibly lead to important consequences in our understanding and perhaps cure of hereditary diseases, cancer (if it is due to mutations), and ageing (if DNA has become 'bad' as the result of somatic mutations) etc.

Speaking about eggs, what might be the future of molecular embryology? The *Xenopus* oocyte system still has a bright future: it is ideal for testing translation of mRNAs injected into the cytoplasm, and transcription of DNA injected into the nucleus. It could potentially become an efficient system of coupled transcription and translation, where the injection of a given DNA (for instance a purified human insulin gene) would be followed by the prolonged synthesis of the corresponding protein; but it seems doubtful this system will compete economically with the bacterial systems discussed below. From a more theoretical viewpoint, it is certain that the still puzzling germinal localizations, morphogenetic gradients and inducers will be much better understood than today. From a practical viewpoint, embryology already plays a decisive role in birth control, a problem which has so many important social, economical, and political implications for mankind, although our knowledge of contraceptives is still in an early stage of development. The separation of spermatozoa bearing the X or Y chromosome, allowing us to obtain females or males at will, should soon become possible, with important practical implications for breeding cattle or poultry. The separation of cells during egg cleavage, followed by re-implantation in the uterus, could supply identical twins at will. In other words, the future of embryology is the *Brave new world* imagined by Aldous Huxley [86] in a remarkable book in which only the keyword DNA is missing.

Will cancer be cured or prevented by vaccination in the next 20 years? Being an optimist, I think so. I do not believe that the large amounts of money and all the research work invested in cancer research have been worthless because so many people still die today from cancer. We know much more about cancer than we did 20 years ago and fundamental work must go on; the cancer battle will not be won by hoping for a miraculous cure (the discovery of a wonder drug), but by patient and intelligent efforts. For the present, it is essential to reduce to a minimum the exposure of man to mutagenic agents (radiations and air, food, and drink pollutants); this, however, is a problem for the politicians and not for the molecular biologists.

Substantial progress can be expected in another field, which has not been touched upon in the present review: studies on the functioning of our central nervous system. This important topic has been left out of the present review because we know next to nothing about the molecular mechanisms of sleep, memory, and thinking. Until there is a major breakthrough in this all-important field, psychology and psychiatry will remain in a state of ignorance comparable to that of physicians treating cases of infectious diseases before Pasteur discovered bacteria.

The origin of life will probably remain a mystery for a long time to come; exploration of the moon and the planet Mars has been disappointing for the 'exo-biologists'. This does not mean that life does not exist on some distant planet, but the astronaut who will discover it is probably not yet born. We could not in any case detect a form of life devoid of photosynthesis, enzymes proteins, nucleic acids, etc. because we cannot imagine it. On the other hand, one can expect progress in our understanding of evolution on our own planet, thanks to the work going on on the synthesis of organic compounds under conditions simulating prebiotic conditions of the earth, and on the evolution of proteins (comparison of changes in amino acid sequences in the same protein isolated from lower and higher organisms) and nucleic acids. This work will eventually lead to deciphering the base sequences of ancestral genes and the codewords of a primitive genetic code.

Genetic engineering (gene manipulation) has become a hotly debated subject. We have already mentioned that it is possible to introduce eukaryotic or prokaryotic genes into the chromosomes of the bacterium *E. coli* after first inserting them into the DNA of an appropriate vector (a phase or a plasmid). The foreign gene will be replicated every 10–15 minutes together with *E. coli*'s own DNA. Improved methods for the selection of the bacteria bearing a foreign DNA material are now being worked out. Since very large numbers of bacteria can easily be grown, it is easy to obtain important amounts of the foreign DNA which has been amplified during bacterial growth. This foreign DNA can be reisolated and submitted to very precise biochemical analysis. This is how we know so much about the fine structure of the ribosomal, 5S, and histone genes. Theoretically (and almost certainly practically) any gene can be inserted into plasmid or phage DNA and then into *E. coli* DNA where it will be replicated as many times as one wishes.

Already *Drosophila*, frog, mouse, and human genes have been inserted, by genetic engineering, into the *E. coli* chromosome, the latter thus becoming a hybrid DNA, carrying eukaryotic sequences as well as bacterial sequences. The bacteria apparently remain perfectly healthy: their DNA polymerase synthesizes the foreign as well as the normal DNA sequences. Ecologists have expressed the fear that such bacteria, which have not followed the proper rules of evolution, might be potentially dangerous; so far, however, this fear has been ill-founded and no harm has come out of bacteria which contain amphibian or mammalian genes.

In fact, man should be proud of such daring experiments where the laws of neo-darwinian evolution (random mutations and natural selection) are violated; they show that evolution is not always the result of blind chance, but that man's intelligence can alter it. Since eukaryotic RNAs are different from *E. coli* rRNA and as this bacterium does not contain measurable amounts of histones, blind chance alone cannot introduce into a bacterium the genes for ribosomal RNAs and histones. Jacques Monod's otherwise excellent book *Le hasard et la nécessité* [87] gives, in my opinion, too pessimistic a view of life because it does not sufficiently stress the capacity given to man to modify nature.

Genetic engineering has tremendous potential; attempts are already being made to introduce, into *E. coli*, the genes for human insulin and those needed for nitrogen fixation. Mass production of human insulin would be a boon for patients of diabetes and mass production of *E. coli* capable of nitrogen fixation would have far-reaching consequences for crop production.

However, genetic engineering, like nuclear physics, could also lead to disaster for mankind. Already *E. coli* bacteria capable of resisting all known antibiotics have been constructed. Aware of the potential dangers of a world epidemic which might take place if such a bacterial strain 'got loose' and out of the control of the laboratories, a number of molecular biologists met together (at the Asilomar Conference) and asked for a moratorium on genetic engineering research. They now regret it, because the *E. coli* strain used for these experiments is not pathogenic. Everybody agrees, however, that further research in genetic engineering should be performed under strictly controlled conditions ensuring security not only for the scientists engaged in the work, but for the whole community. This is the reason why strict guidelines have been proposed for this kind of research by the National Institutes of Health (USA), the British Government, the European Molecular Biology Organization (EMBO), etc. At the time of writing, laws curbing research in the field of genetic engineering are being hotly discussed in the US Senate and House of Representatives. What should of course be absolutely banned is the extension of genetic manipulation to pathogenic strains of bacteria for military or other purposes.

This prospect for the future of molecular biology has been written under the assumption that no major calamities (war, complete economic breakdown, etc.) will fall upon humanity until the end of the present century. But even if such calamities do strike us — and nobody hates them more than I do — one thing will always remain: man's thirst for knowledge will persist, even if it was long ago that Adam and Eve ate the forbidden fruit of the Science of Good and Evil in Paradise.

Acknowledgments
I wish to thank Professor H. Chantrenne for critically reading the manuscript, Professor P. Van Gansen for help in the preparation of the illustration, Dr. J. for improving the English, and Mrs. J. Baltus for typing the text.

References

[1] Descartes, R. *Discours de la méthode*. Jane Maire, Leiden (1637).
[2] Driesch, H. *Z. Wissen. Zool* LIII (1891).
[3] Schleiden, M. J. *Arch. Anat. Physiol. Wiss. Med.* p. 137 (1838).

[4] Schwann, T. Mikroskopische Untersuchungen über die Ubereinstimmungen in der Struktur und dem Wachtum der Tiere und Pflanzen no. 179 (1839).
[5] Claude, A. *Biol. Symposia* **10**, III (1943).
[6] Porter, K. R. *J. Histochem. Cytochem.* **2**, 346 (1954).
[7] Palade, G. E. and Siekevitz, P. *J. Biophys. Biochem. Cytol.* **2**, 171 (1956).
[8] de Duve, C., Pressman, B. C., Gianetto, R. J., Wattiaux, R., and Appelmans, F. *Biochem. J.* **60**, 604 (1955).
[9] Miescher, F. *Hoppe Seyler's Med-Chem. Untersuchungen* **Bd. 4**, Seite 441, A. Hirschwald, Berlin (1871).
[10] Altmann, R. *Archiv für Anatomie und Physiologie*, p. 524 (1889).
[11] Levene, P. A. *Nucleic Acids* Chemical Catalog C°, N.Y. (1931).
[12] Mendel, G. *Experiments in plant hybridisation*. Verh. Naturf. Ver. in Brunn Abkandlungen (1865).
[13] Morgan, T. H. Chromosomes and heredity. *Am. Nat.* **44**, 449 (1910).
[14] Muller, H. J. *J. Genetics* **22**, 299 (1930).
[15] Garrod, A. E. *Lancet* ii 1616 (1902).
[16] Brachet, J. *Enzymologia* **10**, 87 (1941).
[17] Feulgen, R. and Rosenbeck. H. S. *Z. Physiol. Chem.* **124**, 501 (1924).
[18] Caspersson, T. *Cell growth and cell function*. Norton, N.Y. (1950).
[19] Avery, O. T., MacLeod, C. M., and MacCarthy, M. *J. Exptl. Med.* **79**, 137 (1944).
[20] Hershey, A. D. and Chase, M. *J. Gen Physiol.* **36**, 39 (1952).
[21] Garrod, A. E. In *Inborn errors of metabolism*. Oxford University Press, London (1909).
[22] Ephrussi, B. *Quart. Rev. Biol.* **17**, 327 (1942).
[23] Beadle, G. W. and Ephrussi, B. *Genetics* **22**, 76 (1937).
[24] Beadle, G. N. and Tatum, E. L. *Proc. Nat. Acad. Sci. US* **27**, 499 (1941).
[25] Monod, J. In *Enzymes, units of biological structure and function* (Ed. O. H. Gaebler) Vol. 7, p. 28. Academic Press, N.Y. (1956).
[26] Delbrück, M. and Bailey, W. T. *Cold Spring Harbor Sympos. Quant. Biol.* **II**, 33 (1946).
[27] Luria, S. E. *Genetics* **30**, 84 (1945).
[28] Lederberg, E. M. and Lederberg, J. *Genetics* **38**, 51 (1953).
[29] Lwoff, A. *Bacteriol. Rev.* **23**, 109 (1959).
[30] Jacob, F. *Harvey Lectures Ser.* **54** (1960).
[31] Benzer, S. *Proc. Nat. Acad. Sci. US* **41**, 344 (1955).
[32] Brenner, S. *Virology* **3**, 560 (1957).
[33] Cohen, S. S. and Anderson T. F. *J. Exptl. Med.* **84**, 511 (1946).
[34] Wyatt, G. R. and Cohen, S. S. *Biochem. J.* **55**, 774 (1953).
[35] Watson, J. D. and Crick, F. H. C. *Nature* **181**, 737 (1953).
[36] Pertuz, M. F. *Proteins and nucleic acids*. Elsevier, Amsterdam (1962).
[37] Kendrew, J. *The thread of life*. Bell, London (1966).
[38] Ochoa, S. *Experientia* **20**, 57 (1964).
[39] Nirenberg, M. W. and Matthaei, J. H. *Proc. Nat. Acad. Sci. US* **47**, 1580 (1961).
[40] Chargaff, E. *Symposia Soc. Exptl. Biol.* **9**, 32 (1955).
[41] Wilkins, M. H. F. *Cold Spring Harbor Sympos. Quant. Biol.* **21**, 75 (1956).
[42] Franklin, R. and Gosling, R. C. *Nature* **181**, 740 (1953).
[43] Jacob, F. and Wollman, E. L. In *The chemical bases of heredity* (Eds. W. D. MacElroy and B. Glass), p. 468. John Hopkins University Press, Baltimore (1957).
[44] Cairns, J. *J. Mol. Biol.* **6**, 208 (1963).
[45] Pauling, L. *Bull. N.Y. Acad. Med.* **40**, 334 (1964).
[46] Meselson, M. and Stahl, F. W. *Proc. Nat. Acad. Sci. US* **44**, 671 (1958).

[47] Kornberg, A. *The enzymatic synthesis of DNA*. Wiley, London (1961).
[48] Okazaki, R., Okazaki, T., Sakabe, K., Sugimoto, K., Kainuma, R., and Sugino, A. *Cold Spring Harbor Sympos. Quant. Biol.* **33**, 129 (1968).
[49] Jacob, F. and Monod, J. *Symposium of Cytodifferentiation and Macromolecular Synthesis* (Ed. M. Locke), Academic Press, N.Y. (1963).
[50] Chantrenne, H. *Biochim. Biophys. Acta* **53**, 409 (1965).
[51] Gros, F. and Gilbert, W. *Cold Spring Harbor Sympos. Quant. Biol.* **26**, 111 (1961).
[52] Brenner, S., Jacob, F., and Meselson, M. *Nature* **190**, 576 (1961).
[53] Davidson, E. H. *Gene activity in early development*. Academic Press, New York (1968).
[54] Brown, D. D., Wensink, P. C., and Jordan, E. *Proc. Nat. Acad. Sci. US* **68**, 3175 (1971).
[55] Nomura, M. *Cell* **9**, 633 (1976).
[56] Fiers, W., Contreras, R., Duerinck, F., Haegeman, R., Iserantant, D., Merregaert, J., Min Jou, W., Molemans, F., Raemakers, ., Van den Berghe, A., Tolckaert, G., and Ysebaert, M. *Nature* **260**, 500 (1956).
[57] Proudfoot, N. J. *Cell* **10**, 559 (1977).
[58] Thomas, R. *J. Mol. Biol.* **49**, 393 (1970).
[59] Crick, F. *Nature* **234**, 25 (1971).
[60] Paul, J. *Nature* **238**, 444 (1972).
[61] Georgiev, G. P. *J. Theoret. Biol.* **25**, 473 (1969).
[62] Britten, E. H. and Davidson, E. *Science* **165**, 349 (1969).
[63] Op. cit. [53].
[64] Dawid, I. B., Brown, D. D. and Reeder, R. H. (1970). *J. Mol. Biol.* **51**, 341 (1970).
[65] Birnstiel, M. L., Schaffner, W., and Smith, H. O. *Nature* **266**, 603 (1977).
[66] Kedes, L. H. and Birnstiel, M. L. *Nature, New Biol.* **230**, 165 (1971).
[67] Brown, D. and Gurdon, J. B. *Proc. Nat. Acad. Sci. US* **74**, 2064 (1977).
[68] Brown, D. D. and Blackler, A. W. *J. Mol. Biol.* **63**, 75 (1972).
[69] Dawid, I. and Blackler, A. W. *Devel. Biol.* **29**, 152 (1972).
[70] Khorana, H. G. *et al. J. Biol. Chem.* **251**, 565 (1976).
[71] O'Malley, R. W., Sherman, M. R., and Toft, D. O. *Proc. Nat. Acad. Sci. US* **67**, 501 (1970).
[72] Gilmour, R. S. and Paul, J. *Proc. Nat. Acad. Sci. US* **70**, 3440 (1973).
[73] Edström, J. E. and Beermann, W. *J. Cell Biol.* **14**, 371 (1962).
[74] Brachet, J. and Lang, A. (1965). The role of the nucleus and the nucleocytoplasmic interactions in morphogenesis. In *Handbuch der Pflanzenphysiol.* vol. 15, p. 1.
[75] Brachet, J. In *The biological role of ribonucleic acids*, p. 130. Elsevier, Amsterdam (1960).
[76] Hämmerling, J. *Intern. Rev. Cytol.* **2**, 475 (1953).
[77] Brachet, J. and Chantrenne, H. *Nature* **168**, 950 (1951).
[78] Schweiger, H. G. and Berger, S. *Biochim. Biophys. Acta* **77**, 533 (1964).
[79] Gurdon, J. B. *J. Embryol. exp. Morphol.* **20**, 401 (1968).
[80] Fougère, C., Ruiz, F., and Ephrussi, B. *Proc. Nat. Acad. Sci. US* **69**, 330 (1973).
[81] MacCann, J. and Ames, B. N. *Proc. Nat Acad. Sci. US* **73**, 950 (1976).
[82] Harris, H. *Cell fusion*. The Durham Lectures. Clarendon Press, Oxford (1970).
[83] Mintz, B. In *Methods in mammalian embryology*. (Ed. J. C. Daniel). Freeman, San Francisco (1971).
[84] Orgel, L. E. *Nature* **243**, 441 (1973).
[85] Pertuz, M. F. *Nature* **262**, 449 (1976).
[86] Buxley, A. *Brave new world*. Chatto and Windus, London (1932).
[87] Monod, J. *Le hasard et la nécessité*. Le Seuil, Paris (1970).

5

Biologie de l'ensemble de l'organisme: la machinerie immunitaire

G. Mathé

I. La machinerie immunitaire

Introduit dans un organisme, un antigène est reconnu par un très petit nombre de lymphocytes possédant des récepteurs capables de se combiner à lui spécifiquement. Selon la nature, la présentation de cet antigène, des mécanismes différents sont initiés et aboutissent à son élimination en faisant intervenir des effecteurs cellulaires, lymphoïdes et non lymphoïdes, ainsi que leurs produits. Le développement, l'intensité, la durée de cette réponse est sous la dépendance de mécanismes de régulation impliquant d'autres effecteurs. Une réponse immunitaire est donc le résultat d'intéractions complexes entre divers types de cellules et leurs produits, déclenchées par la reconnaissance spécifique de l'antigène.

On distingue deux types d'immunité : l'immunité à médiation humorale, dont l'expression ultime est la synthèse d'anticorps et leur libération dans la circulation et l'immunité à médiation cellulaire, dont les effecteurs sont cellulaires.

5.1 Immunité à mediation humorale et à mediation cellulaire

L'existence des molécules d'anticorps a été découverte en 1890 par Von Behring. Elles sont identifiées ultérieurement comme des globulines sériques et parmi elles comme des immunoglobulines dont les structures primaires et tertiaires ont été élucidées au cours des vingt dernières années.

Une molécule d'immunoglobuline est formée de quatre chaines polypeptidiques, deux chaines légères (L) et deux chaines lourdes (H), réunies par des ponts disulfures. Il existe deux types de chaines L, les chaines λ et κ, la molécule d'immunoglobuline étant constituée soit de deux chaines κ soit de deux chaines λ. On connait actuellement cinq types de chaines lourdes, dénommées γ, μ, a, δ, ϵ. La nature des deux chaines H identiques de la molécule détermine la classe d'immunoglobuline à laquelle elle appartient.

On est donc en présence de cinq classes d'immunoglobulines désignées par les sigles IgG, IgM, IgA, IgD, IgE qui sont représentées en quantités très différentes dans le sérum.

L'anticorps est caractérisé par sa capacité de se combiner spécifiquement avec l'antigène qui a suscité sa formation. Cette spécificité est la propriété de deux sites de combinaisons identiques situés dans les régions variables des chaines H et L aux extrémités NH2. La structure primaire et tertiaire caractéristique dont dépend le site de combinaison de l'antigène se comporte elle-même comme une structure antigénique; les déterminants sont appelés idiotypes et peuvent susciter la formation d'anticorps anti-idiotypes.

La partie COOH terminale des chaines lourdes est une région constante de la molécule. Elle est responsable des propriétés biologiques de l'anticorps après sa combination avec l'antigène. On trouve sur cette partie de la molécule, dite fragment Fc, le site de fixation du complément, complexe de 'protéines sériques' possédant des activités enzymatiques agissant en cascade. Le complément est responsable par exemple de l'attaque de la membrane cellulaire sur laquelle se sont fixés les anticorps spécifiques, ce qui entraînera la lyse cellulaire.

L'immunité à médiation cellulaire ne fait pas intervenir la synthèse d'anticorps. Elle est responsable du rejet de greffes de tissus ou d'organes faites entre individus génétiquement non identiques, de la protection contre les infection virales, bactériennes et contre les parasites intracellulaires et du développement des réactions d'hypersensibilité retardée, c'est-à-dire des réactions inflammatoires cutanées qui se développent au site de l'inoculation de l'antigène chez un sujet sensibilisé par un premier contact avec cet antigène. Les mécanismes effecteurs de ce type d'immunité font intervenir, à côté des lymphocytes réagissant spécifiquement avec l'antigène, des cellules et des produits agissant de façon non spécifique.

Un des grands évènements en immunologie, fut la reconnaissance, il y a une quinzaine d'années, de l'existence de deux classes de lymphocytes, l'une responsable de l'immunité à médiation cellulaire, l'autre de l'immunité à médiation humorale; l'une dépendante de la présence du thymus, l'autre non.

5.2 Les lymphocytes dépendants et indépendants du thymus

Ce sont les expériences de J. F. A. P. Miller en 1962, qui ont démontré le rôle essentiel joué par le thymus dans l'acquisition de l'immunocompétence. Des souris thymectomisées à la naissance sont incapables par la suite de développer une immunité à médiation cellulaire. Une déplétion lymphocytaire est observée dans certaines zones dites pour cette raison, thymo-dépendantes, de la rate et des ganglions lymphoïdes. Les mêmes phénomènes sont décrits chez l'enfant souffrant d'une absence congénitale du thymus (syndrome de Di George).

Parallèlement, les équipes de Warner et de Cooper décrivent, chez les oiseaux, un syndrome inverse à celui provoqué par la thymectomie et se développant après l'ablation d'un organe lymphoïde, la bourse de Fabricius, situé dans la région cloacale: l'immunité à médiation cellulaire n'est pas affectée, mais les taux d'immunoglobulines sériques sont effondrés et les réponses humorales sont inexistantes. La déplétion lymphocytaire dans la rate et les ganglions est située dans des zones particulières dites thymo-indépendantes.

De ces observations et d'études ultérieures, il s'est avéré qu'il existe deux types de lymphocytes selon l'organe lymphoïde primaire dans lequel ils se sont différenciés à partir de cellules souches hématopoïétiques présentes dans la moëlle osseuse, avant de migrer dans les organes lymphoïdes secondaires (rate, ganglions, plaques de Peyer, amygdales, etc.) où ils exercent leurs fonctions immunologiques respectives.

Les lymphocytes dérivés du thymus, ou *lymphocytes T* sont responsables de l'immunité à médiation cellulaire; les lymphocytes se différenciant dans la bourse de Fabricius chez l'oiseau, dans la moëlle osseuse qui apparait comme l'équivalent de la bourse chez les mammifères, appelés *lymphocytes B*, sont les précurseurs des cellules hautement spécialisées dans la synthèse des anticorps, les *plasmocytes*.

Identiques morphologiquement, les lymphocytes T et B peuvent être distingués par la présence de certains marqueurs à la surface cellulaire. Nous citerons ceux particulièrement utilisés pour l'isolement sélectif de l'un ou l'autre type de cellules. Chez la souris, les lymphocytes T expriment un alloantigène dit THY-1 ou θ. Des anticorps dirigés contre cet antigène en présence de complément, détruisent sélectivement les lymphocytes T. Chez l'homme, les lymphocytes T portent des récepteurs fixant les érythrocytes de mouton.

Les lymphocytes B chez toutes les espèces animales sont identifiables par la présence, à leur surface, d'immunoglobulines qui sont responsables de la reconnaissance spécifique de l'antigène. Ces immunoglobulines appartiennent en majorité à la classe des IgD. Des cellules portant des IgM et à la fois des IgM et des IgD sont également détectées. Durant la période embryonnaire, les lymphocytes B portent des IgM.

Un troisième type de cellule lymphoïde a été récemment identifié. Il est caractérisé par l'absence des marqueurs spécifiques des lymphocytes T ou B. Appelées pour cette raison, *cellules nulles*, elles constituent en fait une population très hétérogène de cellules comprenant vraisemblablement des cellules souches hématopoïétiques, des cellules déjà engagées dans la voie de différenciation vers le lymphocyte T (pre-thymocyte) ou vers le lymphocyte B. Certaines de ces cellules nulles portent des récepteurs fixant les immunoglobulines IgG par le fragment Fc de la molécule. Ces récepteurs, dits récepteurs Fc, peuvent étre portés par d'autres cellules, en particulier une sous-population de lymphocytes T, les lymphocytes B, bien que ceci soit maintenant contesté, et les monocytes-macrophages, quatrième type de cellules intervenant dans les réponses immunitaires. Les récepteurs Fc jouent un rôle important dans l'expression de certains mécanismes effecteurs de l'immunité ainsi que dans certains phénomènes de régulation de la réponse immunitaire. Ils seront décrits ultérieurement.

Les *monocytes*, dérivent de la cellule souche pluripotentielle qui donne naissance aux différentes lignées hématopoïétiques: lymphocytaire, granulocytaire, mégacaryocytaire et monocytaire. Issu du promonocyte, précurseur présent dans la moëlle osseuse, le monocyte quitte la circulation pour migrer dans les tissus où il prend le nom de macrophage ou d'histiocyte. Monocytes et macrophages sont doués d'activité phagocytaire. Ils jouent un rôle essentiel dans le développement des réactions immunitaires.

5.3 Développement des reactions immunitaires

La reconnaissance de l'antigène
Les lymphocytes B fixent spécifiquement l'antigène par l'intermédiaire des sites de combinaison des molécules d'immunoglobulines présentes à la surface cellulaire.

La nature des récepteurs pour l'antigène à la surface des lymphocytes T reste encore une énigme. Il est en effet impossible de mettre en évidence des immuno-globulines par les techniques directes utilisant des anticorps anti-immunoglobulines marquées par un élément radioactif ou fluorescent. Certaines équipes ont décrit l'isolement d'immunoglobuline, et particulièrement d'IgM, de la membrane de lymphocytes T, mais ces expériences sont souvent critiquées et il est actuellement admis que les récepteurs pour l'antigène des lymphocytes T n'ont pas la structure d'une immunoglobuline. Néanmoins, des travaux très récents montrent que, pour un antigène donné, ces récepteurs portent les mêmes idiotype que les récepteurs des lymphocytes B et que les anticorps dirigés contre cet antigène, et sont donc partiellement formés d'une partie variable de la molécule d'immunoglobuline.

La proportion de lymphocytes reconnaissant spécifiquement un antigène est très faible, inférieure à 1%, excepté pour les antigène dits d'histocompatibilité qui sont responsables du rejet des greffes pour legquels cette proportion s'élève jusqu'à 5-10% de la population lymphocytaire totale. Cette capacité très anormale à recon-naître les antigènes d'histocompatibilité est maintenant expliquée à la lumière des données très récentes sur le rôle joué par l'ensemble de gènes, dit *complexe majeur d'histocompatibilité*, dans l'initiation et la régulation des résponses immunitaires. Aussi nous allons brièvement décrire ce complexe avant d'aborder les fonctions des différents types de cellules.

Le complexe majeur d'histocompatibilité

A l'origine, il a été reconnu et étudié comme la portion du génome codant pour des antigènes exprimés à la surface des cellules et induisant leur élimination quand elles sont introduites dans un individu génétiquement différent. Chez la souris, le complexe majeur d'histocompatibilité est situé sur le 17ème chromosome et com-prend, entre les deux régions génétiques codant pour les antigènes d'histocompa-tibilité et appelées régions K et D, trois autres régions I, S et G. La *région I*, qui se subdivise actuellement en cinq sous-régions IA, IB, IJ, IE, IC s'est tout d'abord révélée comme renfermant des gènes gouvernant la capacité de répondre à certains antigènes et appelés gènes de réponse immune (gènes Ir). Il s'est rapidement avéré que la région I code également pour des structures antigéniques, appelés *antigènes Ia*, présents à la surface des lymphocytes, des monocytes-macrophages et d'autres cellules. Nous préciserons ultérieurement le rôle joué par les antigènes Ia et les antigènes d'histocompatibilité proprements dits, ou *antigènes H-2*, dans ces réponses immunitaires. Précisons tout de suit que la région S contrôle le taux sérique de certains composants du complément et que la fonction de la région G est peu connue.

5.4 Fonctions des lymphocytes B et T, des monocytes et cellules nulles dans le développement des réactions immunitaires

Rôle des lymphocytes B

Les lymphocytes B stimulés par un antigène prolifèrent de façon clonale et se différencient progressivement en cellules de plus en plus spécialisées pour la synthèse des anticorps, pour aboutir au plasmocyte qui ne se divise plus. Un certain nombre de cellules ne parviennent pas à ce stade final mais reprennent un aspect quiescent. Elles sont devenues des cellules à mémoire et répondront plus vigoureusement et plus rapidement à une nouvelle stimulation.

Les anticorps produits interagissent avec l'antigène donnant naissance à des réactions de précipitation s'il est soluble, d'agglutination, de cytotoxicité (en présence de complément) s'il est porté par une membrane cellulaire, réactions qui aboutissent à son élimination rapide.

Alors que la compartimentalisation en lymphocytes B et lymphocytes T en immunité humorale et immunité à médiation cellulaire venait d'être établie, il est apparu en 1965, que les lymphocytes B exigent l'aide des lymphocytes T pour se différencier en cellules formatrices, d'anticorps, en réponse à une stimulation par la plupart des antigènes naturels. Cette coopération positive entre lymphocytes T et B est la première évidence d'intéractions cellulaires se produisant lors du développement des réactions immunitaires. Néanmoins, certains antigènes, caractérisés par une structure répétitive, ce sont généralement des polymères (polysaccharides et lipo-polysaccharides bactériens, certains produits chimiques de synthèse), peuvent stimuler les lymphocytes B en l'absence de lymphocytes T. Ce sont des antigènes thymo-indépendants et leur capacité de se fixer simultanément à de multiples récepteurs des cellules B, en raison de leur structure, peut expliquer ce phénomène.

Rôle des lymphocytes T

En réponse à des stimuli antigéniques conduisant à des réactions à médiation cellulaire, les lymphocytes T se transforment en grandes cellules, appelées *immuno-blastes*, se divisant activement et caractérisées par un cytoplasme pyroninophile car très riche en ribosomes. Les immunoblastes se différencient ensuite en cellules effectrices exerçant des fonctions différentes selon le type de réponse incriminée.

Des lymphocytes T d'un sujet sensibilisé contre un antigène, tel le bacille tuberculeux, remis *in vitro* en présence de l'antigène sensibilisant (la tuberculine en l'occurence) synthétisent et libèrent dans le milieu des facteurs solubles aux activités biologiques variées. Certains agissent sur les macrophages inhibant leur capacité migratrice, stimulant certaines de leurs activités dont la phagocytose, les attirant par chimiotacticme; d'autres agissent sur les polynucléaires (inhibition de migration, chimiotactisme). Ces médiateurs sont rendus responsables du développement de la réaction inflammatoire d'hypersensibilité retardée qui fait donc intervenir des effecteurs non spécifiques : monocytes, polynucléaires, après intéractions avec des produits de cellules specifiquement impliquées. Ces processus permettent une amplification de la réponse spécifique.

Quand les antigènes sont des constituants de la membrane cellulaire, les immuno-blastes se différencient en cellules effectrices capables de tuer des cibles portant ces mêmes antigènes. Cette activité cytotoxique ne fait pas intervenir le complément et résulte d'un contact direct entre le lymphocyte T cytotoxique et la cellule cible par l'intermédiaire des récepteurs spécifiques pour l'antigène. Quand l'antigène est induit par un virus infectant la cellule, ou par une modification chimique de la membrane, ou résulte d'un phénomène de cancérisation (antigènes associés aux tumeurs), une identité au niveau de la région K ou D du complexe majeur d'histo-compatibilité entre la cellule effectrice et la cellule cible est indispensable pour que la réaction lytique ait lieu. La réponse immunitaire serait en fait induite par, et dirigée contre, un antigène complexe résultant d'une intéraction entre les antigènes d'histocompatibilité H-2 et l'antigène induit par le virus, la modification chimique ou la cancérisation. Cette hypothèse de l'auto-constituant modifié confère enfin aux antigènes d'histocompatibilité un rôle biologique autre que celui d'induire des rejets de greffe. Ils interviennent en particulier dans les défenses contre les virus que ceux-ci soient oncogènes ou non.

A côté de leurs fonctions de cellules effectrices de l'immunité à médiation cellulaire, les lymphocytes T exercent des fonctions régulatrices.

Les lymphocytes T aident les lymphocytes B dans leur réponse aux antigènes dits thymo-dépendants. Les lymphocytes B, pour se différencier en cellules forma-trices d'anticorps auraient besoin de deux signaux : l'un fourni par la fixation de l'antigène sur les récepteurs immunoglobulinemiques l'autre fourni par les *lympho-cytes T 'helper'*. Des données de plus en plus nombreuses tendent à démontrer que les lymphocytes T 'helper' agissent par l'intermédiaire de facteurs diffusibles. De tels facteurs ont été isolés par différents auteurs; ils ont de l'affinité pour l'antigène dont ils sont spécifiques, ils sont constitués en partie par un produit de la région I; de la sous-région IA quand cela a pu être spécifié; ils ne sont pas de nature immunoglobulinemique à l'exception d'un d'entre eux. Ils agissent sur les lympho-cytes B directement ou par l'intermédiaire des macrophages.

Des facteurs produits et agissant de façon non spécifique ont également été décrits. Ils ne renferment ni d'immunoglobuline, ni d'antigènes Ia et agissent en amplifiant la réponse des lymphocytes B plutôt qu'en l'initiant.

Au cours de ces dernières années, il est devenu évident que les lymphocytes T exercent également une activité régulatrice négative sur les réponses à médiation cellulaire et humorale. Cette activité suppressive peut être spécifique de l'antigène et s'effectuer par l'intermédiaire de facteurs solubles. Différent facteurs ont été isolés; ils ne sont pas identiques mais ont souvent une structure semblable à celle des facteurs libérés par les lymphocytes T 'helper' dans les mêmes conditions expérimentales. Certains agissent sur les lymphocytes T 'helper', sur des lympho-cytes T autres que 'helper', sur les macrophages.

Des *lymphocytes T suppresseurs* agissant de façon non spécifique, éventuelle-ment par l'intermédiaire de facteurs solubles, ont également été décrits.

Les lymphocytes T suppresseurs spécifiques ont été impliqués dans le phénomène de la tolérance immunologique aux auto-constituants (une déficience des cellules suppressives entraîne une auto-immunité) et aux constituants étrangers, et dans la non réactivité génétique à certains antigènes. Ils seraient également responsables de situations où l'on observe une inhibition de l'immunité à médiation cellulaire en faveur de l'immunité humorale, l'arrêt de la synthèse d'une classe particulière d'immunoglobulines. Les lymphocytes T suppresseurs agissant de façon non spécifi-que interviennent dans la régulation des réponses à ces antigènes thymo-indépen-dants et sont responsables du phénomène de compétition antigénique : la réponse à un antigène peut inhiber la résponse à un second antigène, différent du premier, mais inoculé peu de temps après.

Les diverses fonctions des lymphocytes T sont exercées par des sous-populations différentes de lymphocytes qui ont pu être distinguées chez le souris gráce à la présence ou l'absence de certains marqueurs de surface: les antigènes Ia, les récepteurs Fc, et des antigènes particuliers dits Ly codés par trois loci Lyl, Ly2, Ly3, chacun d'eux s'exprimant sous deux formes allèliques. Chaque sous-population est caractérisée par un phénotype particulier (Tableau 5.1).

Rôle des macrophages :
Les macrophages interviennent dans la phase d'induction d'une réponse immunitaire. Ils peuvent rendre certains antigènes particulaires plus immunogènes en les digérant partiellement. Les antigènes thymo-dépendants ne peuvent stimuler les différents types de réponse des lymphocytes T (hypersensibilité retardée, fonction helper, fonction suppressive) que s'ils sont présentés par les macrophages en association étroite avec les produits de la région I qui sont exprimés à la surface cellulaire. Il a été montré dans une réponse *in vitro* à un antigène protéinique soluble que l'induction de cellules T 'helper' se fait par l'intermédiaire d'un facteur soluble libéré par les macrophages et formé par des antigènes Ia et un fragment de l'antigène.

Nous avons vu que lors du développement d'une réaction à médiation cellulaire, les lymphocytes T sensibilisés libèrent au contact de l'antigène des médiateurs solubles qui agissent sur les macrophages. Ces cellules subissent une variété de modifications morphologiques et physiologiques, et deviennent des macrophages activés. Leur activité phagocytaire est augmentée. Ils acquièrent une activité bactéricide intense et deviennent capables d'inhiber la prolifération et de tuer des cellules tumorales.

Les macrophages activés peuvent également exercer des fonctions suppressives de façon non spécifique. Ils acquièrent cette fonction également après fixation d'un facteur suppresseur spécifique libéré par les lymphocytes T lors d'une réaction d'hypersensibilité retardée.

Rôle des cellules 'nulles'
Les cellules 'nulles' portant des récepteurs Fc exercent une fonction particulière : celle de détruire des cellules cibles recouvertes d'anticorps dirigés contre des antigènes de la membrane cellulaire. Cette *cytotoxicité à médiation cellulaire dépendante des anticorps*, décrite en 1968 pour la première fois, a été attribuée à des cellules effectrices dénommées *cellules 'K'* (de killer) en raison des difficultés d'identification.

Les cellules K lysent les cellules cibles de façon non spécifique (la spécificité est apportée par l'anticorps). La réaction est induite par la fixation, par leurs récepteurs Fc, du fragment Fc des anticorps liés à la membrane cellulaire de la cible. Les anticorps impliqués apparttiennent à la classe des IgG, mais des données très récentes montrent que des anticorps IgM peuvent également induire la réaction. Dans les deux cas, de faibles quantités d'anticorps sont nécessaires.

L'identification des cellules K fait toujours l'objet de nombreuses controverses

Tableau 5.1 *Caractérisation phenotypique des sous-populations de lymphocytes T*

Sous-population	Antigènes Ly1	Antigènes Ly2	Antigènes Ly3	Antigènes Ia	Récepteur Fc
Lymphocytes T cytotoxiques pour des cellules allogéniques	–	+	+	–	+
pour des cellules tumorales	+	+	–	?	?
Lymphocytes T initiant les réactions d'hypersensibilité retardée	+	–	–	–	?
Lymphocytes T 'helper'	+	–	–	+ autre que la région I-J	–
Lymphocytes T suppresseurs des réactions d'hypersensibilité retardée	+	–	–	?	–
des réponses humorales	–	+	+	+ région I-J	+

car toute cellule possédant des récepteurs Fc est potentiellement capable de lyser une cible recouverte d'anticorps spécifiques. La cytotoxicité à médiation cellulaire dépendante des anticorps est en effet effectuée par des monocytes, des polynucléaires, des plaquettes, des lymphocytes T porteurs de récepteurs Fc, des cellules nulles. Certains auteurs utilisent le terme de cellule K, bien qu'il ne recouvre qu'une fonction. uniquement pour désigner les cellules nulles douées de cette fonction. La situation vient encore de se compliquer dans la mesure où chez l'homme, des lymphocytes T, portant des récepteurs Fc et ayant des récepteurs de très faible affinités, pour les érythrocytes de mouton, semblent en fait recouvrir la population de cellules douées d'activité K et identifiées à tort comme nulles. Ces lymphocytes T lysent des cibles recouvertes d'IgG ou d'IgM.

La complexité des processus qui accompagnent le développement d'une réponse immunitaire menée contre des antigènes simples, bien définis chimiquement, va évidemment se retrouver dans le développement de réactions immunitaires contre des cellules devenues étrangères à un organisme du fait d'un processus de cancérisation.

II. La machinerie immunitaire et la cellule cancéreuse. Les mécanismes de l'immunité tumorale

Certaines observations *in vivo* montrent que, malgré sa croissance progressive, une tumeur suscite chez son hôte des réactions immunologiques susceptibles de freiner son développement. Au court des dix dernières années, la mise au point de techniques *in vitro* a permis de mettre en évidence différents mécanismes immunologiques aboutissant à la destruction des cellules tumorales. Parallèlement, il s'est avéré que d'autres mécanismes permettent aux cellules tumorales d'échapper à cette destruction. Bien qu'il soit souvent difficile d'établir des corrélations entre les phénomènes observés *in vitro* et l'évolution de la tumeur *in vivo*, la notion d'une coexistence de mécanismes de rejet et de protection des cellules tumorales a ouvert de nouvelles perspectives thérapeutiques. L'immunothérapie doit avoir pour objet de stimuler les mécanismes de défense et freiner les mécanismes d'échappement.

5.5 Mise en évidence d'une immunité antitumorale *in vivo*

Chez l'homme:

Les quelques cas indubitables de régression spontanée de tumeurs malignes apportent un argument indirect en faveur de l'existence de mécanismes de défense de l'hôte contre sa tumeur. Cette notion est étayée par l'observation que l'infiltration de certaines tumeurs (particulièrement le cancer du sein) par des cellules lymphoïdes et des macrophages est souvent le signe d'une évolution favorable. Une preuve plus directe de la réactivité immunologique du sujet cancéreux vis-à-vis de sa tumeur est apportée par l'observation d'une réaction d'hypersensibilité retardée en réponse à l'inoculation d'antigènes solubles préparés à partir de cette même tumeur. Rappelons qu'un sujet ne peut développer une réaction d'hypersensibilité retardée à l'injection sous-cutanée d'un antigène que s'il est déjà immunisé contre cet antigène lors d'un contact antérieur. Des corrélations entre la présence d'une hypersensibilité aux antigènes tumoraux et une évolution clinique favorable ont été rapportées chez des patients atteints de leucémies lymphoïdes ou porteurs de certains types de tumeurs solides.

Chez l'animal :
La première preuve expérimentale qu'une tumeur déclenche des réactions immunitaires dirigées contre elle a été fournie par l'observation que des animaux guéris par l'ablation de leur tumeur, avant qu'elle ne donne des métastases à distance, sont devenus capables de rejeter un second inoculum de cette tumeur. Plus convaincante encore est la possibilité de transférer cet état de protection vis-à-vis de la tumeur, chez un animal normal, par l'injection des cellules lymphoïdes de l'animal guéri mais non par l'injection de son sérum, c'est-à-dire d'anticorps.

La notion que, parallèlement aux mécanismes immunologiques responsables de la destruction des cellules tumorales, doivent se développer des mécanismes permettant aux cellules d'échapper à cette agression, trouve une illustraiton parfaite dans les phénomène dit de *l'immunité concomittante*. Quand sa tumeur est encore de faible volume, un animal peut être capable de rejeter un second inoculum de la même tumeur déposé à distance de la tumeur primaire, dont la croissance n'est pas modifiée. A un stade plus avancé du développement de la tumeur primaire, l'immunité concomittante peut devenir non spécifique, c'est-à-dire qu'elle entraïne le rejet d'une seconde tumeur antigéniquement différent de la première. Quand le volume tumoral est trop important, le phénomène disparait : les deux tumeurs, antigéniquement identiques ou différentes, se développent indépendamment l'une de l'autre.

Les fonctions respectives des différents éléments (cellules et leurs produits) de la machinerie immunitaire dans les mécanismes aboutissant soit à la destruction, soit à la protection, des cellules tumorales ont été mises en évidence et analysées en majeure partie grâce à des études *in vitro*.

5.6 Mecanismes impliqués dans la destruction des cellules tumorales

Rôle des lymphocytes dépendants du thymus (lymphocytes T) :
C'est à partir de données expérimentales montrant que les lymphocytes T d'animaux ayant rejeté une greffe d'organe ou de peau, sont capables de lyser *in vitro* des cellules portant les mêmes antigènes d'histocompatibilité que le greffon, que Burnet a postulé dans sa *théorie de l'immunosurveillance* que les lymphocytes T sont responsables de l'élimination des cellules malignes apparaissant dans un organisme.

Les techniques mises au point pour l'étude de phénomène de rejet de greffes incompatibles ont été utilisées pour analyser l'action des lymphocytes T d'animaux ou de patients cancéreux, sur la survie des cellules tumorales *in vitro*.

Des lymphocytes T, capables de tuer spécifiquement les cellules tumorales dans une réaction exigeant un contact direct entre le lymphocyte et la cellule cible, sont présent dans les organes lymphoïdes de l'animal ou du patient cancéreux et également dans sa tumeur. Chez l'animal, dans certains systèmes tumoraux, une disparition des cellules T cytotoxiques dans les organes lymphoïdes a été observée quand la tumeur atteint un volume trop important. Elles réapparaissent après l'ablation de la tumeur. Chez l'homme, pour des raisons techniques, la preuve directe de la participation des lymphocytes T dans les réactions de cytotoxicité vis-à-vis de cellules tumorales autologues ou allogéniques, n'a pas encore été apportée.

La réaction de cytotoxicité exige au départ une reconnaissance spécifique par les lymphocytes T des antigènes présents à la surface cellulaire contre lesquels ils ont été sensibilisés *in vivo*. Il est difficile à l'heure actuelle de déterminer quels sont les antigènes tumoraux impliqués : sont-ce les antigènes spécifiques de tumeur, et parmi eux les antigènes de transplantations associés aux tumeurs, ou bien des antigènes de transplantation associés aux tumeurs, ou bien des antigènes embryonnaires réapparaissant sur les cellules tumorales, ou encore des antigènes complexes résultant d'une interaction entre les antigènes d'histocompatibilité et les antigènes induits par un virus oncogène ou un carcinogène chimique?

Il est difficile d'établir quel est le rôle *in vivo* des lymphocytes T cytotoxiques dans le rejet d'une tumeur. Le transfert adoptif d'une immunité antitumorale par les lymphocytes d'un animal guéri est beaucoup moins efficace quand les lymphocytes T ont été éliminés de la population. Il a été également possible de prévenir le développement de certaines tumeurs fortement antigéniques en injectant les cellules tumorales mélangées à des lymphocytes T spécifiquement cytotoxiques obtenus par une immunisation *in vitro* contre les antigènes tumoraux.

Les lymphocytes T peuvent intervenir dans le rejet des tumeurs par un autre mécanisme que la lymphocytotoxicité. Les lymphocytes T d'un animal ou d'un sujet cancéreux, incubés *in vitro* avec des extraits acellulaires de la tumeur, libèrent les médiateurs solubles dont la production est considérée comme une manifestation de l'immunité à médiation cellulaire et constitue donc un moyen de détection de l'existence de ce type d'immunité chez le sujet cancéreux. Les études sont encore trop récentes pour que des corrélations entre la production de médiateurs solubles *in vitro* et l'évolution de la tumeur *in vivo* soient établies.

Certains médiateurs, et particulièrement le facteur inhibiteur de la migration des leucocytes, peuvent être détectés dans le sérum des sujets cancéreux et on essaie actuellement d'établir si leur présence peut constituer un facteur de pronostic. Les médiateurs peuvent exercer un effet antitumoral par plusieurs mécanismes : certains sont directement toxiques pour les cellules tumorales, d'autres peuvent agir en amplifiant des réactions inflammatoires et favorisant ainsi l'accumulation de cellules capables de détruire les cellules tumorales, d'autres encore agissent sur les macrophages et les rendent capables de détruire les cellules tumorales. (Ces phénomènes sont décrits ci-dessous.) Une preuve directe du rôle de ces médiateurs dans le rejet des tumeurs est apportée par l'obtention de régressions tumorales chez l'animal et chez l'homme par l'injection locale de milieu ou de sérum renfermant ces médiateurs.

Rôle des lymphocytes indépendants du thymus (lymphocytes B) :
Des anticorps spécifiquement cytotoxiques, c'est-à-dire capables de tuer *in vitro* les cellules tumorales en présence de complément (ensemble de protéines présentes dans le sérum et responsable de l'attaque de la membrane cellulaire après que l'anticorps se soit fixé spécifiquement sur les antigènes de surface), ont été détectés dans le sérum d'animaux porteurs de tumeurs fortement antigéniques, généralement

après l'apparition des lymphocytes T cytotoxiques dans les organes lymphoïdes, et parfois seulement après l'exérèse tumorale.

L'observation la plus intéressante faite chez l'homme concerne le lymphosarcome de Burkitt dont l'agent étiologique serait le virus d'Epstein-Barr. La présence d'anticorps cytotoxiques dirigés contre les antigènes de membrane induits par le virus serait un signe pronostic de rechute de la maladie.

Les cellules tumorales sont plus ou moins sensibles à la lyse par les anticorps et le complément pour des raisons encore non définies. Cependant, cette susceptibilité peut être augmentée après traitement avec certaines drogues utilisées en chimiothérapie anticancéreuse.

Les observations suggérant un rôle éventuel des anticorps cytotoxiques dans le rejet des tumeurs sont excessivement limitées. L'administration d'anticorps cytotoxiques provenant d'animaux hyper-immunisés, et injectés à des animaux porteurs de tumeurs ont pu, dans quelques cas, défavoriser la croissance tumorale.

Ces dernières années, il a été montré que les anticorps peuvent collaborer avec des cellules définies par leur fonction, celle de lyser des cellules cibles recouvertes d'anticorps dirigés contre des antigènes de la surface cellulaire. Les cellules responsables de cette cytotoxicité à médiation cellulaire dépendante des anticorps sont désignées sous le nom de cellules K (K de killer).

Rôle des cellules K :

Les cellules K sont caractérisées par la présence, à leur surface, de récepteurs pour le fragment Fc des immunoglobulines, dans la majorité des cas de type IgG, et exceptionnellement de type IgM. La double fixation de l'anticorps, d'une part spécifique à la surface de la cellule cible, d'autre part non spécifique à la surface de la cellule K provoque l'activation de cette dernière qui lyse alors la cellule cible. Cette réaction réclame de très faibles quantités d'anticorps et sa spécifité ne dépend que de celle de l'anticorps.

La majorité des travaux effectués jusqu'à présent sur les cellules K avaient pour but leur identification. A côté des monocytes, des polynucléaires, des plaquettes et surtout des cellules nulles qui peuvent exercer cette cytotoxicité à médiation cellulaire anticorps dépendante, une dernière population cellulaire mise en évidence chez l'homme caractérisée par la présence à sa surface cellulaire de récepteurs pour les érythrocytes de mouton de très faible affinité, et qui pourrait être des lymphocytes T immatures, prend une importance de plus en plus grande dans les phénomènes de cytotoxicité dirigés contre des cellules tumorales. Il se dégage en effet que les cellules cibles ont une préférence, mal expliquée encore, pour certaines cellules effectrices. Les érythrocytes humains recouverts d'anticorps anti Rh sont ainsi presque exclusivement lysés par des monocytes.

L'étude des corrélations entre l'activité des cellules K et le développement tumoral est encore limitée. Chez la souris porteuse d'un fibrosarcome chimio-induit, l'activité des cellules K, testée sur des cellules cibles non tumorales, augmente avec le volume tumoral et retourne à des valeurs normales après éxérèse de la tumeur. Dans le modèle du sarcome induit par le virus de Moloney, pour lequel des régressions spontanées sont observées, une corrélation entre le taux sérique d'anticorps cytotoxiques et l'activité des cellules K, testée sur les cellules tumorales du sarcome, a été mise en évidence. Les deux types de cytotoxicité sont très élevés chez les animaux dont la tumeur régresse mais restent très faibles chez ceux dont la tumeur progresse.

Les observations cliniques sont encore très rares. Certains patients porteurs de carcinomes de la vessie, de mélanome, de cancer du sein, ont dans leur sérum des anticorps qui augmentent l'activité cytotoxique de leurs propres lymphocytes contre les cellules tumorales autochtones ou allogéniques. Des diminutions de l'activité des cellules K, dirigée contre des cibles non tumorales ont été rapportées chez certains patients atteints de leucémie ou porteurs de tumeurs solides.

Rôle des macrophages :

Celui-ci a été démontré par l'observation que des animaux infectés par divers parasites intracellulaires présentent une résistance accrue au développement de tumeurs spontanées ou greffées. Les macrophages des animaux infectés transfèrent cette résistance à des animaux non infectés et ils sont capables de freiner la prolifération ou de tuer des cellules tumorales, *in vitro*, de façon non spécifique. Ce sont des *macrophages activés*. Des phénomènes semblables peuvent être induits par le traitement avec certains adjuvants de l'immunité (BCG, *Corynebacterium parvum*, etc.).

Des macrophages provenant d'animaux immunisés contre une tumeur peuvent tuer spécifiquement ces mêmes cellules tumorales *in vitro*. Des macrophages d'animaux normaux peuvent devenir spécifiquement cytotoxiques après incubation avec des cellules lymphoïdes immunes. Ils sont dits armés et cet armement a été imputé, soit à la fixation d'un facteur spécifique libéré par les lymphocytes T (specific macrophage-arming factor or SMAF), soit à des anticorps cytophiles, l'armement aboutissant en fait dans ce cas à une cytotoxicité cellulaire dépendante des anticorps.

Rôle des cellules spontanément cytotoxiques, ou cellules NK :

Le développement des tests de cytotoxicité *in vitro* a abouti à l'observation que les lymphocytes d'animaux ou de sujets normaux sont capables de lyser une grande variété de cellules tumorales allogéniques ou hétérologues, en l'absence d'anticorps et de complément. On a donc été amené à admettre l'existence d'une résistance naturelle vis-à-vis des cellules tumorales, faisant intervenir des cellules appelées naturellement cytotoxiques (natural killer : NK).

L'identification des *cellules NK* a fait l'objet de nombreux travaux ces cinq dernières années. Les cellules NK ont été retrouvées dans la population de cellules nulles actives dans la cytotoxicité cellulaire dépendante des anticorps et très récemment dans la population particulière de lymphocytes T douée de cette même activité.

La cytotoxicité naturelle s'exerce contre la plupart des cellules tumorales maintenues en lignée mais la nature des structures reconnues à la surface des cellules cibles fait l'objet de nombreuses controverses. Chez la Souris, on a impliqué les antigènes dont l'expression est associée aux virus endogènes de type C. Certains faits expérimentaux suggèrent que des anticorps naturels, spécifiques pour des antigènes de tissu ou pour des antigènes associés aux tumeurs sont fixés à la surface des cellules NK par des récepteurs Fc dont l'affinité pour les anticorps serait beaucoup plus faible que celle des récepteurs portés par les cellules K.

Les cellules NK sont détectées chez les animaux et les patients cancéreux et les cytotoxicités imputées aux lymphocytes T cytotoxiques sont en fait généralement dues également aux cellules NK. Parmi les rares corrélations établies entre les cellules NK et l'évolution tumorale, citons l'observation d'une cytotoxicité naturelle plus élevée chez les femmes atteintes d'un cancer mammaire, dont la maladie progresse lentement que chez les femmes dont la maladie progresse rapidement et que chez les sujets normaux.

Chez l'animal, l'augmentation importante de l'activité des cellules NK observée chez la souris 'nue' est rendue responsable de l'efficacité des défenses antitumorales de cet animal, malgré l'absence de thymus.

La corrélation entre la sensibilité à l'action des cellules NK de différentes lignées de cellules tumorales, induites par le virus de Moloney, et la résistance des animaux à une greffe de ces mêmes cellules, apporte un argument convaincant en faveur de l'efficacité des cellules NK dans le contrôle de la croissance tumorale.

5.7 Les mecanismes d'échappement a l'immunité anti tumorale

La machinerie immunitaire est donc capable de détruire des cellules tumorales par différents mécanismes impliquant les lymphocytes T, les lymphocytes B, les cellules K, les cellules NK et les macrophages. Le développement tumoral résule donc de l'altération de ces mécanismes.

Rôle des cellules suppressives :
La responsabilité des lymphocytes T suppresseurs dans les mécanismes d'échappement a été suspectée après l'observation que, chez la Souris, la thymectomie à l'âge adulte permet une croissance plus lente de certaines tumeurs et diminue la fréquence de certaines tumeurs spontanées. Par contre; une accélération de la croissance tumorale et une augmentation de la fréquence des métastases a été induite par l'addition de lymphocytes T d'animaux porteurs de la tumeurs aux cellules tumorales au moment de la greffe. Dans le cas d'un sarcome chimio-induit de la souris, cet effet a pu être attribué à l'action spécifique de lymphocytes T suppresseurs agissant par l'intermédiaire d'un facteur soluble qui a été isolé. Il est spécifique, ne renferme pas d'immunolglobulines mais un produit codé par la région I-J du complexe majeur d'histocompatibilité. La libération de ce facteur est inhibée par l'ablation de la tumeur.

A côté de cette démonstration du rôle des lymphocytes T suppresseurs agissant de façon spécifique, de nombreuses observations montrent la présence de cellules suppressives non spécifiques dans les organes lymphoïdes d'animaux porteurs de tumeurs induites par des virus ou des carcinogènes chimiques. Ces cellules ont généralement été identifiées comme des macrophages, probablement activés, mais leur rôle *in vivo* sur la croissance tumorale reste à démontrer bien que la disparition des cellules suppressives ait été observée après l'ablation de la tumeur.

Chez l'homme, let états d'immuno-déficience accompagnant le développement de certains cancers ont été parfois attribués à la présence de cellules suppressives non spécifiques.

Rôle des anticorps et des facteurs bloquants :
Hellström et Hellström, en 1969, rapportent que les sérums de souris porteuses de sarcomes induits par le virus de Moloney préviennent spécifiquement la lyse des cellules sacromateuses par des lymphocytes immuns. Cette observation a soulevé un intérêt énorme et on y a vu l'explication de la croissance progressive des tumeurs par blocage des effecteurs cytotoxiques. Ces *facteurs sériques dits bloquants*, sont détectés dans le sérum d'animaux porteurs de tumeurs viro-ou chimio-induites, ou de tumeurs spontanées, et, dans le sérum de patients cancéreux. Des facteurs bloquants ont également été détachés de la surface des cellules tumorales.

L'étude de la nature de ces facteurs bloquants et de leurs mécanismes d'action ont fait l'objet de nombreux travaux. Il s'est révélé que l'activité bloquante est due essentiellement à des complexes formés par les anticorps spécifiques et l'antigène tumoral libéré sous forme soluble dans la circulation. Ils agissent, soit en masquant les antigènes de la cellule cible prévenant l'action cytotoxique des lymphocytes T ou des cellules K, soit en bloquant, (a) par la partie antigène, les récepteurs des lymphocytes T; (b) par la partie anticorps, les récepteurs Fc des cellules K. L'antigène soluble libre peut également être bloquant en se fixant sur les récepteurs des lymphocytes T, ou en neutralisant les anticorps intervenant dans l'activité des cellules K.

Les mécanismes d'action des facteurs bloquants ne sont pas encore élucidés. La fixation sur les récepteurs des lymphocytes T préviendrait la reconnaissance ultérieure de l'antigène. Récemment, il a été isolé du sérum de souris porteuses d'un fibrosarcome, un facteur glycoprotéique qui ne semble être ni un anticorps, ni de l'antigène soluble, et qui rend compte de l'activité bloquante du sérum. Ce facteur pourrait agir en activant des lymphocytes T suppresseurs.

Les facteurs bloquants sont mis en évidence par des test *in vitro*, mais leur présence semble liée à l'évolution de la tumeur *in vivo*. Chez l'homme et chez l'animal, ils disparaissent, en effet, après l'éxérèse tumorale quand la tumeur régresse spontanément ou sous l'influence d'un traitement. Une preuve plus directe de leur intervention dans les mécanismes d'échappement est apportée par la possibilité d'obtenir une facilitation de la croissance tumorale par l'injection de sérum ayant une activité bloquante à une dose suffisante pour retrouver cette activité dans le sérum des animaux traités.

Le sérum d'animaux ou de patients dont la tumeur a régressé devient capable de neutraliser l'activité des facteurs bloquants. Cette activité débloquante est attribuée à des anticorps spécifiques qui agissent en dissociant les complexes antigène–anticorps bloquants fixés sur les cellules cytotoxiques. Des injections de sérum débloquant a pu, dans certains cas, entraîner des régressions tumorales chez l'animal.

Nous avons décrit les mécanismes principaux qui peuvent conduire à l'élimination des cellules tumorales et à leur protection contre cette destruction immunologique. Certains d'entre eux sont induits spécifiquement par les antigènes associés aux tumeurs. D'autres s'exercent de façon non spécifique, c'est le cas des macrophages activés, des cellules NK et, si on fournit les anticorps, des cellules K. Dans la mesure où il apparait de plus en plus probable que les tumeurs spontanées sont peu ou pas antigèniques, il faut s'intéresser aux moyens d'amplifier les mécanismes de rejet et de freiner le développement des mécanismes d'échappement faisant intervenir des effecteurs non spécifiques.

III   L'immunosurveillance des cancers

Ehrlich suggéra, en 1908, que l'une des fonctions du système immunitaire est de prévenir la formation des tumeurs. Cette idée a été reprise par Thomas en 1959 puis par Burnet, en 1970, sous la forme d'une théorie générale, selon laquelle la fonction majeure des mécanismes immunologiques chez les mammifères est de reconnaitre et d'éliminer les cellules anormales qui apparaissent à la suite de mutations somatiques ou de processus èquivalents.

Ce concept de l'immunosurveillance renferme les postulats suivants ; (1) les cellules tumorales apparaissent avec une fréquence élevée; (2) ces cellules portent des antigènes nouveaux;(3) elles sont, en conséquence, reconnues comme étrangères par l'hôte et éliminées par un mécanisme immunologique identique à celui responsable du rejet des allogreffes, avant que la tumeur ne soit cliniquement détectable.

Les arguments les plus convaincants en faveur de l'immunosurveillance ont été apportés par l'analyse de la résistance aux tumeurs induites par des virus cancérogènes. Ces tumeurs possèdent des antigènes de surface spécifiques du virus, et certains d'entre eux fonctionnent comme des antigènes de transplantation.

La fréquence d'apparition des tumeurs induites par les virus cancérogènes, et particulièrement les virus à ADN, est augmentée chez les animaux immunodéprimés par une thymectomie néonatale, une irradiation ou l'administration de sérum anti-lymphocytaire, chez l'animal nouveau-né et chez les souris dépourvues congénitalement de thymus (souris 'nues'). **Les lymphocytes T semblent jouer un rôle essentiel dans le développement de la résistance aux tumeurs induites par des virus, mais des mécanismes immunologiques indépendants des lymphocytes T sont également parfois impliqués.**

Les adversaires de la théorie de l'immunosurveillance interprètent la résistance aux tumeurs induites par les virus comme un reflet de l'intensité de l'immunité anti-virale. En fait, il est possible de susciter une réaction immunitaire contre les antigènes de transplantation associés aux tumeurs en l'absence de toute production de virus. **Cependant, la capacité de reconnaitre ces antigènes est sous la dépendance de gènes** liés au complexe majeur d'histocompatibilité, du moins en ce qui concerne le virus de la leucémie de Marek chez la poule.

Les mécanismes de défenses contre les tumeurs induites par des cancérogènes chimiques et contre les tumeurs spontanées donnent matière à de nombreuses controverses. Un certain nombre de tumeurs induits par des cancérogènes chimiques sont caractérisées individuellement par des antigènes de surface associés aux tumeurs et induisent des réactions de rejet chez des hôtes autochtones ou syngéniques. Ces deux dernières années, les preuves se sont accumulées en faveur de la faible ou de la non-immunogénicité des tumeurs spontanées. La théorie de l'immunosurveillance explique le développement des tumeurs par une sélection de cellules tumorales ayant une antigénicité ou immunogénicité faible ou des cellules résistantes aux mécanismes effecteurs de l'immunité antitumorale. Cette hypothèse est infirmée par le fait que les cellules transformées *in vitro* par des cancérogènes chimiques, donc en dehors de toute pression sélective de nature immunologique, ont une antigénicité très variable.

Les analyses de l'influence de l'état immunitaire de l'hôte sur le développement des tumeurs induites par des cancérogènes chimiques et des tumeurs spontanées ont souvent mené à des interprétations contradictoires. La fréquence des tumeurs spontanées est très augmentée chez les sujets souffrant de déficits immunitaires primaires ainsi que chez les patients ayant reçu des traitements immunosuppressifs pour des transplantations d'organe. Ces observations semblent en faveur de l'efficacité de l'immunosurveillance, mais il est difficile d'expliquer, à la lumière du concept, pourquoi la majorité des tumeurs sont de type lymphoïde. Certains faits expérimentaux suggèrent que l'assocation d'une immunodépression et d'une stimulation antigénique chronique (infections, greffon) favorise la prolifération incontrôlée des cellules lymphoïdes. Cependant, certaines formes de lèpre s'accompagnent d'un effondrement de l'immunité à médiation cellulaire sans qu'on observe une augmentation de la fréquence des tumeurs lymphoïdes. Expérimentalement, il n'est pas observé d'augmentation de la fréquence des tumeurs chimio-induites ou des tumeurs spontanées dans les sites immunologiquement protéges (cerveau, poche jugale du hamster), mais on connait mal le mécanisme de protection. Les observations faites chez les souris 'nues' sont souvent difficiles à interpréter. Les animaux éléves dans des conditions normales ne parviennent pas à l'âge auquel apparaissent les tumeurs spontanées chez la souris normale. Les animaux maintenus dans un milieu exempt de germes ont une durée de vie normale. La fréquence des tumeurs chimio-induites et des tumeurs spontanées, dans ces conditions, n'est pas différente de celle observée chez la souris normale, exception faite pour les tumeurs lymphoïdes dont la fréquence est augmentée chez la souris 'nue' âgée.

Les études menées chez les souris 'nues' ont permis de mettre en évidence l'efficacité de mécanismes immunologiques 'nues' indépendants les lymphocytes T dans l'élimination des cellules tumorales. Ces souris possèdent des anticorps naturels capables de détruire des cellules tumorales en présence de complément. Elles peuvent synthétiser des quantités suffisantes d'anticorps IgG spécifiques pour permettre l'action des cellules K dans une cytotoxicité à médiation cellulaire dépendante des anticorps. L'activité des cellules naturellement cytotoxiques (cellules NK) pour certaines cellules tumorales est plus élevée que chez les animaux normaux. Les macrophages sont naturellement fortement activés et, par là même, ils peuvent détruire efficacement et non spécifiquement des cellules tumorales.

Parallélement au concept de l'immunosurveillance, s'est développée la théorie de l'immunostimulation défendue par Prehn, selon laquelle une réponse à médiation cellulaire est nécessaire pour que se développe une tumeur maligne. La plupart des faits expérimentaux étayant cette théorie peuvent s'expliquer par l'action de cellules T suppressives.

En conclusion, le concept de l'immunosurveillance, tel qu'il a été formulé à l'origine, ne peut s'appliquer qu'à un nombre restreint de tumeurs expérimentales induites par des virus. Il semble raisonnable, pour tous les autres types de tumeurs expérimentales ou spontanées, d'élargir ce concept plutôt que de le rejeter sans appel. Tout d'abord, il ressort d'observations sur certaines lignées de cellules dites 'lymphoblastoïdes' humaines infectées par le virus d'Epstein–Barr, et sur l'action de cancérogènes chimiques sur des cultures de cellules, que la transformation maligne, définie par la capacité de donner naissance à une tumeur *in vivo*, est un évènement rare. D'autres mécanismes immunologiques que ceux faisant intervenir les lymphocytes T peuvent jouer en rôle essentiel dans la surveillance, et ceci d'autant plus efficacement que certains d'entre eux, impliquant les macrophages activés et peut-être les cellules NK, n'exigent pas que les cellules tumorales portent des antigènes spécifiques de tumeur. Des facteurs génétiques semblent également intervenir : gènes de susceptibilité à certains virus, gènes intervenant dans la capacité de répondre à certains antigènes, etc. Il serait plus exact de parler de surveillance plutôt que d'immunosurveillance qui présume trop de la spécificité des mécanisms impliqués.

6

Astrophysics

Bengt Strömgren

6.1 Introduction

In 1908 G. E. Hale published a book, *The study of stellar evolution*, that was to exert a significant influence on the development of astrophysics. Its opening sentence is of interest, even today:

It is not too much to say that the attitude of scientific investigators toward research has undergone a radical change since the publication of the *Origin of Species*. This is true not only of biological research, but to some degree in the domain of the physical sciences. Investigators who were formerly content to study isolated phenomena, with little regard to their larger relationships, have been led to take a wider view.

Fifty years later, at the beginning of the period 1958–78 that is under consideration here, research in astrophysics was indeed aiming at the solution of problems of the *evolution* of celestial bodies over millions or billions of years, as well as the exploration of the properties of these bodies as observed during the present epoch.

At the time when Hale's book was published, some progress in the study of evolution of the stars had been achieved through attempts to put together data concerning stellar properties referring to the same epoch in order to form a scheme of evolution. Within the following two decades, considerable progress in this direction was made through the work of E. Hertzsprung, H. N. Russell, and A. S. Eddington. The situation of the astrophysicist trying to understand problems of stellar evolution was then sometimes compared to that of the botanist studying a forest and putting together the evidence observed at a given time in order to obtain a picture of growth and evolution.

Fifty years later, the astrophysicist concerned with stellar evolution had at his disposal much more powerful tools. It had become possible to follow, through calculation, chemical and physical changes with time that take place in stellar interiors and to investigate the consequences of such changes for the total structure and observable appearance of the stars. The properties of interstellar matter in our galaxy were being explored using observational techniques of both optical and radio astronomy, and the interaction of stars and interstellar matter was under study. Investigations of the structure and evolution of our own galaxy, as well as neighbouring galaxies, were under way.

In the sections that follow, we consider progress during the period 1958–78 in a number of areas of astrophysics. The emphasis is on the astrophysics of our galaxy and its principal components, stars and interstellar matter; the solar system is dealt with here very briefly, and problems of extragalactic astrophysics only in outline.

6.2 Structure and evolution of stars

Massive, luminous early-type stars radiate so much energy per second into surrounding space that it can be concluded with certainty that their lifetimes as luminous stars must be a good deal less than 100 million years. This follows from a comparison of energy output per second with the available nuclear energy. On the other hand, the age of our galaxy, as estimated from the age of its oldest stars, is somewhat over 10×10^9 years, and it is known that general conditions in our galaxy have changed very little over the last 100 million years. It follows that massive, luminous early-type stars must have been formed out of galactic interstellar matter with properties such as are observed in the present epoch.

In certain galactic clusters, massive, luminous early-type stars are found together with much less luminous, less massive stars of later spectral types. The stars in a cluster must be of approximately the same age, and it can thus be concluded that under conditions as observed at present the formation of stars out of interstellar matter takes place in such a way that their masses cover the whole range from about 50 solar masses to 0.1 solar mass, and possibly down to one-hundredth of a solar mass.

We shall return to questions concerning the formation of stars out of interstellar matter in §§ 6.3 and 6.5. Here we consider the evolution of celestial bodies that have passed the earliest stages of contraction and have become opaque star-like bodies in quasi-hydrostatic equilibrium. Detailed investigations have shown that, in this stage of evolution, the temperature throughout the star is so low that no appreciable release of nuclear energy can take place. The stars contract, and thereby release gravitational energy. The rate of contraction is such that the corresponding net release of energy is equal to the loss of energy by radiation into surrounding space. Energy transfer through the star is by convection. In this phase, the convective zone extends all the way from the atmosphere to the centre of the star. As a consequence, the stellar matter in the interior is throughly mixed, and a chemically homogeneous star should result, even if inhomogeneities existed in a previous phase of evolution of the body. In this phase, a star of solar mass has a surface temperature of about 3000 K. At the beginning of quasi-hydrostatic contraction, the luminosity is more than 100 times the present luminosity of the sun, i.e. the star appears as a red giant. In the course of a relatively very short time, about a million years, the luminosity decreases to a value only a few times higher than that of the sun, while the surface temperature increases to about 5000 K, not much different from that of the sun. After about 30 million years, the temperature in the central regions reaches several million degrees, and it is now high enough for release of nuclear energy to take place. The internal structure of the star readjusts itself to a situation where the energy lost by radiation into space is completely balanced by release of nuclear energy through conversion of hydrogen into helium in the central regions of the star, i.e. there is no further gravitational contraction. The star is thus secularly stable and has entered the so-called main-sequence phase of evolution. The period of readjustment lasts another 30 million years. The phase preceding the main-sequence phase has a total duration that is less than one-hundredth that of the lifetime in the main-sequence phase.

For stars more massive than the sun, the evolution through the contraction phase follows a similar pattern. However, for a star of 4 solar masses, the duration of the contraction phase is more than 10 times shorter. On the other hand, for a star with a mass 0.1 that of the sun the contraction period lasts about 10 times longer than for the sun, but it is still of much shorter duration than the main-sequence phase. For stars with masses less than about 0.07 solar mass, the central temperature never becomes high enough for significant energy release through nuclear processes to take place, and the star continues to contract, albeit at a very slow rate because its rate of radiation into space is very low.

We turn now to the important main-sequence phase of stellar evolution. By 1958, questions of the structure and evolution of main-sequence stars were well understood. During the next decade investigations of the physical properties of stellar interiors were combined with extensive computer-aided calculations of evolutionary sequences for various stellar masses. This work led to a quantitative description of internal structure and evolution during the main-sequence phase.

Stars of about 1 solar mass have an outer convective zone, beginning at the bottom of the atmosphere and extending to a depth equal to about one-tenth of the solar radius. Inside this zone the matter is in radiative equilibrium, i.e. energy transport is by photons, a net flux of radiation being driven outward by a temperature gradient. In the present sun the central temperature is close to 15×10^6 K, while the density is somewhat over 10^5 kg m^{-3}. The matter is a highly ionized plasma, and in spite of the high density this plasma behaves very nearly like a perfect gas. Energy production is through conversion of hydrogen into helium. It is practically limited to a central region with a radius equal to one-fifth of the solar radius. Outside the energy-generating core the temperature is below 7×10^6 K, too low for the nuclear processes to proceed at a high enough rate for significant release of nuclear energy.

Throughout the main-sequence phase, the relative content of hydrogen in the energy-producing central region decreases, while there is a corresponding increase in the relative helium content. Since there is no mixing of the matter by convection, the relative hydrogen and helium are 'frozen in', i.e. a chemical gradient results, the relative hydrogen content being lowest at the centre. At the present phase of evolution of the sun, corresponding to its age of 4.5×10^9 years, the relative hydrogen content in the outer part is practically equal to its original value, a little over 70 per cent by weight. In the centre, however, the relative hydrogen content has been reduced to about one-half of the original value, with a corresponding increase in helium content. The gradients in the hydrogen and helium contents cause a change in the entire structure of the star. Through the main-sequence phase, central temperature and density increase somewhat with age, and radius and luminosity also increase.

For stars with masses larger than about 1.5 times the mass of the sun, the picture is different. The deep outer convective zone disappears, while a convective core is present. Outside this core the matter is in radiative equilibrium. Nearly all of the release of nuclear energy, which is still caused by conversion of hydrogen into helium, takes place in the convective core. As a consequence, the relative hydrogen content is the same throughout the core, but with increasing stellar age the hydrogen content becomes less within the core than in the surrounding region of the star. Computer-aided calculations have shown that the energy-producing convective core shrinks as the star grows older. As the core recedes, a transition zone is formed through which the relative hydrogen content increases from the lower value in the core to the initial value found in the outer regions where energy-production has never occurred. For a star of 10 solar masses the central temperature is somewhat higher than in the sun, being about 30×10^6 K at the beginning of the main-sequence phase, and increasing somewhat with stellar age. The central density, on the other hand, is an order of magnitude lower than for the sun, and increases slightly with age during the main-sequence phase. The rate of energy production through conversion of hydrogen to helium increases very steeply with temperature, and this explains why a doubling of the temperature (from 15×10^7 to 30×10^6 K) suffices to increase energy production by a factor of about 10 000. Throughout the main-sequence phase, and for the whole relevant mass range, the total energy release in the central regions is equal to the luminosity of the star, i.e. the energy radiated by the star into surrounding space. Since the luminosity is much higher for the more massive stars — it increases almost as the fourth power of the mass — the rate of conversion of hydrogen into helium is much higher. This means that the duration of the main-sequence phase is much shorter for massive stars than for the solar-type stars.

For stars in the range 10–50 solar masses, the problems of structure and evolution are more complicated, particularly with regard to the transition zone between the convective core and the outer mantle, which retains the original chemical composition. Furthermore, for stars of about 20 solar masses or more, conditions are such that during the main-sequence phase there is an appreciable mass loss through streaming from the surface into surrounding space. These very luminous, massive stars with lifetimes on the main sequence of only a few million years, although relatively rare, are of great interest to the astrophysicist. In recent years, a combination of theory and observation has led to considerable progress with regard to the problems they present.

For main-sequence stars with masses that are appreciably lower than that of the sun, the most important fact is that their radiation into surrounding space is relatively very low. For a mass equal to one-quarter of a solar mass, the luminosity is more than one hundred times smaller than for the sun, and for a main-sequence star of 0.1 solar mass the luminosity is less than a thousandth of the solar value. This means that the stars in question convert hydrogen into helium at a very low rate. In other words, their evolutionary changes during the whole period of existence of our galaxy are practically negligible. Only when the value of the mass is fairly close to the solar value do appreciable evolutionary changes occur. This fact is of considerable importance, for the majority of the main-sequence stars have masses that are a good deal smaller than the sun's mass. We shall discuss this point later.

For all main-sequence stars with masses larger than about 0.7 solar masses, an appreciable, continuing reduction of the relative hydrogen content occurs in those parts of the stellar interior where the temperature is high enough for energy release through nuclear processes to take place. This takes place over a time period equal in length to the age of our galaxy ($10-15 \times 10^9$ years). All these stars will reach a point during that period when the nuclear fuel that is sufficiently hot and available is used up. For large masses this happens quickly: for masses larger than 2 solar masses within 1000 million years, for those larger than 5 solar masses within 100 million years, and for those larger than 10 solar masses within 20 million years.

It is clear that the epoch when this happens marks a critical point in the evolution of the star: it is the end of the main-sequence phase. Drastic changes in the internal constitution now take place, and the star gradually becomes a red giant, characterized by high luminosity, large radius and relatively low surface temperature.

At the end of the 1950s this phase of stellar evolution was partially understood. Extensive investigation, based on new physical information about the properties of high-temperature stellar matter and very effectively supported by large-scale computer-aided calculations, have led to a much more complete and detailed picture.

Consider the evolution of a star with a mass equal to that of the sun. When the main-sequence phase has come to an end, the star as a whole will expand. However, the central regions will contract and will also reach higher temperatures than those occurring during the main-sequence phase. As a consequence, layers outside the helium core that was formed during the main-sequence phase, i.e. layers that still contain a large fraction of hydrogen, will now become hot enough for conversion of hydrogen to helium to take place at an appreciable rate. This means that release of nuclear energy now takes place in a hydrogen-rich shell just outside the helium core. The hydrogen in this shell will gradually become exhausted, as happened during the main-sequence phase in the core region. However, the internal structure will change in such a way that sufficiently high temperatures are obtained in regions still further away from the centre where hydrogen is present. Thus the energy-producing shell will gradually move outward, and the helium core will grow. The internal structure continues to change, the radius increasing with stellar age, so that the mean density of the star decreases while the density in the inner regions increases, as does the temperature there. These gradual changes indicate an evolution of the star into the red-giant stage.

Two important developments take place in the latter part of the life of a star of solar mass in the red-giant stage. As already mentioned, the central-region temperature increases with age, and when it reaches about 200×10^6 K in the helium core, conversion of helium into carbon occurs at an appreciable rate. The nuclear processes in question were understood before 1958 and further investigated subsequently.

Two helium nuclei (i.e. α-particles) collide and form an unstable nucleus of atomic weight 8, and when the temperature is high enough, capture of a third helium nucleus will occur before fission for an appreciable fraction of the unstable nuclei, thereby forming a stable carbon nucleus. This so-called triple-alpha process leads to very substantial release of nuclear energy in the helium core; in other words, the star has a new source of energy when it reaches this stage. Under these circumstances, a star of solar mass is, however, unstable. This obtains because the nuclear-energy release takes place in a high-temperature plasma where the free electrons form a degenerate gas in which the pressure is practically independent of temperature. When this is the case, a temperature-control mechanism such as exists in the main-sequence phase, and keeps nuclear-energy production at a stable level, is not present. The result is a very rapid increase of nuclear-energy production, a thermal runaway causing a so-called helium flash. As a consequence, the whole internal structure of the star changes rapidly. The radius of the star becomes much smaller, the luminosity somewhat smaller, the degeneracy of the electron gas in the helium core is lifted, and the star can now burn its helium into carbon under stable conditions. What happened during the main-sequence phase is thereafter repeated under different circumstances. The helium supply in the core gradually becomes completely exhausted, and energy production moves to a surrounding helium shell, the star as a whole moving into the red-giant region again; i.e. the mean density of the star as a whole decreases, while the density of the central regions increases. The second evolution into the red-giant stage has been studied in recent years through extensive computer calculations, and a number of difficult problems connected with the development of both hydrogen-rich and helium-rich energy-producing shells, and the temporary occurrence of thermal instabilities have been clarified.

The second development that takes place in the latter part of the red-giant phase is the occurrence of very appreciable loss of mass by the star through streaming of matter from the stellar surface into surrounding space. That this is the case has been concluded on the basis of three observations.

(1) A number of red-giant stars have been found, through spectroscopic investigation and infrared photometry of grains in a circumstellar shell, to be surrounded by expanding shells of matter. Estimates of the magnitude of the mass flow, combined with theoretical calculations of the duration of the evolutionary phase in question, show that the mass loss amounts to an appreciable fraction of star's mass.

(2) Investigations of the important class of celestial objects called planetary nebulae strongly suggest that an object of this kind, consisting of a relatively dense star of high surface temperature surrounded by a luminous expanding shell of appreciable mass, originated in a red-giant star that split into a dense star containing the red-giant core, and a surrounding shell that gradually (over some 30 000 years) dissipated into surrounding space through its expansion. The dense star contracts still further and, with the complete exhaustion of its nuclear energy supply, becomes a so-called white dwarf of very low luminosity, extremely small radius (of the order of one-hundredth of a solar radius) and very high mean density (almost 10^9 kg m^{-3}).

(3) A few galactic clusters, among them the Hyades cluster, are known to contain white dwarfs, i.e. stars that are the end-product of an evolution such as is described above. Now it can be assumed that all the stars in a galactic cluster are of nearly the same age, and the age in question can be fairly accurately deduced from the observed state of evolution of the most luminous and most massive cluster stars. The age being known, it can be estimated what the original mass must have been of

those stars that have been through their evolution and become white dwarfs. This mass must exceed a certain minimum, otherwise evolution would have been too slow for the star to reach the white-dwarf stage. A comparison of the original mass with the actual white-dwarf mass, typically 0.6 solar mass (always less than the so-called Chandrasekhar limit of 1.4 solar masses), shows that a very substantial mass loss must have taken place.

Quantitative theoretical calculations of the rate of mass loss from red giants have not yet been carried out, but the phenomenon is understood in a general way. It is connected with the fact that the surface gravity of red giants is quite low in comparison with the surface gravity of main-sequence stars. The existence in red giants of deep outer convective zones is also relevant.

We have described the evolution of a star of solar mass. Numerical calculations of evolutionary sequences have shown that the general pattern of evolution is very much the same for the initial mass range from 0.7 to 2 or 2.5 solar masses. It is quite possible that stars up to 3 or 4 solar masses follow much the same pattern, in particular that the end product also is a white dwarf. In fact, the galactic-cluster argument just mentioned leads to this conclusion, and similar investigations of the components of certain binary stars, assumed to be of nearly equal age, point in the same direction. The explanation would be that mass loss in the red-giant stage for these more massive stars is so large that they are forced to follow an evolutionary path similar to that of stars starting with somewhat lower masses.

An important point must be noticed in considering the stars with masses in the range from about 2 to 3 or 4 solar masses. Here the evolution that takes the star into the red-giant region after the end of the main-sequence phase is a very rapid one. While for the solar-mass star, where this evolution takes place quite gradually, intermediate-stage stars are well represented among actual stars, for instance in clusters, this is not so for those of larger mass. Between the main-sequence stars and the red giants there is a distinct gap, often referred to as the Hertzsprung gap. The distribution of luminosities and surface temperatures observed for the stars of the Hyades cluster illustrates the point. Here the most luminous stars fall into two widely separated groups, namely, main-sequence stars of spectral type A and red giants of spectral type K.

We now turn to the question of the post-main-sequence evolution of more massive stars. Consider stars that have masses in the range 4–8 solar masses. When they reach the red-giant stage after a very rapid transition from the main-sequence, such as just described, they are characteristically different from the lower-mass stars, in that no helium flash takes place. The reason for this is that the conversion of helium into carbon in the stellar core now takes place at temperatures and pressures such that the electron gas is nondegenerate, i.e. there is no thermal runaway, but stable burning of helium does occur. Under these circumstances, helium is gradually used up and the structure changes in the sense that core temperatures and densities become still higher. A point is reached, at temperatures of about 300×10^6 K and densities near 2×10^{12} kg m^{-3} when carbon nuclei react with each other and form heavier nuclei, such as ^{23}Na, ^{20}Ne, and ^{24}Mg. A fraction of the carbon nuclei have been built up to oxygen nuclei through capture of a helium nucleus. Fusion of two oxygen nuclei yields ^{28}Si and other heavier nuclei.

It has been possible to follow the evolution of stars with initial masses in the range 4-8 solar masses with reasonable accuracy up to the point just considered, when carbon nuclei react with each other. This has been achieved through extensive numerical calculations that simulate the physical and chemical changes taking place in the stellar interior, a large number of successive relatively small-time steps being taken in the numerical process. Of great importance has been the availability of nuclear-physics input. This concerns reaction cross-sections for the relevant nuclear processes as well as, very importantly, neutrino emission rates as a function of temperature and density of the stellar matter. At this stage of evolution, the emitted neutrinos have energies and corresponding cross-sections of interaction with stellar matter low enough for neutrinos to travel straight through the star into surrounding space after emission. This leads to a very considerable energy loss for the matter in the stellar core, with important consequences for the evolution of the star.

When the stage of carbon–carbon reactions has been reached for stars 4-8 times more massive than the sun, the density (2×10^{12} kg m^{-3}) is so high in comparison with the temperature (3×10^8 k) that the electron gas is degenerate. This means that there is again a thermal runaway, a carbon flash, and this time a much more violent event than the helium flash described above. It should be emphasized that the situation encountered here leads to questions that have not yet been answered satisfactorily through theoretical calculations. It appears that there are two possibilities: either the star explodes in such a way that all the matter dissipates into space, or else the outer part of the star explodes into space while a core remains as a remnant star. In both cases, the outside observer should see a supernova explosion, and in the second case a remnant in the form of an extremely dense star should be observed near the center of an expanding nebula, formed by the matter streaming with very high velocity order (of magnitude 10^3-10^4 km s^{-1}) into surrounding inter-stellar space.

At this point we depart from the description of results of theoretical calculations and turn to observation. One of the most remarkable discoveries made since 1958 is that of extremely dense neutron stars, observed by radio astronomers as so-called pulsars. The pulsars stand out as radio point sources that show a periodic variation of the emitted intensity, with periods in some cases considerably shorter than one second, in other cases of several seconds. The great majority of the periods are in the range 0.2-2 s. The periods are remarkably constant, but very accurate observations have revealed the existence of extremely small secular changes, and also jumps in the period. The fact that the periods are short and very constant suggests that the radio emission has its origin in a rotating star of very small radius, much smaller than even the radius of a white dwarf. Detailed examination has definitely established that pulsars are stars with radii of the order of 10 km rotating with periods of the order of a second and endowed with strong magnetic fields on the surface. The radio emission is caused by the combination of rotation and a magnetic field.

In some cases it has been possible to estimate the mass of a pulsar because it is a member of a binary star, and it appears that pulsar masses are of the order of 1 solar mass, and that they are somewhat larger than typical white-dwarf masses. From the known radius and mass, it is found that the average density of the stellar matter in a pulsar is of the order of 10^{17} kg m^{-3}. Theoretical calculations concerning the properties of matter of extremely high densities then lead to the conclusion that the larger fraction of the interior of a pulsar consists of matter in the form of neutrons: hence the term 'neutron star' to describe these objects.

Somewhat more than 300 pulsars have been discovered. Here we shall describe results concerning the pulsar found at the centre of the so-called Crab nebula, because the results have a particularly important bearing on the question of supernova explosions.

The Crab nebula is a galactic object located at a distance of about 2000 parsecs. It has an angular extension of 6 minutes of arc, and its linear size is thus a few parsecs. Detailed measurements have shown that the nebular matter is expanding in all directions, away from a well-defined centre, with a velocity of about 1000 km s^{-1}. If it is assumed that the expansion has taken place at a constant rate, then the nebula originated in an explosion that took place a little less than 800 years ago. Records of new stars show that a very bright nova appeared at the position in the sky of the Crab nebula in the year A.D. 1054. From the estimated brightness of the new star and its distance, it is concluded that it was a supernova. There is no doubt that the phenomena are connected. The observed expansion rate is, in fact, compatible with an orgin in the year 1054, if it is assumed that a relatively small acceleration has occurred.

At the centre of the expanding Crab nebula, a pulsar emitting with a period of 0.03 s has been found, and also an optical counterpart whose light varies with the same period as found for the pulsar. There is no doubt that this is a neutron star. The result of the supernova explosion in 1054 was thus an expanding nebula and a neutron star remnant.

Statistical investigations concerning the rate of supernova explosions in our galaxy, and the rate of formation of neutron stars, lend further support to the conclusion that the phenomena are connected. The statistics are not, however, very accurate, and the detailed comparison of the phenomena is complicated by the fact that there are at least two types of supernova explosions. The greater part of the supernovae appear to have been connected with explosions in massive stars, but in the case of the particularly well studied Crab nebula, there is uncertainty about the mass of the exploding star. Thus, although a very important connection has been established, there is still no definite answer to the question of the ultimate fate of stars in the range of 4–8 solar masses.

We finally consider the most massive stars, those of 8–50 solar masses. Like the stars of 4–8 solar masses, they exhaust hydrogen in the central regions, very rapidly move into the red-giant region, then pass through a helium-burning phase, which is thermally stable, and reach the stage of evolution where the central temperature is so high (about 700×10^6 K) that carbon burning takes place. For stars in this mass range, however, there is a very important difference: the density at this stage is much lower (of the order of 10^8 kg m^{-3}); this means that the electron gas is non-degenerate so that there is no thermal runaway. In other words, the carbon-burning phase is now passed without an explosion occurring, and the star can develop further.

The phase of evolution that follows is of great interest to the astrophysicist, and during the past few years considerable progress has been made in its exploration. During this phase, the temperature and the density in the central regions of the star continue to increase, and as a consequence a large number of nuclear reactions become important. To follow the evolution in detail is a computational task of great magnitude. Certain general features of the evolutionary changes emerge, and we shall describe the most significant ones.

As a typical and important example consider the evolution of a star of 20 solar masses. In the earlier stages of evolution a helium core of 8 solar masses develops, and in this core burning of helium to carbon (and to some extent to oxygen) occurs. When the temperature reaches about 400×10^6 K, the burning of carbon to heavier nuclei begins; as explained above, this takes place at a density (about 10^8 kg m^{-3}) that is too low for thermal runaway to occur, so that further evolution proceeds through the stage of stable carbon burning.

The density of the core is now so high compared to the average density of the star, that the core evolution is only weakly coupled to evolutionary changes in the outer mantle; the core, so to speak, lives its own life. This simplifies the numerical calculations of the evolutionary changes.

What happens is that the carbon fuel in the central regions is exhausted, and carbon burning, with energy release, is shifted to a surrounding shell. A neon core is produced, and grows as the surrounding carbon-burning shells are successively exhausted. As the neon core grows, its central temperature and density increase, with the consequence that neon burning sets in and nuclei of still heavier elements are formed. The play continues with core-fuel exhaustion, shell burning, core growth with increase of central temperature and density, and finally ignition of new, heavier nuclei continues. A build-up of elements from silicon to iron and nickel takes place. However, as has long been known, this is where build-up with release of energy stops, and no appreciable further thermonuclear energy generation is possible.

We have followed the core of the star of 20 solar masses to a point where the central temperature is 8×10^9 K and the density is 4×10^{12} kg m^{-3}. What was originally a helium core of 8 solar masses has become an iron core of about 1.4 solar masses surrounded by successive shells consisting mainly of silicon, oxygen, neon, and carbon. At this point an extremely pronounced change in conditions takes place.

The change is triggered by a process of capture of electrons by nuclei, which now – at the high densities that we are considering – takes place at a very great rate. The effective adiabatic constant of the plasma is thereby reduced below the critical value of $\frac{4}{3}$, with the consequence that the iron core becomes dynamically unstable and undergoes violent contraction.

The violent contraction, or collapse, of the central part of the core leads to an increase of the central density from the value just mentioned, 4×10^{12} kg m^{-3}, to

1.5×10^{16} kg m^{-3} while the central temperature rises to 1×10^{11} K. As a result of these changes of the physical parameters the nuclear composition is radically altered. The iron nuclei are broken down by photoionization processes, first to a-particles and then to protons and neutrons. Since electron capture takes place at an extremely rapid rate, the end result is *neutronization*. At a density of 1.5×10^{16} kg m^{-3} more than one-half of the weight of the matter consists of neutrons, with a-particles contributing less than one-third, and heavier nuclei less than one-sixth of the weight.

All this happens in about *one second*. Violent as the collapse is, the speed of contraction is yet more than an order of magnitude slower than that corresponding to free fall forward the centre. In other words, there are opposing pressure gradients. The whole scheme here outlined has been followed using computer-aided calculations of the solutions of the appropriate hydrodynamic equations.

We have mentioned that neutrino emission is an important process during the later phases of red-giant evolution. During the collapse phase, the electron capture just referred to is accompanied by emission of neutrinos, typically with energies of about 8 MeV. Contrary to what was the case during the earlier phases the densities are so high that the neutrinos do not travel unimpeded through the matter in the core, but are trapped. Most of the neutrinos are emitted during the last 10–20 ms of the collapse to the central density of 1.5×10^{16} kg m^{-3}, and although they will ultimately largely escape into the surrounding mantle, the trapping keeps them in the inner part of the core for times that are more than an order of magnitude greater than the collapse time mentioned.

The release of energy in the form of neutrinos is of enormous magnitude. When the density reaches 1.5×10^{16} kg m^{-3}, the neutrino luminosity is almost 10^{46} Js^{-1}, and the total energy of the neutrinos emitted during the one-second collapse is about 10^{44} J. For comparison, the luminosity of the sun is 4×10^{26} J s^{-1}, that of all the stars in our galaxy put together about 2×10^{37} J s^{-1}, while the total energy which the star of 20 solar masses considered has emitted during the whole of its previous history is about 10^{46} J.

We come now to a very significant question: will the star explode as a supernova as a consequence of the collapse of the central part of its core?

Theoretical calculations of the further hydrodynamic evolution of star have not yet yielded a definite answer to the question, but two very important clues have been found.

(1) The numerical calculations based on the hydrodynamic equations show that the mass flow in the stellar core changes character very soon after the density has reached the value 1.5×10^{16} kg m^{-3}. When the density reaches 10^{18} kg m^{-3}, i.e. a few times nuclear density, the contraction stops and the implosion 'bounces' to become an explosion. Strong outward accelerations occur not only in the core but also in the mantle.

(2) As a consequence of the enormously high neutrino luminosity of the inner part of the core, there will be a flow of neutrinos through the outer parts of the star. The neutrino energies are so high, and have correspondingly high cross-sections for interaction with matter, that very large amounts of energy and outward momentum are communicated to the mantle.

In view of the observational results concerning supernova explosions and neutron stars, which were described before, it is reasonable to conclude that the result of the core collapse in the 20 solar-mass star is indeed a supernova explosion resulting in the formation of a neutron star remnant with a mass of about 1–2 solar masses and an expanding envelope containing the greater part of the mass, which will ultimately mix with surrounding interstellar matter. Attempts to demonstrate through theoretical calculations that this is what happens have not yet been successful, but the theoretical experience gathered in this field does not exclude the possibility of such demonstration through future work.

One more question must be considered in this context: is it possible that a remnant with properties different from that of a neutron star could be formed under certain circumstances, for example after a supernova explosion of a still more massive star?

Theoretical calculations pertaining to the structure of neutron stars show that there is an upper limit to the mass of stable neutron-star configurations. This upper limit, which is not precisely known, is of the order of 3 or 4 solar masses. When the mass limit is exceeded, the remnant will continue to contract and become a so-called 'black hole'. The surface gravity of the black hole is so high that radiation cannot escape to the surroundings, whereas surrounding matter is subject to the gravitational attraction of the black hole.

The conclusion concerning an upper limit to the mass of stable neutron stars has been reached by assuming the validity of the predictions of the general theory of relativity for the matter densities in question, which clearly go far beyond the range for which experimental tests of the gravitational theory have been possible.

Theoretical work has not yet given an answer to the question of whether or not remnants with masses larger than the quoted upper limit could be formed as the result of evolution of large stellar masses. Nor has a definite answer yet been derived from observation.

If it could be established by observation that binary star systems exist in which one component has a mass exceeding the limit for stable neutron star masses, and if such components were found by observation not to emit radiation even approaching that expected according to the observed mass, then it could be concluded that the evidence indicated the presence of a black hole. Interesting candidate binary systems are under investigation, but there are as yet no certain conclusions. We shall return to the general question of the possible existence of black holes in § 6.5, in connection with the discussion of an active nucleus of our galaxy.

Summarizing the discussion of stellar evolution, we can see that the overall trend is one of *contraction*. The protostars are very tenuous stellar objects that contract to the main-sequence stars. Here contraction is halted, because nuclear energy is released through conversion of hydrogen into helium; so effectively, indeed, that the star spends by far the greatest part of its lifetime as a lumious object in this phase.

After the exhaustion of the available hydrogen fuel, the star moves into the red-giant region; however in the central regions there is contraction, with the result that temperatures and densities rise and new nuclear fuels become available. As these in turn are exhausted, fresh contraction takes place, and in massive stars this continues until neutronization occurs. A neutron-star remnant of extremely high density is formed as a result of a supernova explosion. In the case of less massive stars, the end-product is a dense white-dwarf remnant, which is the result of an evolution that is characterized by mass loss through gentle outflow of mass from the stellar surface.

Evolution through contraction from interstellar matter to massive protostar to neutron-star remnant corresponds to a density increase of almost forty orders of magnitude. This implies radical changes in the physics of the matter, and it is clear that extensive knowledge of basic physics is necessary for calculations in theoretical astrophysics.

The evolutionary contraction of stellar matter is, as described above, accompanied by a buildup from hydrogen nuclei of heavier nuclei, from helium to iron. Some of the processed matter is returned to interstellar space. This causes a secular enrichment of galactic interstellar matter in terms of heavier elements. In other words a great mechanism operates in our galaxy in that large numbers of stars are formed out of interstellar matter, evolve, and return enriched matter to interstellar space. In consequence, later generations of stars, formed out of this enriched interstellar matter, will have a higher content of heavier elements. We shall return to this in § 6.5 when considering the evolution of our galaxy.

Using the results of investigations of stellar evolution we can now describe certain important features of this mechanism of enrichment of galactic matter.

First of all, it is clear that a large fraction of the mass that has gone into stars through star formation from interstellar matter has not contributed at all to the enrichment process during the lifetime of our galaxy. This is because stars that are formed with masses smaller than about 0.7 solar mass evolve so slowly that they have not evolved beyond the main-sequence stage during the $10-15 \times 10^9$ years in question.

Next, consider the mass range 0.7 to about 4 solar masses, where the end-product of evolution is a white dwarf. Here, nearly all the matter from which build-up to heavier elements has taken place is locked up in a white dwarf, while that part of the stellar matter which streamed back into interstellar space was not enriched, or enriched only to a small extent. Thus the stars formed in this mass range do not contribute much to the enrichment process.

For stars in the mass range 4–8 solar masses, the situation may be very much the same. As we have seen, it is not unlikely that the end-product here is a neutron-star remnant. If this is the case, then most of the processed matter containing heavier elements is again locked up in a remnant, and consequently the contribution to enrichment of interstellar matter is small.

For the stars with masses greater than about 8 solar masses, the situation is different. Taking the star of 20 solar masses that was considered above, only a relatively small fraction of the matter in the 8-solar-mass helium core, which was built up to heavier elements during the evolution, is locked up in a neutron-star remnant, the mass of which is only 1–2 solar masses. A very considerable amount of enriched matter is blown into surrounding space through the supernova explosion and ultimately mixes with the existing interstellar matter.

Since only massive stars contribute to the enrichment process, its efficiency is relatively low. For each solar mass of interstellar matter that contracts into stars, about 0.4 or 0.5 solar mass stream back without having been processed, while another 0.4 or 0.5 solar mass are locked up in low-mass stars that are still on the main-sequence, in white dwarfs or in neutron stars. The amount of heavy-element matter produced in massive stars and returned to interstellar space is perhaps 0.02 solar mass. The corresponding very rough efficiency estimate is thus 4 per cent.

At the present stage of evolution of our galaxy, most of the very tenuous matter originally present has become stars. The rough efficiency estimate just referred to is compatible with the observational result that the present-phase heavy-element content of our galaxy amounts to a few per cent. We shall return to this in § 6.5.

As a final comment on the enrichment mechanism, let us emphasize that the efficiency of the mechanism is proportional to the fraction which the contributing massive stars represent in the mass distribution of the stars formed out of interstellar matter. It is quite conceivable that the fraction in question differed from its present value during earlier phases of evolution of our galaxy.

Having considered important results obtained after 1958 in the field of stellar structure and stellar evolution, we shall briefly refer to problems of solar physics that are relevant to the problems discussed.

The sun is the only star for which it is possible to study surface details that are small compared to the radius. Such studies have led to very extensive knowledge of characteristic features in the directly visible outer layers of the sun, namely the granulated photosphere, sun spots, chromosphere, flares, prominences, and the corona.

By the 1950s, solar research in the far ultraviolet had already been carried out with the help of rockets. Since 1958 observations from satellites have yielded important new results in the far ultraviolet and X-ray regions. These results are of particular value in the study of the properties of the tenuous, high-temperature (of order of magnitude 1×10^6 K) coronal matter.

Ground-based solar spectroscopy has provided new information concerning the patterns of motion in the visible solar layers and the structure of the magnetic field in these layers. The phenomena are presumably not of direct importance in the discussion of large-scale structure, energy balance, and evolution of solar-type stars. Their study nevertheless leads to important general insights into problems of cosmic hydrodynamics and magnetohydrodynamics, which may prove valuable in future attacks on these problems as they relate to stellar interiors.

Finally, we mention observations aimed at detecting the emission of neutrinos from the sun. The conversion of hydrogen into helium in the central regions of the sun is accompanied by neutrino emission. A simple calculation based on the rate of energy production and the number of neutrinos emitted per unit energy produced leads to the conclusion that the number of solar neutrinos reaching the earth is close to 10^{15} m^{-2} s^{-1}. Because of the very small cross-section for interaction of the solar neutrinos with laboratory matter it is very difficult to detect a stream of neutrinos of this magnitude. However, a radiochemical method has been developed (the Brookhaven solar neutrino experiment) and pushed to a level of sensitivity that should be sufficient, or almost sufficient, for the detection of the solar neutrino current. The experiment uses as a detector element the stable isotope chlorine-37 (in perchloroethylene, C_2Cl_4) which, upon capture of a neutrino of minimum energy 0.81 MeV, emits a negative electron and is transformed into argon-37, a radioactive element with a half-life of 35 days. In the experiment, a tank containing about 600 tonnes of C_2Cl_4 is placed in a rock cavity 1500 m underground to reduce the background effect produced by cosmic rays. The neutrino-produced ^{37}Ar can be removed and recovered with high efficiency, and the number of ^{37}Ar atoms produced over long observing periods can be measured.

The number of ^{37}Ar atoms produced is a sensitive function of the neutrino energy. This means that it is not sufficient to calculate the total number of solar neutrinos arriving at the surface of the earth per square centimetre per second, but the energy distribution must also be evaluated. The conversion of hydrogen to helium in the solar interior takes place through a fairly complicated chain of reactions, starting with a proton–proton reaction that produces ^2H. The energies of the emitted neutrinos range from zero to 14 MeV, and those with higher energies are the most effective in producing ^{37}Ar in the Brookhaven experiment. These high-energy neutrinos are emitted when, in one of the branches in the proton–proton chain of reactions boron-8 is transformed through positron emission into beryllium-8, which then splits into two ^4He atoms. However, the fraction of proton–proton captures that lead to the $^8B \to {}^8Be$ reactions is a sensitive function of temperature. It follows that the predicted effectivity with which the solar neutrinos produce ^{37}Ar in the Brookhaven experiment depends in a sensitive way on the theoretical solar model that gives the temperature as a function of distance from the solar centre. In other words, the Brookhaven experiment is a very important test of theoretical calculations of structure and of nuclear process rates for the central region of the sun.

So far the results of the neutrino experiment have not been conclusive. Calculations of the ^{37}Ar production rate by solar neutrions based on the generally accepted standard solar model lead to values that are at least twice as high as an upper limit to this rate deduced from the experimental results.

Considerable efforts have been made to see whether acceptable changes in the solar model might eliminate the discrepancy, but no definite conclusions have been reached. One possibility that has been discussed is to change the chemical composition of the solar-interior matter assumed in the theoretical calculations of temperatures in the central regions. A reduction of the relative content Z of the elements heavier than hydrogen and helium, together with a reduction of the relative helium content Y, does lead to model predictions that are compatible with the experimental results. However, it is difficult to reconcile these assumptions with conclusions based upon direct spectrophotometric determinations of Z that are valid for the outer convective mantle of the sun. This is so because of the theoretical inference that thorough mixing of solar-interior matter took place during the early phase of solar evolution.

A possibility remains that the results of this very difficult experiment might ultimately come out in such a way that there is no irreconcilable conflict with theoretical calculations. However, should a firmly established discrepancy result, then there would be serious doubts about the validity of certain sectors of stellar-interior calculations.

6.3 Interstellar matter

By 1958 a large amount of information had been gathered concerning the physical and chemical properties of interstellar matter. The interstellar regions where hydrogen is ionized – the so-called H II regions – had been studied through detection and spectrophotometry of emission lines, particularly the Balmer lines of hydrogen. The regions where interstellar hydrogen is non-ionized – the H I regions – were investigated using the extensive material from observations of radio emission in the 21-cm line of the neutral hydrogen atom. Work on the absorption lines due to interstellar atoms and ions, particularly on the resonance lines of neutral sodium, neutral calcium, and singly ionized calcium, observable in the spectra of early-type stars, gave further valuable information on H I regions. The interstellar grains had been studied through measurements of the absorption and scattering effects that they produce on the radiation from stars as it passes from the source to the observer. Particular attention had been given to the wavelength dependence of absorption and scattering by interstellar grains (an effect that causes reddening of the light of stars observed through the grain medium), as well as to the polarization of starlight caused by aligned, non-spherical interstellar grains. The observations had made it possible to construct theoretical models describing the properties of interstellar grains and to estimate their relative contribution to the density of interstellar matter.

Since 1958, great progress in the investigation of interstellar matter in our galaxy was achieved through radio-astronomical observations and through spectroscopic observations from satellites. The radio-astronomical measures pertain to continuum emission at centimetre wavelengths, to emission in high-level recombination lines of hydrogen, both from H II regions, to combined absorption–emission observations of the 21-cm line of hydrogen (H I regions), and finally to observations of lines in the centimetre and millimetre region belonging to a large number of interstellar molecules (H I regions). The spectroscopic observations from satellites have yielded new information on absorption by interstellar atoms, ions and molecules in spectral lines located in the far-ultraviolet region (100–300 nm) that is not observable from the ground. Observations of the Lyman-alpha line (at 121.6 nm) of neutral hydrogen in absorption, and of absorption lines due to molecular hydrogen are of particular importance.

Here we shall summarize the new results with particular reference to their importance in the discussion of the problems of interaction between stars and interstellar matter, and thereby for the discussion of the evolution of our galaxy.

Through radio astronomical measures in the 21-cm line of atomic hydrogen it has been possible to identify typical interstellar clouds consisting mainly of neutral hydrogen (H I regions) and to evaluate their densities and temperatures. Particularly valuable information has been derived from observations in which a distant extra-galactic radio source of very small angular diameter is observed through a galactic interstellar cloud relatively near by that appears with a larger angular diameter. In this case both absorption and emission by the near-by interstellar cloud can be determined, which makes it possible to find the number of hydrogen atoms per square centimetre in a column along the line of sight through the interstellar cloud, as well as the cloud temperature. The observations are preferably made for lines of sight that make a not very small angle with the galactic plane (larger than 10-20°). In this way one can make sure that only one, or a very few, interstellar clouds are intersected by the line of sight in question. Since there is a significant, although very small, spread in the wavelengths of the lines owing to the differing line-of-sight velocities of the interstellar clouds (Doppler effect), it is possible to derive data pertaining to a single interstellar cloud when indeed only a very few clouds are intersected.

It has been found that a typical cloud has a density of 20×10^6 hydrogen atoms per m³, a temperature around 100 K a diameter of 5 parsecs (1 parsec = 30×10^{15} m = 3 light-years), and a mass equal to 30 solar masses. The actual values scatter over a moderate range around the typical values.

The intensity of the 21-cm emission of such clouds has been found to be closely correlated with observed reddening of the radiation from stars seen through the clouds. The reddening is produced by interstellar grains located in the hydrogen clouds, which absorb and scatter radiation. The attenuation of the radiation by the grains increases with diminishing wavelength, and this causes the reddening referred to. The wavelength dependence in question has been studied over a considerable range of spectrum, and this has made it possible to determine the nature and size distribution of the grains. A typical interstellar grain has a diameter of 0.2-0.3 μm and consists of a silicon core surrounded by a mantle formed by a frozen substance containing molecules and radicals built largely of the atoms hydrogen, carbon, nitrogen, and oxygen. The emission and absorption properties of the grains are such that the grain temperature is lower than that of the surrounding interstellar gas, being around 10 K.

From the observations of the degree of reddening produced by the interstellar grains it has been possible to deduce their total mass. In a typical cloud of the type considered, the mass of the grains is 100-200 times smaller than the total mass of hydrogen. However, it should be noticed that the grains consist largely of elements heavier than hydrogen and helium, such as carbon, nitrogen, oxygen, and silicon. In § 6.2 we referred to the fact that these heavier elements have a total abundance in the sun measured by $Z = 0.02$ (2 per cent by weight).

The attenuation of passing radiation produced by absorption and scattering due to the grains in a typical interstellar cloud, such as that just described, is about 10 per cent (at 550 nm wavelength). We shall refer to this kind of cloud as an Ambarzumian-type cloud. A different type of interstellar cloud which reduces the light passing through it much more will now be discussed. This is the classical dark cloud.

Astrographic plates that record large numbers of faint stars show that in the Milky Way, i.e. in directions making a small angle with the galactic plane, there occur areas that stand out with respect to their relative emptiness of stars. It has been established that this phenomenon is caused by extinction of light in dark clouds, rather than by variations in the space density of stars. Investigations carried out at different wavelengths show that the dark clouds in question not only attenuate the starlight that passes through them, but also cause a reddening of the light. In the case of the classical dark clouds, the attenuation is very pronounced, the starlight that passes through being reduced by factors up to several hundred. This should be compared with the modest reduction (by 10 per cent) caused by the Ambarzumian-type interstellar clouds that we first considered. The effect of a cloud that causes a 10 per cent light reduction is not clearly noticable in star counts.

The area in the sky which is covered by a classical dark cloud varies within wide limits, from a small fraction of a square degree to tens, or in a few cases hundreds of square degrees. The very great majority of classical dark clouds are, however, small, covering areas less than 0.1 square degree, i.e. their dimensions are smaller than about $\frac{1}{200}$th of their distances.

The small dark clouds found from inspection of astrographic plates, particularly those taken for the *Palomar sky atlas*, are at distances of less than about 1000 parsecs. For larger distances the number of foreground stars which are not obscured by the dark cloud in question is relatively so large that the fluctuation in the star number becomes too small for certain identification of the cloud. Somewhat over a thousand small dark clouds have been identified and catalogued so far. An estimate of the number of such clouds per cubic parsec shows that they are less numerous than the small interstellar clouds that we considered first, by a factor between 10 and 100. The diameter of a small dark cloud is typically about 1 parsec, i.e. the volume is about 100 times smaller than for the Ambarzumian-type cloud. However, the density of grains that cause the attenuation and reddening of the star light passing through is much higher (by a factor of about 100) for the small dark cloud, as can be deduced from the fact that the light reduction factor is much larger in spite of the shorter path traversed. Thus the total mass of the grains in a small dark cloud is about the same as in the much less dense, but larger Ambarzumian-type clouds.

It is now important to determine whether or not the ratio of the total mass of hydrogen in the cloud to the mass of the grains is the same for the two types of clouds. Now, radio-astronomical investigation of the intensity of the hydrogen 21-cm line has shown that the small dark clouds do not stand out at all. This is partly due to the effect of self-absorption in the 21-cm line. The main cause, however, is that hydrogen is largely in molecular form in the small dark clouds.

Theoretical investigations have shown that the equilibrium between H and H_2 in dense clouds containing grains is strongly shifted in the direction of H_2. A considerable fraction of the hydrogen atoms that hit the cold grain surfaces form H_2 molecules, which then return to surrounding space. In the dense clouds the effect of dissociation processes caused by the general radiation from the star field is greatly reduced, partly through self-absorption in the relevant H_2 molecular lines, partly through attenuation of the radiation by the grains.

The prediction that the equilibrium between H and H_2 is strongly shifted toward

H_2 in dense clouds has been fully confirmed by direct observations by the Orbiting Astronomical Observatory (OAO), carried by the *Copernicus* satellite.

With the OAO it has been possible to observe absorption lines produced by inter-stellar matter in the spectra of early-type stars which are seen through the matter in question. This is a type of observation that has long been possible from the ground for the resonance lines of Ca I, Ca II, and Na I. The extremely important difference is, however, that from the OAO the spectrum can be observed in the far ultraviolet down to about 100 nm, whereas the terrestrial atmosphere cuts off radiation below 300 nm. This has meant that resonance lines of a large number of atoms and ions could be studied from the OAO. In particular, absorption lines of molecular hydrogen are observable, and also the Lyman-alpha resonance line of atomic hydrogen at 121.6 nm.

The OAO observations have indeed shown that in tenuous interstellar matter only very small fractions of H_2 are present, whereas in dense interstellar clouds H_2 dominates very strongly over H. These findings suggest that the ratio of the total mass of hydrogen to the total mass of grains has about the same value in dense interstellar clouds and in much more tenuous regions of interstellar space.

Further important progress has been made through the study, by radio-astronomical methods, of emission by molecular carbon monoxide. During the past 20 years the antennae and receivers used in ratio astronomy for short wavelengths have been gradually improved, and as a result much work has been carried out in the millimetre range. This work includes investigation of the emission by the inter-stellar molecule carbon monoxide at 2.6 mm (rotational transition $J = 1 \rightarrow 0$). Important observations have been made pertaining to the $^{12}C^{16}O$ as well as the $^{13}C^{16}O$ line (at 115.2 and 110.2 GHz, respectively; isotope ratio C^{12}/C^{13} about 100).

The carbon monoxide emission line at 2.6 mm has been found to be an excellent indicator of the density of hydrogen. Tenuous interstellar regions do not show this emission line, while denser regions, such as the small dark clouds, invariably show it.

Investigations of a number of small dark clouds based on observations of $^{12}C^{16}O$ and $^{13}C^{16}O$ emission lines have led to determinations of the $^{13}C^{16}O$ density and the cloud temperature. While self-absorption is strong for the $^{12}C^{16}O$ line, so that the intensity of emission yields the cloud temperature, this is not the case for $^{13}C^{16}O$; hence reliable determination of the number of $^{13}C^{16}O$ per square centimetre of a column along a line of sight passing through the small dense cloud is possible. The $^{13}C^{16}O$ densities can be converted to H_2 densities with reasonable accuracy (within a factor of about 2). Thus it has been found that the ratio of total hydrogen mass to total grain mass is about the same in the small dark clouds as in the Ambarzumian-type cloud, which we considered at the beginning of this section. This is an important result. The kinetic temperature of the gas in the small dark clouds is low, about 10 K.

The foregoing adds up to the following comparison between the more tenuous Ambarzumian-type clouds and the small dark clouds. The former are more numer-ous, the density (per cubic parsec) being between 10 and 100 times higher. Their volume is typically about 100 times larger, while the density of both hydrogen and grains is lower by a factor of about 100. This means that typical hydrogen densities for the tenuous clouds are 20×10^6 atoms per m^3; those for the small dark clouds are a few thousand million per m^3. The masses of the two types of clouds are about equal: 20-50 solar masses. Temperatures for the tenuous Ambarzumian-type clouds are typically about 100 K, ranging from somewhat less than fifty to a few hundred degrees. The small dark clouds appear to be fairly uniform with respect to the kinetic temperature of the gas, with values close to 10 K.

When the distribution of angular diameters of small dark nebulae is examined, it is seen that the number increases by a factor of more than two as one goes from the range 10-13 minutes of arc to 7-10 minutes of arc, and it decreases again for smaller diameters. In this connection it should be mentioned that it is observationally difficult to identify dark nebulae with diameters of only a few minutes of arc because the number of background stars on record even in, say, the *Palomar sky atlas* is quite small, and statistical fluctuations of the numbers are consequently large. However, quite a few dark nebulae of this kind have been found, particularly where they are seen against a background of emission nebulosity. These objects are referred to as globules. Diameters range from a few hundredths to a few tenths of a parsec. For globules with diameters of about 0.5 parsec, masses have been found similar to those of the typical small dark nebulae considered above. The corresponding globule densities are an order of magnitude higher, and exceed 10^{10} hydrogen atoms (in molecular form) per m^3.

In our description of the properties of interstellar matter, we now come to an important third component, namely the large dark clouds. We have already referred to the fact that photographs of the Milky Way, as reproduced, for example, in the *Palomar sky atlas*, reveal dark clouds covering tens or hundreds of square degrees in the sky. A typical linear diameter of a large dark cloud is 50-100 parsecs. As already emphasized, these large dark clouds are much less numerous than the small dark clouds. In fact, within 700 parsecs (about 2000 light-years), only five very large dark clouds — or rather cloud complexes — are known, but together they cover an area of about 2000 square degrees of the sky. The corresponding number of Ambarzumian-type clouds in the volume in question is probably close to 10^5, while it is a few thousand for the small dark clouds.

Investigations of the attenuation of transmitted light and of carbon monoxide emission show that the density of hydrogen as well as of grains in a large dark cloud is similar to that found for the Ambarzumian-type clouds, i.e. about 20×10^6 hydrogen atoms per m^3 (mostly in molecular form). This contrasts with the small dark clouds where the densities are a few thousand million per m^3. However, with diameters of the order of 100 parsecs, the total masses are quite high: typically between 10^4 and 10^5 solar masses. The kinetic gas temperatures are around 20 K.

Referring to the values quoted for the number of clouds per cubic parsec for the three types of clouds considered (Ambarzumian-type clouds, small dark clouds, and large dark clouds), and their respective typical masses, we see that most of the mass is contributed by the first category; the large dark clouds follow, with a total that is perhaps 2-3 times smaller; and the small dark clouds together contribute only a few per cent of the interstellar mass.

The filling factor for the Ambarzumian-type clouds is 0.02–0.03. It is smaller for the large dark clouds and very much less for the small dark clouds. This means that most of the interstellar space considered is intercloud. The important question then arises whether appreciable amounts of matter are indeed present to form an intercloud medium.

There are admittedly still significant uncertainties in regard to this question. It is, however, possible to set a fairly reliable upper limit to the density of matter between clouds. This has been possible on the basis of 21-cm hydrogen-line studies, studies of hydrogen emission lines (Balmer lines and Lyman-alpha) as well as investigations of interstellar reddening. It appears that a density of 2×10^5 hydrogen atoms per m^3, or 3×10^{-28} kg m^{-3}, is an upper limit, at least in the part of our galaxy that lies within 500 parsecs.

This means that there is a strong density contrast, at least 100 to 1, between Ambarzumian-type clouds and the large dark clouds on the one hand, and the intercloud medium on the other. In the case of the small dark clouds the contrast is even stronger (10^4 to 1).

The temperature of the intercloud medium is several thousand degrees. This is significant, because it means that the pressure is comparable to that in the Ambarzumian-type clouds and the large dark clouds.

The OAO observations of interstellar absorption lines have revealed the presence of O VI in interstellar space. Regions where O VI ions occur must contain tenuous gas of very high temperature (10^5–10^6 K). It appears, however, that O VI is limited to a fairly small fraction of intercloud space. It is possible, although as yet not at all certain, that the O VI regions are either close to very hot stars, or else are regions disturbed by supernova explosions.

We now consider problems concerning the influence of radiation from the stars upon the surrounding interstellar matter.

First, we shall examine a specific example that illustrates some cases of importance. At a distance of about 900 parsecs in the direction of the constellation Monoceros in the Milky Way, there is a large dark cloud which covers an area in the sky of about 10 square degrees. The linear diameter is close to 50 parsecs. Investigations of the degree of light attenuation caused by the grains in the cloud, and investigations of the carbon monoxide emission, indicate that the cloud consists largely of H$_2$ molecules and grains in about the standard proportion and that the total mass is close to 10^5 solar masses. A large number of stars are embedded in the cloud, and among them are several stars of spectral class B. These stars are surrounded by reflection nebulae with diameters of a few minutes of arc or, in linear measure, about 1 parsec. The stars in question illuminate grains in the nebula, and these reflect starlight, as can be concluded from the observation that the luminous nebulae show a spectrum identical to that of the illuminating star. Two of the stars are B1 stars of relatively high surface temperature. Their radiation causes the hydrogen molecules in the near surroundings to dissociate and the H atoms to become ionized, i.e. an H II region is created. The stellar radiation that ionizes hydrogen is in the part of the spectrum that is below the Lyman limit at 91.1 nm. For the two B stars in question, the intensity of the radiation below 91.1 nm is high enough for appreciable ionization to take place in the surrounding part of the nebula. In other words, the stars have a high enough atmospheric temperature to create an H II region in the immediate surroundings. The diameter is about 2 parsecs, and the density in the H II region is about 20×10^6 hydrogen atoms and 20×10^6 free electrons per m^3.

The Monoceros cloud has a core in which the density of matter is much higher than in the surrounding nebula. This is shown by the fact that the carbon monoxide emission has a pronounced peak in a region about 10 minutes of arc in diameter (corresponding to a linear diameter between 2 and 3 parsecs), and also through emission by the carbon sulphide molecule. The density in the core is a few times 10^{10} hydrogen atoms (in molecular form) per m^3, and the total mass in the core amounts to a few thousand solar masses.

Embedded in the core is a very small region, about 0.1 parsec in diameter, that has been found to emit strongly in the radio continuum at both 3 cm and 6 cm, and is also measurable at wavelengths of 11 and 21 cm. From the intensity distribution in the continuum, it can be concluded that the emission comes from a plasma of ionized hydrogen – an H II region. The emission mechanism is a well-known one: quantum jumps in the kinetic energy of free electrons moving in the Coulomb fields of protons (free-free transitions). This interpretation is confirmed through the observation, from the same small region, of the hydrogen 109α recombination line (of wavelength close to 6.0 cm), which is emitted when α hydrogen ion captures an electron into the quantum state $n = 110$ and a transition occurs to the state $n = 109$ as the electron cascades down to the ground state. Analysis of the intensities in the 109α line and in the radio emission continuum show that the temperature of the H II region is close to 8000 K, while the density is about 10^{10} H II ions (i.e. protons) and as many free electrons per m^3. The total mass of the H II region is small, about a tenth of a solar mass. The small dense H II region is surrounded by an H II envelope (of diameter 1 parsec), where the density is nearly 100 times lower; the mass contained in the envelope is about 1 solar mass. The H II region is quite invisible in the optical part of the spectrum because of the obscuration by the very dense molecular hydrogen and grain medium of the core in which it is embedded.

There can be no doubt that the ionization of hydrogen in the H II region, located in the dense core of the dark nebula, is caused by radiation from a star with high surface temperature; considerably higher, in fact, than that of the B1 stars that ionize a surrounding medium of thousandfold lower density. The conclusion is that the ionizing radiation comes from a young (i.e. newly formed) hot, luminous O star, with a mass probably between 10 and 20 solar masses.

Another example will illustrate the influence of stellar radiation upon the surrounding interstellar matter. At a distance of about 1600 parsecs, also in the constellation Monoceros (but not connected with the object just described), is an emission nebula called the Rosette nebula. Its diameter is about $1°.5$; in linear measure about 40 parsecs. A cluster of stars is embedded in the Rosette nebula, among them a few massive, luminous O stars of high surface temperature. The optical spectrum of the emission nebula contains the Balmer lines of hydrogen, and from their intensity as well as from the intensity of 10-cm radio emission (free–free transitions), it has been concluded that the Rosette nebula is an H II region with a density of about 15×10^6 H II ions and as many free electrons per m^3. The density decreases toward the centre as well as towards the edge of the nebula, so that most of the nebular matter is contained in a wide shell around the ionizing O stars.

The total mass of the H II region is about 10^4 solar masses. The whole Rosette nebula H II region is embedded in a large dark cloud, containing neutral hydrogen atoms, hydrogen molecules, and grains, with a mass about equal to 10^5 solar masses. In this respect, there is a similarity with the object described above, but in the former case there is a dense core containing molecular hydrogen and grains, with a very compact H II region, completely obscured in the optical spectrum.

The comparison strongly suggests that the two H II regions are successive stages of evolution: when one or a few O stars have been formed out of dense interstellar matter, they will ionize the surroundings, creating an H II region with a temperature of several thousand degrees. As a consequence of the high temperature, the pressure in the compact H II region will be much higher than that in the surrounding neutral matter. Under these circumstances the matter in the H II region will expand, pushing the surrounding neutral matter before it, thus forming a shock wave. At the same time, ionization produced by the radiation of the O stars will cause the mass of ionized hydrogen to grow. In this way a much larger H II region will result. The grains will largely follow the outward motion of the gas, and in addition there will be some destruction of grains by the stellar radiation. Thus the compact H II region, completely obscured in the optical wavelengths at first, will expand and will also become optically visible.

This picture of evolution has been confirmed through observations of H II regions that are in stages intermediate between the two described. Further observations that clarify the picture have been made in the infrared region of the spectrum. In the infrared the obscuration by the grains is reduced, and emission by the grains, heated by stellar radiation, has been observed.

It has been possible to follow the evolution from compact to greatly expanded H II regions through theoretical calculations. Observation and theory have led to a good understanding of these processes, which are clearly an important part of star formation from interstellar matter.

As mentioned, the Rosette nebula contains a whole cluster of stars; several similar cases have been found. An essential observation is that the process leads to the conversion of large amounts of interstellar mass into stars, in many cases 10^3 to 10^4 solar masses. We shall return to this point, but first we shall refer to other effects of interaction between stellar radiation and interstellar radiation.

Consider the effect upon an Ambrazumian-type cloud of the radiation from a massive, luminous O star with high intensity in the region of the spectrum below the Lyman limit at 91.1 nm. The Ambarzumian-type cloud, according to observations, is an H I region, and it is easily computed that the region is highly opaque to radiation below 91.1 nm, the optical depth here being of the order of a thousand. This means that ionizing radiation from an O star will be checked by the attenuation of an Ambarzumian cloud, and that only when the O star is fairly close to the cloud, or even embedded in the cloud, would appreciable ionization of the hydrogen take place. Now there are within 1000 parsecs of the sun only a few dozen O stars, and this means that only a relatively small fraction of the Ambarzumian clouds will be affected by radiation from the stars in question. The corresponding number of B stars is much higher, but here the intensity in wavelengths below 91.1 nm is so low that their total effect is a good deal smaller than that of the O stars. On the other hand, calculations show that dissociation of H_2 molecules by stellar radiation is effective for Ambarzumian-type clouds, so that hydrogen should be present mostly in atomic form. This prediction agrees with observation.

The small dark clouds and the large dark clouds are even less affected by hydrogen-ionizing radiation. Only O and early B stars actually embedded in the clouds will produce H II regions, in the way described earlier.

The situation with regard to ionization of the intercloud medium by O and B stars is less clear. If the density is as low as 10^5 H atom per m^3, then an O O7 star can ionize hydrogen at distances of more than 200 parsecs from the star if the intercloud region is not shielded from the O star by an intervening Ambarzumian-type cloud (which is, however, generally the case for distance of more than 200 parsecs). Even the more numerous B stars have an appreciable ionizing effect. However, should the intercloud medium have a density of 3×10^5 H atoms per m^3, then the ionized volumes would be reduced by a considerable factor (about 10).

Finally we shall consider the case where one very massive O star or a group of O stars have dispersed and completely ionized the hydrogen–grain medium out of which they were formed, and are now radiating into the surrounding space. We shall illustrate this case with an example. In the direction of the constellation Orion, at a distance of between 400 and 500 parsecs, there is a huge complex of stars and interstellar matter. The region in question is bounded by a nebular feature, long known from astrophotographs, named the Barnard Loop after its discoverer. The area inside the loop, about $14° \times 10°$ (corresponding to a maximum diameter of about 100 parsecs for the region), contains an association of stars, all relatively young (ages less than 20 million years), with a total mass of approximately 10^4 solar masses. Among these young association stars there are a few O stars and hundreds of B stars, and it is characteristic that stars of smaller masses (e.g. of solar masses) have not yet reached the main sequence in their evolution (cf. § 6.2). Furthermore, in this region there are two large molecular clouds containing H_2 and grains and having masses about equal to 10^5 solar masses. The molecular clouds contain cores where the density is between 10^{10} and 10^{11} H atoms per m^3 in molecular form, and also have compact H II regions, obscured in the optical part of the spectrum, which are ionized by a newly formed O or early B star. Particularly remarkable is a very strongly emitting H II region located in the larger and southernmost of the two molecular clouds. This is the Orion nebula, an emission nebula that has been known for a long time. The Orion nebula H II region contains near its center a cluster of stars, and one of the cluster members is a massive O star. The radiation field in the region of the spectrum below 91.1 nm is dominated by the contribution from this star, and it is this star that causes the ionization of the Orion nebula. In this case the hydrogen–grain matter has been dispersed to the extent that hydrogen is ionized right up to the boundary of the concentration. Here the grains cause only moderate attenuation of the radiation of the O star, and the same is the case for the optical emission by the Orion nebula. In fact, the ionizing radiation from the O star penetrates into the surroundings and affects the tenuous interstellar matter inside the Barnard loop, except in rather small sections that are shielded from the O star by the two large molecular clouds. Thus, the whole region of interstellar space inside the Barnard loop forms an H II region with a maximum diameter of about 100 parsecs. Outside the molecular clouds the density of hydrogen is about $1-2 \times 10^6$ hydrogen ions and as many free electrons per m^3. This is not much larger than the average density in these regions of interstellar space. The interstellar matter here is presumably a mixture of matter dispersed as a consequence of star formation and matter that resided there before the event.

Outside the Barnard loop, neutral hydrogen is present, forming an H I shell around the large H II region.

The three examples considered, i.e. the two clouds in Monoceros and the Orion complex, illustrate important aspects of the interaction of stars and interstellar matter. Several similar star–cloud configurations have been investigated in detail, and this material forms a good basis for study of the mechanisms of formation of stars out of interstellar matter.

Using these observations, the following question can be posed: are the three types of clouds dynamically stable?

Consider first the typical Ambarzumian-type cloud. With the known values for density, temperature, and total mass as input, it is found that self-gravitation is not sufficient to keep the cloud matter together. If located in a vacuum, an Ambarzumian-type cloud would expand and gradually disperse. However, the pressure of the very tenuous, but much hotter, intercloud medium acts to slow down the expansion and may even cause the Ambarzumian cloud to be stable.

Of other outside influences on an Ambarzumian-type cloud, we have mentioned the − relatively rare − case in which an O star passes close by and ionizes the cloud, and thereby disperses it. Ambarzumian-type clouds will also collide, at time-intervals of the order of 10–30 million years. Theoretical calculations have clarified some of the consequence of collisions. In some cases there will be partial heating and dispersal of cloud material; compression, and perhaps even build-up to larger clouds, may also occur.

In the typical small dark cloud, the radius is smaller and the temperature lower, while the mass is about the same as for the Ambarzumian cloud. This means that the effect of self-gravitation is much stronger and may stabilize the cloud, or even force the cloud to collapse. The very dense globules are indeed dynamically unstable and collapse, large increases in density occurring over periods of the order of a few million years.

The pressure from a surrounding intercloud medium with a density of 2×10^5 H atom per m^3 has little effect on a small dark cloud, whereas the larger pressure in an H II region, such as that inside the Barnard loop, might presumably trigger collapse of a small dark cloud.

Finally, for the large dark clouds, density, temperature, and total mass are such that self-gravitation will overcome the tendency toward dispersal through thermal motion, or possible existing turbulent motion. In the case of these clouds, it is known from observation that dense cores will form and that collapse in high-density regions will lead to star formation. It is also known from observation that star formation, if it includes the formation of massive O stars, can lead to dispersal of the cloud matter, in some cases to ultimate destruction of the cloud.

It is clearly of importance to understand the processes that lead to the compression of parts of a large dark cloud, with star formation following. One possible mechanism is known from observation, namely, the compression in shock waves produced by ionization of hydrogen in the cloud by an O star, as described above. Another mechanism, also suggested by observation, is compression through the effect of a supernova explosion. We shall return to this question in § 6.5, where we shall briefly discuss still another mechanism, namely, compression of interstellar matter in shock waves caused when interstellar matter flows through the gravitational field in a spiral arm of a galaxy.

There are still many uncertainties concerning the mechanisms through which the observed space distribution of interstellar matter — the three types of clouds plus intercloud matter — originates and is maintained over long intervals of time. However, as we have seen, some of the processes through which compression of interstellar matter into dense regions, and ultimately into stars, is occurring have been discovered and clarified through the combination of observation and theoretical calculation.

We have already referred to the fact that interstellar grains are largely composed of elements heavier than hydrogen and helium, and that the ratio of grain mass to hydrogen mass is such that the relative content of the heavier elements, counting only the grain contribution, is smaller than that found in the sun and solar-type stars, but not very much smaller. How about the relative abundance of the heavier elements in the interstellar gas?

As already mentioned, measurements of absorption lines produced by interstellar gas and observed in the spectra of early-type stars (where the relevant stellar absorption lines are weak) have been carried out both from the ground and, most importantly, from space with the help of the OAO carried by the *Copernicus* satellite.

We have here a parallel to the classical laboratory experiment in which absorption of a gas column is observed by placing the column between a light source and the observer's spectrograph. In the interstellar gas practically all atoms and ions are in the lowest stationary state, the ground state, and the observable absorption lines are limited to those absorbed by the ground state. Here the OAO observations have very much extended the scope of investigations, because absorption lines at wavelengths from 100 nm to 300 nm have now become observable.

We have already referred to results obtained for H (Lyman-alpha) and H_2. For the determination of relative abundances of elements heavier than hydrogen and helium in the interstellar gas, it is very important that absorption lines have been observed for a considerable number of elements, not only for the neutral atoms, but also for ions. Thus, the relative abundance of an atomic species can in several cases be found through adding up abundances (derived from absorption line strengths) of the relevant atoms and ions, rather than depending on theoretical calculations of the distribution over the stages of ionization, which are sometimes not very accurate.

Through investigations of this type, it has been found that the total heavy-element content in the interstellar gas is not very different from that in the grains, but is in fact a little larger. The gas-and-grain content of elements heavier than hydrogen and helium adds up to about 1 per cent (by mass) of the hydrogen content, comparable to, but somewhat smaller than, the corresponding value of 2 per cent for the sun ($Z = 0.02$). It remains to be seen whether or not continued observation will lead to a value closer to the solar value. In particular, there is the possibility that appreciable amounts of heavier-element interstellar matter might be present in molecules not yet observed.

A point of interest is the fact revealed by this kind of observation that certain of the heavier elements, for example calcium, aluminium, and titanium, are strongly depleted in the interstellar gas, presumably because the great majority of the atoms of these species are bound in interstellar grains.

We return to the question of interstellar molecules in the gas phase. More than thirty such molecules have been discovered through observation of their emission lines in the centimetre and millimetre ranges of the radio spectrum. In most cases they are combinations of hydrogen, carbon, nitrogen, and oxygen: from simple diatomic molecules like CO, OH, and CN, to molecules such as ethanol, CH_3CH_2OH. The molecules SiO, SO, SO_2, and SiS have also been identified.

Theoretical investigations of the processes by which molecules are formed in interstellar space, and of the way in which equilibrium distributions are reached, have not yet led to many definite results. It is not clear whether reactions in the gas phase or reactions in the grains, followed by evaporation, are the most effective in establishing the equilibrium distribution.

With regard to possibly undiscovered molecules, it should be noted that the chances of discovery are reduced for molecules with particularly low transition probabilities of observable lines.

Absorption lines of helium as produced in H I regions are not observable because their wavelengths are much shorter than the OAO limit of about 100 nm. However, the observed optical spectra of several H II regions contain emission lines of helium. Analysis based on measured emission-line strengths for hydrogen and helium has led to a determination of the abundance ratio of helium to hydrogen. The result is 1 to 10 in terms of numbers of atoms, corresponding to a ratio of masses equal to 0.4. The relative weights of hydrogen, helium, and all the heavier elements in interstellar space are therefore close to 70, 30, and 1 percent respectively. As mentioned above, it is possible that further observations may lead to a revision of the last value to about 2 per cent.

It must be emphasized that the results concerning interstellar matter described so far refer to a rather restricted part of the galaxy, the region within about 2000 parsecs from the sun. Presumably the picture derived from observations pertaining to this region is valid in a general way for a whole ring round the centre of the galaxy, namely, the parts of the galactic disc located between 7000 and 11 000 parsecs from the centre. (The distance of the sun from the centre is taken to be 9000 parsecs.)

Interstellar matter in the disc that lies closer to the centre than 7000 parsecs is difficult to observe in the optical and rocket-observed ultraviolet part of the spectrum because of very strong attenuation of the radiation by interstellar grains along the long object–observer paths in question. Here observations in the radio spectrum, and also in the infrared part of the spectrum, are of great importance, the attenuation being negligible for the former and much less pronounced for the latter. Significant results have already been obtained.

Observations of the radio spectrum in the centimetre range have led to the identification of H II regions in the part of the galactic disc between 4000 and 7000 parsecs from the centre. We have already discussed the relevant emission mechanisms, namely, continuous-spectrum emission through free–free transitions, and transitions between stationary states of H with very high radial quantum numbers (e.g. $100 \rightarrow 109$). It has been established that the disc-ring located between 4000 and 7000 parsecs from the galactic centre contains many giant H II regions, each with stronger emission and larger total mass than even the Orion complex discussed earlier. The many giant H II regions in the 4000–7000-parsec ring, together with some giant H II regions located at larger distances from the centre, outline a system of spiral arms in our galaxy. We shall return briefly to this question in § 6.5.

In the 4000–7000-parsec ring, many compact H II regions of the type described above have been located and studied through radio-astronomical observations in the centimetre range. These observations reveal strong star-formation activity. Observations of CO lines at 2.6 mm have shown that there occur in the 4000–7000-parsec ring considerable numbers of large molecular clouds: an important result. The conclusion has been reached that the average density of interstellar matter in the 4000–7000-parsec ring is larger than in the 7000–11000-parsec ring, and that star formation in the former ring goes on at a higher rate than in the latter.

In the part of the galactic disc closer than 4000 parsecs to the centre, the average density of interstellar matter appears to be low, except close to the centre (within 1000 parsecs), where there is a high density. The rate of star formation, too, seems to be lower in the region of low average density of interstellar matter between 1000 and 4000 parsecs from the centre.

In the region within 100 parsecs of the galactic centre, the density of stellar and interstellar matter is very high. Within 1 parsec of the centre it is extremely high, more than a million times greater than in the 7000–11 000-parsec ring. In § 6.5 we shall briefly return to problems of the central region – the galactic nucleus.

6.4 Astrophysics of the solar system

Since 1958 great progress has been made in the study of the moon and the planets. The development of space astronomy meant that large amounts of new information became available as a basis for studies of the physics and chemistry of the solid bodies in question as well as their surrounding atmospheres.

Other areas in which very significant progress has been made since 1958 are the astrophysics of comets and the quantitative analysis of the chemical composition of meteorites.

We shall refer only to a few points that are particularly relevant to the principal topic of this chapter.

The great progress achieved in the physics and chemistry of the moon is of particular importance to studies of the evolution of the solar system, because the lunar surface has been practically undisturbed over billions of years by processes of the kind that have so completely changed the outer layers of the earth. The clues obtained from the lunar studies, together with those derived from observations of planets and their satellite systems, comets, and meteorites, form the basis for attempts at achieving an inductive, theoretical reconstruction of the history of the solar system. Such additions to knowledge will contribute significantly to our understanding of the processes of evolution in the galaxy.

Direct and most valuable contributions to the studies of the physics and chemistry of the sun and the stars have come from the development of refined methods of quantitative chemical analysis of meteoritic material. The relative chemical abundances of non-volatile elements, which have been determined for material from carbonaceous chondrites, supplement in important ways the results from quantitative spectro-analysis of the solar and stellar atmospheres. In fact, the cosmic abundance tables that form the basis for many discussions of the physics of stars and interstellar matter have been derived through a combination of the two sets of data mentioned. The meteorite data have been of particular importance for establishing isotope ratios.

6.5 Structure and evolution of our galaxy

By the late 1950s the main features characterizing the distribution of stars and interstellar matter in our galaxy were known: the central bulge, containing at its centre a dense nucleus, the flattened disc rotating around an axis through the centre and at right-angles to the disc, and finally the nearly spherical halo, with matter at density relatively low.

It was also known that the disc contains spiral arms which stand out through a higher density of young stars and interstellar matter. During the years 1958–78, much information pertaining to the spiral arms has been added. In § 6.3 we mentioned the investigations of giant H II regions and of the associations and clusters containing young stars which are connected with these H II regions. We have referred to the result that these complexes are concentrated in and outline a structure of spiral arms in the disc, and that they are particularly valuable as spiral-arm tracers in the part of the disc that is closer to the galactic centre than the sun. In the outer regions of the galaxy, the complexes in question are still valuable tracers. Here, however, determinations of the distribution of neutral hydrogen, as revealed by observations of the 21-cm emission, yield even more important results concerning the spiral structure.

A significant feature of our galaxy, connected with spiral-arm structure, is its so-called non-thermal radio emission. It was well known before 1958 that cosmic rays consisting of atomic nuclei, particularly protons; and free electrons moving with relativistic velocities, are an important component in our galaxy. It was also known that the galactic disc possesses an interstellar magnetic field, possibly extending into the galactic halo, and that this field guides the cosmic-ray particles, thus holding them within the galaxy. Finally, a very important step was taken when it was realized that as cosmic-ray electrons move with relativistic velocities in the galactic magnetic field, they will emit observable synchrotron radiation. Theory as well as observation had shown that this galactic non-thermal synchrotron radiation has a frequency distribution quite different from the galactic thermal radiation, emitted from H II regions through free–free transitions of electrons, as described in § 6.3. In the radio spectrum, the non-thermal synchrotron radiation falls off rapidly with increasing frequency, whereas the thermal radiation from H II regions has an approximately flat spectrum.

Since the late 1950s knowledge of these galactic components has increased considerably. The properties of the galactic magnetic field have been investigated through observations based on Zeeman-splitting of emission lines in the radio spectrum, through observations of Faraday rotation of polarized radio emission from extragalactic sources, and through studies of the influence of the galactic magnetic field upon interstellar grains, an influence that causes partial alignment of elongated grains and consequent polarization effects in the attenuation of optical radiation passing through the interstellar grain medium. Although a definite picture has not yet emerged, certain important features of the galactic magnetic field have become known.

The galactic magnetic field appears to have a regular component several times stronger within the spiral arms than in the interarm regions and runs parallel to the relevant spiral arm. Superimposed on this regular field is an isotropically random field. In the neighbourhood of the sun, the regular component has a strength of a few microgauss.

The galactic non-thermal synchrotron emission dominates over the galactic thermal emission in the radio spectrum between 300 and 70 cm, whereas thermal radiation from H II regions dominates in the centimetre and millimetre ranges. We mention in passing that another type of thermal emission may play a dominating role in the submillimetre and far-infrared part of the spectrum, namely the thermal emission from dense clouds of interstellar grains, heated by luminous neighbouring stars, embedded in the cloud (cf. § 6.3).

The theory of synchrotron radiation shows that when the magnetic field is of the order of a few microgauss, radio emission in the radio spectrum between 300 and 70 cm will be produced mostly by cosmic-ray electrons with energies in the range 10^3–10^4 MeV. The intensity and energy distribution of cosmic-ray electrons in the galactic neighbourhood are approximately known (within perhaps a factor of 2) from measurements on the earth. This knowledge provides an important input to the theory of galactic synchrotron radiation. Comparison of theoretical calculations with the observed intensity distribution over the sky show general agreement, and in particular they confirm the assumed picture of strengthening of the galactic magnetic field in the spiral arms.

As for the cosmic-ray component in our galaxy, very important questions about its origin and the mechanism of its containment remain unsolved, although the last 20 years has seen great progress in the knowledge of relative abundances and energy distributions for the cosmic-ray particles, as well as in the study of interaction between cosmic rays and interstellar matter. Important steps have been made in examining the possibility that the cosmic rays originate in supernova explosions, but here agreement between theory and observation has not yet been achieved.

We return to problems of the galactic spiral arms. Since the late 1950s, a gravitational theory of the spiral-arm distribution of interstellar matter and young stars was developed. According to this theory, the spirals are not material arms but travelling waves of density condensation of interstellar and stellar matter. The spiral-arm configuration shows rigid rotation around the axis through the galactic centre-perpendicular to the plane, such that the wave pattern lags behind the rotating interstellar and stellar matter. At the distance of the sun from the centre, where the matter rotates with a velocity of about 250 km s^{-1}, this lag of the spiral wave configuration amounts to somewhat more than 100 km s^{-1}.

We shall here consider one aspect of the theory of galactic spiral arms that is particularly important in the discussion of the astrophysics of stars and interstellar matter, namely its predictions concerning the streaming of interstellar matter through the lagging spiral wave-pattern. Analysis of the hydrodynamics of this streaming motion has shown that shock-waves are created by the gravitational forces connected with the wave-pattern of spiral arms, and that the consequence is a compression of matter in interstellar clouds as they pass through spiral arms. This is the compression mechanism that was referred to in § 5.3, where the mechanisms of compression near interstellar hydrogen ionization fronts and through supernova explosions were also mentioned. The theory gives an explanation of the phenomenon of concentration of newly formed stars to the spiral arms of the 'grand design'. At the same time, the possibility of some star-formation outside the 'grand design' is left open. The effect of secondary waves of star formation, triggered by the birth of massive O stars in a primary wave, is an important part of the picture.

Discussion of phenomena near the galactic centre has a significant place when progress in astrophysics over the last 20 years is considered. By the late 1950s it was known that there exist in a ring near the galactic plane between 1000 and 4000 parsecs from the centre (where the average density of interstellar matter is low — cf. § 5.3), hydrogen masses that move away from the centre with velocities of the order of 100 km s^{-1}. Further investigations have shown that, in a ring between 250 and 800 parsecs from the centre, the existing hydrogen masses are quiescent and rotate around the centre with approximately circular velocities. Analysis based on the observed velocities gives important clues concerning the gravitational field and the mass distribution around the centre. Within 250 parsecs of the centre dense molecular clouds have been shown by radio-astronomical observations of CO to be moving outward with velocities between 100 and 200 km s^{-1}. At about 40 parsecs from the centre, observations pertaining to compact H II regions indicate the presence of masses of ionized hydrogen, likewise ejected from the central region.

Within 0.5 parsec from the galactic centre, there is a core containing both stars and interstellar matter. In the optical part of the spectrum the core is invisible because of very strong attenuation of the light by interstellar grains, but the core has been investigated in the infrared, particularly at 2.2 and 10 μm. The observations show that the core contains a mixture of stars and interstellar matter. The 10 μm radiation comes from grains heated by the radiation from neighbouring stars. The total mass in the core within 0.5 parsec is several million solar masses, so that the density of matter here is about 10^7 times that in the galactic neighbourhood of the sun.

Recent observations of a Ne II line at 12.8 μm pertaining to the core have indicated a velocity spread of 500 km s^{-1} in the gas core, suggesting violent expulsion of matter from a centre.

Observations of continuous radio emission in the centimetre range, using very-long-baseline interferometry, have indicated that there is within the core a very strong concentration of emission in a region that is only 0.001 second of arc, or only 5×10^{-5} parsec in diameter. The total mass within this tiny region is presumably a few million solar masses.

All this information adds up to a picture of an active nucleus of our galaxy. At intervals of the order of a million years, matter appears to be expelled from a central engine, with kinetic energies totalling 10^{48} J. This engine has tentatively been identified with the mass concentration just described, a mass concentration which might possibly be a black hole.

Since 1958, efforts to unravel the past history of the galaxy have largely confirmed the scheme of evolution of our galaxy that had already been developed at that time, leading however to more detailed information in many respects.

According to this scheme the galaxy, when it first separated out as an independent unit, consisted of gaseous matter. The chemical composition was hydrogen and helium, approximately in the same proportion (10 to 1) as observed at the present time, with an extremely small admixture of other elements. In the earliest phase of evolution the gaseous mass filled a volume corresponding to that now occupied by the galactic halo. The distribution of gaseous matter within this volume was non-uniform, with clouds of much higher than average density. Star-formation began in these clouds, and produced the first generation of stars, the low-mass fraction of which has survived to the present day as halo stars (including globular cluster stars) with, characteristically, a very low content of heavy elements. These are the so-called extreme population II stars.

During the following phases, a collapse of the largely gaseous matter toward a centre (occupied in the present epoch by the central bulge stars) took place. At the same time, collisions between clouds led to a dissipation of turbulent kinetic energy, with the result that the shape of the galaxy was forced to correspond to the total rotational energy. A flattened, still largely gaseous, galactic disc resulted.

Star-formation continued in the central bulge and the galactic disc, and was followed by evolution of stars resulting in stellar remnants as well as matter returning to interstellar space with an increased content of heavy elements, as described in § 6.2. In other words, the great mechanism of gradual enrichment of interstellar matter operated and caused successive generations of stars to be born, with heavy-element contents increasing with the age of the galaxy.

A generation of stars, the so-called intermediate population II, with heavy element contents much larger than those of extreme population II stars and occupying a space with rather pronounced flattening toward the galactic plane, was formed only a few billion years after extreme population II. Gradually conditions changed into those of the present epoch, with population I stars of approximately solar composition being formed in a disc of very pronounced flattening.

As a result of the continuing process of star-formation, the galactic matter was transformed to its present form, 96 per cent matter in stars, and 4 per cent interstellar matter, the whole process taking place during an interval of time estimated to be $12-15 \times 10^9$ years.

The tools of research used in the investigations that have led to this picture of evolution of our galaxy were developed on the basis of the theory of stellar structure and evolution, described in § 6.2, as well as the theory of stellar atmospheres and quantitative determination of atmospheric chemical composition. Knowledge of stellar ages and chemical compositions, combined with knowledge of stellar galactic orbits, all this for large numbers of stars, made possible the analysis that revealed the consecutive stages of galactic evolution.

The history of our galaxy has thus been followed back in time to an early epoch, where the present discussion can be linked to another: the formation of galaxies as independent units in the expanding universe.

6.6 Morphology and physics of galaxies

Studies of the morphology of galaxies require instruments of great radiation-gathering power and high angular resolution (1 second of arc or better). Since 1958, research has been furthered through the addition to the 5-metre telescope on Mount Palomar (in operation since 1948) of a number of effective 3–4-metre telescopes in both the northern and the southern hemispheres. In the radio region, big antenna-arrays (synthesis radio telescopes) have made possible intensity mapping with an angular resolution about equal to that obtained with large optical telescopes, while very-long-baseline interferometry (with antenna separations of many thousand kilometres) has yielded resolutions of better than one-hundredth of a second of arc. These instruments have been particularly applied to studies of the active nuclei of galaxies, including quasars.

Systematic studies of radiation distribution and velocity distribution, as well as of variation of chemical composition with position within galaxies, have been undertaken for spiral galaxies, elliptical galaxies, and irregular galaxies. Theoretical models describing the evolution of different types of galaxies have been developed, and their predictions compared with observation.

Great progress was made in the detailed intensity mapping of radio galaxies. Theoretical investigations have led to at least a partial understanding of the phenomena based on models of ejection of relativistic electrons from active nuclei, with radio emission of the synchrotron type (cf. § 6.5).

The discovery and intensive investigations of quasars opened a new chapter in astrophysics, in which the behaviour of matter under extreme conditions of concentrated energy release in an active galactic nucleus is studied. A picture has gradually emerged, according to which a vast array of observed phenomena are interpreted as caused by activity in nuclei of galaxies. Our galaxy has a nucleus with comparatively mild activity, whereas strong radio galaxies, Seyfert galaxies, and a large fraction of, if not all quasars are galaxies with extremely active nuclei.

6.7 Spatial distribution of galaxies

Large observing programmes have been carried out for the study of distances and radial velocities relative to our galaxy for galaxies within about 10×10^6 parsecs (30×10^6 light-years). These investigations have yielded important results on the average numbers per unit volume of various types of galaxies, and on inhomogeneities in the spatial distribution, in particular on the role of galaxy clusters and galaxy grouping. Furthermore, they have allowed determinations of the Hubble constant that measures the rate of expansion of the universe, and also estimates of the properties of the Hubble flow, i.e. its degree of uniformity.

For greater distances, information is less detailed, but highly important for the discussion of cosmological problems. The range of classical optical astronomy in galaxy studies has been extended beyond 1000×10^6 parsecs, and studies of strong radio galaxies and quasars have yielded data on the properties of the universe out to distances many times further. A picture emerges of a highly uniform and isotropic distribution of matter when densities are averaged over distances of 100×10^6 parsecs. This picture has received confirmation through the discovery and investigation of the microwave background, the so-called 3-degree radiation. At the greatest distances, reached using strong radio galaxies and quasars as tracers, there are good indications for departure from universal homogeneity. These are of particular importance in cosmology, because the distances in question correspond to look-back times approaching the age of the universe in its present phase, as computed from the observed Hubble expansion constant and the present average density of matter.

The available information on the large-scale properties of the universe has been utilized as a basis for theories of the formation of galaxies as independent units in the expanding universe. A partial understanding has been obtained of the formation of cosmic units of the size of galaxies, and of the properties of their present-phase spatial distribution, as dominated by clusters and groupings of galaxies.

6.8 Conclusions and future trends

In the period 1958-78, observational astronomy developed greatly, particularly through the introduction of new electronic techniques in classical optical astronomy, through developments in infrared astronomy and in radio astronomy, and through the impact of space astronomy, which opened up new possibilities of observation in the far ultraviolet and X-ray wavelengths.

Advances in theoretical astrophysics have been made possible through the vast input of new observational data, through the development of new numerical techniques utilizing electronic computers, and from the great strides made in theoretical and experimental physics.

We can expect that this trend will continue during the next twenty years. A big space telescope, of very high angular resolution, may go into operation, and its findings may be combined with those resulting from still bigger ground-based telescopes. Very large and effective new synthesis radio telescopes may also be built.

Progress in each of the areas discussed above has led to the formulation of new important questions that will be tackled during the next twenty years. These should lead to advances in many fields, including such very important ones as the late phases of stellar evolution, the early history of the evolution of galaxies, understanding of the different types of galaxies, the physics of extremely dense bodies (i.e. neutron stars and, possibly, black holes), the physics of active galactic nuclei, radio galaxies and quasars, and finally, basic questions of cosmology.

References

The *Annual Review of Astronomy and Astrophsics* (Annual Reviews Inc.) contains review articles on many of the subjects discussed in this chapter. The series of Symposium Volumes sponsored by the International Astronomical Union (D. Reidel Publishing Company) contains a large number of articles on these subjects. Specific references are given below.

Stellar evolution
Bahcall, J. N. and Sears, R. L. (1972). Solar neutrinos. *Annual Review of Astronomy and Astrophysics* 10, 25.
Cameron, A. G. W. (1970). Neutron stars. *Annual Review of Astronomy and Astrophysics* 8, 179.
Hayashi, C. (1966). Evolution of protostars. *Annual Review of Astronomy and Astrophysics* 4, 171.
IAU Symposium No. 42 (1971). *White dwarfs* (Ed. W. J. Luyten).
— No. 53 (1974). *Physics of dense matter* (Ed. C. J. Hansen).
— No. 66 (1974). *Late stages of stellar evolution* (Ed. R. J. Tayler, co-Ed. J. E. Hesser).
Iben, Jr., I. (1967). Stellar evolution within and off the main sequence. *Annual Review of Astronomy and Astrophysics* 5, 571.
— (1974). Post main sequence evolution of single stars. *Annual Review of Astronomy and Astrophysics* 12, 215.
Schramm, D. N. and Arnett, W. D. (1973). *Explosive nucleosynthesis, Proceedings of Conference, Austin, Texas, 1973.* University of Texas Press, Austin.
— (Ed.) (1976). Supernovae. *Proc. Special IAU Section, Grenoble.* D. Reidel Publishing Company.
Weidemann, V. (1968). White dwarfs. *Annual Review of Astronomy and Astrophysics* 6, 351.
Woltjer, L. (1972). Supernova remnants. *Annual Review of Astronomy and Astrophysics* 10, 129.

Interstellar matter and star formation
Aannestad, P. A. and Purcell, E. M. (1973). Interstellar grains. *Annual Review of Astronomy and Astrophysics* 11, 309.

Castellani, V. and Gratton, L. (1974). General properties of the diffuse interstellar matter, Proceedings of the Second European Regional Meeting in Astronomy, Trieste 1974. *Memorie della Società Astronomica Italiana,* **45**, N 1/2, Section A.

Heiles, C. (1976). The interstellar magnetic field. *Annual Review of Astronomy and Astrophysics* **14**, 1.

IAU Symposium No. 52 (1973). *Interstellar dust and related topics* (Ed. J. M. Greenberg and H. C. van de Hulst).

— No. 75 (1977). *Star formation* (Ed. T. de Jong and A. Maeder).

Salpeter, E. E. (1977). Formation and destruction of dust grains. *Annual Review of Astronomy and Astrophysics* **15**, 267.

Spitzer, Jr., L. and Jenkins, E. B. (1975). Ultraviolet studies of the interstellar gas. *Annual Review of Astronomy and Astrophysics* **13**, 133.

Zuckerman, B. and Palmer, P. (1974). Radio radiation from interstellar molecules. *Annual Review of Astronomy and Astrophysics* **12**, 279.

Structure and evolution of our galaxy

IAU Symposium No. 69 (1974). *Galactic radio astronomy* (Ed. F. J. Kerr and S. C. Simonson).

Lin, C. C. (1971). Theory of spiral structure, *IAU Highlights of Astronomy*, Vol. 2, p. 58 (Ed. C. de Jager). D. Reidel Publishing Company.

Oort, J. H. (1977). The galactic center. *Annual Review of Astronomy and Astrophysics* **15**, 295.

Toomre, A. (1977). Theories of spiral structure. *Annual Review of Astronomy and Astrophysics* **15**, 437.

Morphology, physics and spatial distribution of galaxies

Audouze, J. and Tinsley, B. M. (1976). Chemical evolution of galaxies. *Annual Review of Astronomy and Astrophysics* **14**, 43.

van den Bergh, S. (1975). Stellar populations in galaxies. *Annual Review of Astronomy and Astrophysics* **13**, 217.

IAU Symposium No. 44 (1972). *External galaxies and quasi-stellar objects* (Ed. D. S. Evans, ass. by D. Wills and B. J. Wills).

— No. 58 (1974). *The formation and dynamics of galaxies* (Ed. J.R. Shakeshaft).

— No. 63 (1974). *Confrontation of cosmological theories with observational data* (Ed. M. S. Longair).

van der Kruit, P. C. and Allen, R. J. (1976). The radio continuum morphology of spiral galaxies. *Annual Review of Astronomy and Astrophysics* **14**, 417.

De Young, D. S. (1976). Extended extragalactic radio sources. *Annual Review of Astronomy and Astrophysics* **14**, 447.

7

Mathematics

Mark Kac

7.1 Introduction

Unlike disciplines with empirical backgrounds, mathematics lacks central problems that are clearly defined and universally agreed upon. As a result, the development of mathematics proceeds along a number of seemingly unrelated fronts, which tends to present a picture of fragmentation and division. Adding to the difficulty of evaluating its present state and of (guardedly!) predicting its future, is the fact that during the past few decades mathematics became increasingly isolated from its sister disciplines, and as a result of turning inward there was a marked increase in the level of abstraction and a reinforcement of the ever-present trend to greater and greater generality.

Ironically, the isolationist and introspective trend in mathematics coincided with a veritable explosion in demand for its 'services' to an ever-widening circle of 'users'. There is now hardly a corner of human activity which has not for better or for worse been affected by mathematics. So much so, in fact, that one may speak with considerable justification of a 'mathematization' of culture.

Ever since Plato's invectives against Eudoxus and Archytas there have been tensions between the two main trends of mathematical creativity: the abstract ('unembodied objects of pure intelligence') and the concrete ('means of sustaining experimentally, to the satisfaction of the senses, conclusions too intricate for proof by words or diagrams'[1]). The tensions between the two trends (more easily recognized under the names of *pure* and *applied*) persist to this day, aggravated by the emerging power of the computer and complicated by socio-economic factors. Thus Professor Marshall Stone, writing in 1961 asserted that 'while several important changes have taken place since 1900 in our conception of mathematics or in our points of view concerning it, the one which truly involves a revolution in ideas is the discovery that mathematics is entirely independent of the physical word' [1]. A few lines later Professor Stone elaborates:

---

[1] I am referring here to the following passage in Plutarch's *Life of Marcellus*: 'Eudoxus and Archytas had been the first originators of this far-famed and highly prized art of mechanics, which they employed as an elegant illustration of geometrical truths, and as means of sustaining experimentally, to the satisfaction of the senses, conclusions too intricate for proof by words and diagrams.... But what of Plato's indignation at it, and his invectives against it as mere corruption and annihilation of the one good in geometry, which was thus shamefully turning its back upon the unembodied objects of pure intelligence to recur to sensation and to ask help (not to be obtained without base supervisions and depravation) from matter....'

When we stop to compare the mathematics of today with mathematics as it was at the close of the nineteenth century we may well be amazed to see how rapidly our mathematical knowledge has grown in quantity and in complexity, but we should also not fail to observe how closely this development has been involved with an emphasis upon abstraction and an increasing concern with the perception and analysis of broad mathematical patterns. Indeed, upon closer examination we see that this new orientation, made possible only by the divorce of mathematics from its applications, has been the true source of its tremendous vitality and growth during the present century.

No one will deny that mathematics can reach ultimate heights in splendid isolation. One does not even have to look to our century for striking examples. Surely Galois theory, culminating in the proof that algebraic equations of degree 5 and higher are, in general, not solvable in terms of radicals, is gloriously convincing.

There are even examples, again taken from the past, of mathematical developments which, having begun with a concern for 'unembodied objects of pure intelligence', have ultimately become instruments of deeper understanding of the physical universe. I have in mind especially the General Theory of Relativity, which in its mathematical apparatus depends crucially on differential geometry, a branch of mathematics which owes its inception and much of its early growth to non-Euclidean geometry. There is, however, an implication in Professor Stone's thesis that isolation is also *necessary* for survival and growth of mathematics, and this I find totally unacceptable.

On a number of occasions, I made reference to a wartime cartoon depicting two chemists surveying ruefully a small pile of sand amidst complicated looking pieces of equipment. The caption read: 'Nobody really wanted a dehydrated elephant, but it is nice to see what can be done'. Forced isolation of mathematics, though it may result in many spectacular achievements, will also proliferate dehydrated elephants to a point where the price may be too high to pay.

With all this as background, I shall attempt to give a glimpse of some of the prevailing themes, trends, and tensions in contemporary mathematics. The picture will be fragmentary and incomplete, but this I am afraid is unavoidable. The limitations of one's competence and the very nature of modern mathematics are too formidable to overcome.

7.2 Scaling the heights

In an article 'American mathematics from 1940 to the day before yesterday' [2] six authors selected (according to certain clearly stated rules and with due apologies for biases and omissions) ten theorems which they considered to have been pinnacles of achievement of American mathematics in the period from 1940 to 24 Janaury 1976, i.e. the day Professor Halmost delivered an abbreviated version of the paper as an invited address before a meeting of the Mathematical Association of America. Since much of the material depended significantly on the work of non-Americans (underscoring the truly international character of mathematics) it is safe to say that the selection can be taken as representative of the achievements of mathematics the world over. The authors by one of their rules included only 'pure' mathematics, and there is little doubt that their picture of this part of mathematical landscape is photographically accurate.

All the theorems confirm, with vengeance I may add, that 'mathematics is entirely independent of the physical world', thus lending support to Professor Stone's neo-Platonist view of advantages of isolation. On the other hand, the very fact that it required a collaboration of *six* very able and high accomplished mathematicians to merely explain to their fellow practitioners the backgrounds and formulations of the ten most spectacular achievements of their subject is in itself a telling commentary on the state of the art. Of the ten, I have chosen three for a brief presentation. My choice is dictated by the belief that these are less 'internal' to mathematics than the others and hence perhaps of greater general interest.

The first of the three is the independence of the continuum hypothesis from the standard (Zermelo–Fraenkel) axions of set theory, and it represents a solution to the first on the famed list of unsolved problems posed by Hilbert in 1900. The question goes back to Georg Cantor, the creator of set theory. Cantor defined two sets to be of equal cardinality ('being equally numerous') if a one-to-one correspondence between the elements of the two sets could be established. A set A is said to be of greater cardinality than a set B if there is a *subset* of A of equal cardinality with B, while the cardinality of A itself is not equal to that of B. In other words, there is a subset of A whose elements can be put into one-to-one correspondence with those of B, but there is no such one-to-one correspondence between all the elements of A with those of B.

For finite sets, equal cardinality simply means that the sets contain the same number of elements. For infinite sets, the situation is more subtle since a subset of an inifinite set can have the same cardinality as the set itself. For example, the set of even integers, though clearly a subset of the set of all integers, is of the same cardinality as the larger set, since a one-to-one correspondence between the elements of the two can be established as:

$$
\begin{array}{ccccc}
2 & 4 & 6 & 8 & 10 \quad \dots \\
\updownarrow & \updownarrow & \updownarrow & \updownarrow & \updownarrow \\
1 & 2 & 3 & 4 & 5 \quad \dots.
\end{array}
$$

The set of (positive) integers is the 'smallest' infinite set, i.e. its only subsets of smaller cardinality are finite. All sets of the same cardinality as the set of integers are called denumberable. It was shown by Cantor that all rational numbers (i.e. numbers of the form p/q, with p and q integers without a common divisor) form a denumerable set. Much more surprisingly, Cantor has also shown that the set of all real numbers is not denumerable and hence of greater cardinality than that of the set of integers. The proof is simple and is so well known that I reproduce it only with profound apologies to the overwhelming majority of my readers. Since I want to comment on it later on, I thought it best to have it in front of one's eyes rather than rely upon one's memory.

Every real number can be uniquely written as a non-terminating decimal (this requires, for example, writing $6/5 = 12/10 = 1.2$ in the equivalent way as $1.1999\dots$).

If real numbers could be written as a sequence (which is another way of saying that if the set of real numbers were denumerable), we could have a tableau

$$c_{11} \cdot c_{12} c_{13} c_{14} \cdots$$
$$c_{21} \cdot c_{22} c_{23} c_{24} \cdots$$
$$c_{31} \cdot c_{32} c_{33} c_{34} \cdots$$
$$\vdots \quad \vdots \quad \vdots \quad \vdots$$

containing all of them. But the number

$$d_1 \cdot d_2 d_3 \cdots ,$$

where $d_1 \neq c_1$, $d_2 \neq c_{22}$, $d_3 \neq c_{33}$, etc. is clearly different from all elements of the tableau and thus we have a contradiction which means that the set of all real numbers is of *greater* cardinality than the set of integers.

The continuum hypothesis of Cantor is the assertion that the set of real numbers is the *next* largest after the set of integers or, in other words, there are no sets whose cardinality is greater than that of the set of integers and less than that of the set of real numbers.

There is a historical analogy between the fate of the continuum hypothesis and Euclid's fifth postulate. Cantor believed his hypothesis to be nearly self-evident. As doubt arose, an enormous amount of work was undertaken in constructing statements logically equivalent to Cantor's hypothesis or its negation in the search for one so bizarre that an acceptance or rejectionof the hypothesis could be forced. How very much like attempts of Sacheri and his fellow toilers in their search for a self-evidently rejectable consequence of a negation of the fifth polstulate!

In both cases the labour was in vain. For, as Paul J. Cohen showed in 1963, paralleling the feat of Bolyai and Lobachevsky, one can keep or reject the continuum hypothesis without any fear of logical contradiction. Here the analogy ends. For as non-Euclidean geometries became part of the mainstream of mathematics and lived to influence profoundly our perception of the physical world, the non-Cantorian set theories have not, so far at least, had any discernable effect on the development of mathematics, let alone on our views of the physical universe.

Is it possible that the strange non-Cantorian creations could become frameworks for some future physical theories? Personally I doubt that this will ever be the case. My main reason for feeling doubtful is that the independence of the continuum hypothesis can be restated as an assertion that there are models (realizations) of the set of real numbers of arbitrarily high cardinality. In other words, the usual collection of axioms which describe the properties of real numbers are logically consistent with the *set* of real numbers being of arbitrarily high cardinality (higher, of course, than the cardinality of integers), or in yet other words, there are non-Cantorian models of the set of reals.

The set of reals has been a troublesome concept from the beginning. Cantor's striking proof that it is not denumerable led immediately to a paradoxical conclusion. Since the set of numbers which can be defined in a finite number of words can be easily shown to be denumerable, it followed that there must be real numbers which *cannot* be defined in a finite number of words. Can one sensibly speak of objects definable in a sentence of infinite length? The great Henri Poincaré thought that the answer should be an emphatic one. 'N'envisagez jamais que des objets susceptibles d'être définis dans un nombre fini des mots' he proclaimed, and there is a whole school of logic (the intuitionists) which lives by this dictum. So it all boils down to whether we have found a proper axiomatic description of reals, and this question appears closer to philosophy than to physics. Nevertheless, one cannot tell and there are serious mathematicians (S. M. Ulam is one) who would disagree with me.

One final remark. Cantor's diagonal construction, which is at the heart of his proof that the set of reals is non-denumerable, has to be treated with care, for it can be turned into a source of paradoxes. The following (a variant of a pardox of Jules Richard) is a good illustration.

We say that a function f defined for all integers 1, 2, 3, ... and assuming only integer values is *computable* if for every integer n there is a *prescription* consisting of a finite number of words which allows one to calculate $f(n)$ in a finite number of steps. One can easily show that the computable functions form a denumerable set, and they can thus be arranged into a sequence

$$f_1, f_2, f_3, \ldots.$$

Consider now the function g defined by the formula

$$g(n) = f_n(n) + 1.$$

Now $g(n)$ is clearly computable and yet since $g(1) \neq f_1(1)$, $g(2) \neq f_2(2)$, etc. it is not included in our sequence: a contradiction!

The source of the paradox is subtle, and it has to do with the fact that we are asking whether g is a member of a set in terms of which g itself is defined. That such circularity is fraught with danger has been known since antiquity, certainly since Epimenides the Cretan proclaimed that all Cretans lie.

Let me now turn to my second illustration. It again involves the solution of a problem of long standing and one which also appeared (as No. 10) on Hilbert's list of 1900. The problem concerns diophantine equations, i.e. polynomial equations with integer coefficients for which we seek only integer solutions. The simplest is

$$ax + by = 1,$$

(a, b, x, and y being positive or negative integers) and its solution was already known to the Greeks. The next simplest is the so-called Pell's equation

$$x^2 - Dy^2 = 1,$$

where D is a positive integer. The complete theory of this equation has been known for over two hundred years.

In both of these cases there are simple algorithms for generating all integer solutions. In the linear case it is related to the so-called Euclid algorithm; in the quadratic case the algorithm is to expand $\sqrt{D}$ in a continued fraction:

$$\sqrt{D} = a_0 + \cfrac{1}{a_1 + \cfrac{1}{a_2 + \ldots}}$$

Consider the continued fraction

$$\frac{p_n}{q_n} = a_0 + \cfrac{1}{a_1 + \cfrac{1}{a_2 + \cfrac{\displaystyle \cdot^{\cdot^{\cdot}}}{\cfrac{1}{a_n}}}}$$

A solution to Pell's equation can then be obtained simply from the p_ns and q_ns.

What is an algorithm? Without entering into technicalities it is sufficient to identify algorithms with computer programs. It is quite easy to program a computer to generate solutions of the linear and quadratic diophantine equations, but for equations of higher degrees and involving more than two variables the situation is clearly much more complicated.

Hilbert's tenth problem was to prove (or disprove) that, given a (polynomial) diophantine equation, an algorithm can be found to generate all of its solutions. The answer is in the negative, i.e. there are diophantine equations for which no algorithm to generate all of its solutions can be found.

The theorem is purely existential, i.e. it merely asserts that certain objects *exist* without providing us with a *concrete* prescription of how to exhibit them. In fact, the Cantor diagonal construction is used in the final stage of the proof, thus making it non-constructive in the usual sense of the word.

'Impractical' as the negative solution of Hilbert's tenth problem is, it was not without influence on the rapidly growing discipline of computer science. It brought the power of modern mathematical logic to bear upon questions of computability, and it brought these seemingly mundane questions closer to foundations of mathematics and even to philosophy.

So far my illustrations were from the realm of foundations. My third (and last) is from the heart of the mathematical superstructure. The actual theorem chosen by Halmos *et al.* is the so-called index theorem of Atiyah and Singer, which is rather technical. It has to do with certain elliptic (differential) operators defined on manifolds, and it culminates in proving that an integer defined in terms of topological properties of the manifold is equal to another integer defined in terms of the operator. Although the motivation for the Atiyah–Singer index theorem came from algebraic geometry (the theory of Riemann surfaces to be more specific) the theorem is one of a class of theorems connecting differential equations and topology, whose roots can also be traced to physics. More importantly, the physical background can be turned into a potent mathematical weapon, thus reminding us of the words of Poincaré: 'La Physique ne nous donne seulement l'occasion de résoudre des problèmes ..., elle nous fait présenter la solution.'

Let me very sketchily and in a greatly oversimplified manner try to describe what is involved. In ordinary Euclidean space — let us say the plane — heat conduction is described in appropriate units by the differential equation

$$\frac{\partial P}{\partial t} = \Delta P = \text{div (grad } P) = \frac{\partial^2 P}{\partial x^2} + \frac{\partial^2 P}{\partial y^2},$$

where $P(x,y,t)$ is the temperature at (x,y) at time t.

If we consider a bounded region whose boundary is kept at temperature 0, then starting with an arbitrary temperature distribution $P(x,y;0)$ at $t=0$ we shall observe an approach to 0 of the form

$$P(x,y;t) = \sum_{n=1}^{\infty} c_n(x,y) \exp(-\lambda_n t),$$

where the λ_ns are positive numbers. The terms $\lambda_1, \lambda_2, \ldots$ are called the eigenvalues of the Laplacian Δ, and it turns out that

$$\Delta c_n + \lambda_n c_n = 0,$$

with $c_n = 0$ on the boundary of the region. (If the region were a membrane held fixed along its boundary, the λ_n would be the squares of frequencies of fundamental tones which the membrane was capable of producing.)

Let us now consider the same heat-conduction problem but on a smooth, closed, two-dimensional surface (e.g. the surface of a sphere of radius R).

The differential equation is still of the form

$$\frac{\partial P}{\partial t} = \text{div (grad } P) = \Delta P,$$

but Δ is now called the Laplace–Beltram operator (it can be expressed explicitly in terms of curvilinear coordinates) and there is no boundary, so that $\lambda_0 = 0$ and an arbitrary temperature distribution will approach the uniform distribution on the surface.

It is a natural question to ask what information about the surface can be obtained from the λ_ns. The answer, striking in its simplicity and beauty, is contained in the following formula of McKean and Singer [3] valid for small t:

$$\sum_{n=0}^{\infty} \exp^{(-\lambda_n t)} = \frac{S}{4\pi t} + \frac{1}{6} E + \text{terms which approach 0 as } t \to 0.$$

Here S is the area of our surface and E the so-called Euler–Poincaré characteristic, a topological invariant which can be defined in an elementary way as follows:

Consider a polyhedron inscribed in the surface and let F, E, and V be respectively the numbers of faces, edges, and vertices of this polyhedron. The Euler–Pointcaré characteristic is then given by the formula

$$E = F - E + V.$$

It does not in fact matter that the polyhedron is inscribed; it suffices that it be *continuously deformable* into the surface. For a sphere $E = 2$ and for a torus $E = 0$.

The theorem connecting the eigenvalues of the Laplacian and the Euler–Poincaré characteristic has been generalized (to manifolds of arbitrary even dimension) and extended in many directions.

Perhaps if one compares the circles of ideas involved in this problem and in the Cantor problem discussed at the beginning of this section one might gain a measure of appreciation of the unbelievable breadth of mathematical problems. One might also wonder as to how the existence of non-Cantorian systems could in any way affect the theory of heat conduction or the eigenvalues of the Laplacian.

7.3 Finding common ground

I have already mentioned that during the past few decades mathematics has largely developed in isolation. Whether as a result of a revolution, as Professor Stone would have us believe, or simply because internally generated problems provided a sufficiently exciting intellectual fare, the fact remains that mathematics all but severed its ties with sister disciplines.

Even in the past much of the development of mathematics was mostly internal and wholly autonomous. But the lines of communication between mathematics in the one hand, and physics and astronomy on the other were open, so that a healthy flow of ideas in both directions could be maintained.

Thus when in about 1914 Einstein struggled with formulating field equations of general relativity, he soon found out that a mathematical apparatus that he needed was already in existence in the form of so-called Ricci calculus, an outgrowth of Riemann's inspired *Habilitationsvortrag* 'Ueber die Hypothesen welche der Geometrie zugrunde liegen'. Similarly when the new quantum mechanics was being created in the 1920s and 1930s, some of the necessary conceptual and technical mathematical background was again available in the form of the theory of Hilbert spaces, although much of the theory had to be developed further to fit the needs of the new physics.

The Ricci calculus, which is a branch of differential geometry, is directly traceable to non-Euclidean geometries. Its development was internally motivated and constituted a natural generalization and extension of the work of Gauss and Riemann. Riemann expected that his geometry could some day intervene in problems of physics and astronomy, but niether he nor his followers could have imagined the spectacular and dramatic way in which it ultimately did. The theory of Hilbert spaces on the other hand, evolved from problems of nineteenth-century physics and culminated in the miracle of unifying atomic spectra and fundamental tones of membranes and strings.

The interweaving of mathematical and physical ideas in general relativity and in quantum mechanics belongs to the most glorious chapter of history of natural sciences. It is therefore truly surprising that the partners in these great adventures of human mind gradually came to a parting of ways.

It is not that mathematics abandoned either differential geometry or the theory of Hilbert spaces. Quite to the contrary, both continued to thrive. It is just that the preoccupations, the problematics, and the directions of mathematics and physics diverged to such an extent that the resulting separation seemed irreversible. Fortunately we may be witnessing a change in the trend and, though it is too early to tell, we may in fact be in the midst of a period in which ideas which have originated in vastly disparate contexts are beginning to show signs of yet another miraculous confluence. Interestingly enough, the mathematical ideas have come from a far-reaching extension of those underlying Riemannian geometry, itself an inspired extension of the theory of curved surfaces. It is difficult without considerable technical preparation to give an honest account of what is really involved, but I will try to give the reader a little taste of the brew.

First though, a quick glance at the theory of surfaces, illustrated on the example of the surface on a sphere of radius R in ordinary three-dimensional Euclidean space.

Introducing polar coordinates (with ϕ the latitude and θ the longitude) we have

$$x = R \sin \phi \cos \theta \, ;$$

$$y = R \sin \phi \sin \theta \, ;$$

$$z = R \cos \phi.$$

The Euclidean distance ds between two infinitesimal close points is (from Pythagoras' theorem)

$$ds^2 = dx^2 + dy^2 + dz^2.$$

If the points are constrained to lie on the surface of our sphere, one has

$$ds^2 = R^2 \, (d\phi^2 + \sin^2\phi d\theta^2).$$

At each point P of the surface of our sphere we consider the plane tangent to it, and a vector $\vec{v}(P)$ in that plane which changes smoothly as P moves among the surface. How do we define the differential of $\vec{v}(P)$ as P changes infinitesimally? One would be inclined to merely take the difference $\vec{v}(P') - \vec{v}(P)$ (P' and P infinitesimally close). But this difference would also have a component in the direction perpendicular (normal) to the surface of the sphere, and one takes the view that a flat creature confined to live on that surface would be incapable of detecting changes in the normal direction. We thus define the differential D as the tangential component of the usual differential d, and call it the *covariant differential*. This covariant differential leads to the *covariant derivative* by the formula

$$\frac{D\vec{v}}{ds} \, ,$$

ds being the differential distance between P and P'.

If P moves along a curve in such a way that the covariant derivative of $\vec{v}$ is zero, we say that $\vec{v}$ is undergoing *parallel displacement*. If $\vec{v}(P)$ is parallel displaced along a closed curve it may not come back to its original value. In such a case the surface is curved, and one can define (and describe) its curvature in terms of the changes suffered in parallel displacement.

All this (in a different terminology) was already known to Gauss. Gauss made one more observation, which he recognized as being of crucial importance. (The observation is embodied in a theorem he called Theorema Egregium.) The observation was that the concept of curvature is *intrinsic*, i.e. it can be defined without any reference to the Euclidean space in which the surface is imbedded. In other words, the hypothetical flat creature could define curvature without knowing that there is a world in the direction normal to the surface on which it lives.

Riemann took the next step by replacing surfaces by more general manifolds, which could be coordinatized (at least locally) and endowed with a metric

$$ds^2 = \Sigma g_{ij} dx^i dx^j$$

(by analogy with such formulae for surfaces) which *need not* have been 'inherited' from an Euclidean metric in the way

$$ds^2 = R^2(d\phi^2 + \sin^2\phi d\theta^2)$$

was 'inherited' from

$$ds^2 = dx^2 + dy^2 + dz^2 .$$

The Italian geometers developed this idea further by extending the concepts of covariant differentiation, parallel displacement, and curvature. Unlike in the case of surfaces in Euclidean three space, in which curvature could be described by scalars, the higher-dimensional cases required tensors for a full description of how the space is curved. By the time Einstein needed a theory of curved spaces, it was there for him to use.[1] Riemannian geometry, rooted as it was in the theory of curved surfaces, made essential use of tangent spaces and defined covariant differentiation and parallel displacement by analogy with that theory. The underlying skeleton is the space (manifold) M with a space T_x (tangent space) attached to every point x of M and a rule (connection) which tells us how to perform parallel translations on elements of T_x as x moves in M.

I have just given a vague and imperfect description of a fibre bundle which became the stepping-stone to the next development of differential geometry. Tangent spaces (fibres) can be replaced by more complicated structures not necessarily related to the space to which they are attached. Similarly, connections need not be defined in terms of metric properties of the underlying space. To be sure some restrictions have to be imposed, and these are dictated by a degree of adherence to the parental Riemannian geometry.

But the newly gained freedom is great — so great in fact, that the familiar theory

---

[1] Riemannian geometry was developed for the case of *definite metric*, i.e. the case in which $\Sigma g_{ij} dx^i dx^j$ is positive definite (in other words for locally Euclidean spaces). In relativity theory the space (or rather space–time) is locally Minkowskian, i.e. by a suitable change of local coordinates $\Sigma g_{ij} dx^i dx^j$ can be transformed into

$$(dy^1)^2 - (dy^2)^2 - (dy^3)^2 - (dy^4)^2 .$$

of electromagnetism can be couched in geometric terms. Vector potential intervenes in the definition of covariant differentiation and the electromagnetic field tensor becomes the curvature tensor in the new formulation. The much studied Yang–Mills fields also fit neatly into the new scheme, and results of direct physical relevance are uncovered in the intricate web of abstract mathematical speculations.

Let me conclude with a simple illustration of the interplay between geometrical and physical ideas, which I have taken from the excellent notes by Professor Sidney Coleman ('Classical lumps and their quantum descendents') containing his lectures delivered at the 1975 International School of Subnuclear Physics 'Ettore Majoranana'.

The underlying manifold is the ordinary plane, and to each of its points $\vec{r}$ we attach a two-dimensional (plane) real vector potential $\vec{A}(\vec{r})$ and a complex number (scalar field) $\phi(\vec{r})$.

We now define the *covariant gradient* of ϕ by the formula

$$D\phi = \nabla\phi + ie\vec{A}\phi \,,$$

where $\nabla\phi$ is the ordinary gradient and e a real scalar.

Now let

$$U(\phi) = \frac{\lambda}{2}(\phi\phi^* - a^2)^2 = \frac{\lambda}{2}(|\phi|^2 - a^2)$$

and define an energy integral by the formula

$$E = \iint d\sigma \,\{(\text{curl}\,A)\cdot(\text{curl}\,A) + D\phi\cdot D\phi^* + U(\phi)\} \,,$$

the integral being taken over the whole plane. The question is whether there are fields ϕ and $\vec{A}$ which yield finite E.

If we were to set

$$\phi(\vec{r}) = \psi(\vec{r})\exp(i\alpha(\vec{r}))$$

(i.e. change the phase of ϕ by α) and if we were to set

$$\vec{A}(\vec{r}) = \frac{1}{e}\nabla\alpha(\vec{r}) \,,$$

then the integrand of E above would become much simpler, namely

$$\nabla\psi\cdot\nabla\psi^* + U(\psi) \,.$$

But then we could not make the integral

$$\iint d\sigma\,\{\nabla\psi.\nabla\psi + U(\psi)\}$$

over the whole plane finite, except in the trivial case in which for large $r = |\vec{r}|$, $\psi(\vec{r}) = \psi(r,\theta)$ would have no angular dependence, i.e.

$$\lim_{r\to\infty}\psi(r,\theta) = \text{const}.$$

There are however choices of 'gauge fields' $\vec{A}(\vec{r})$ for which the original integral for E is finite.

To see this, note first of all that as $r\to\infty$ $U(\phi)$ must approach zero, i.e. $|\phi|^2$ must approach a^2 and hence

$$\lim_{r \to \infty} \phi\,(r,\theta) = a \, \exp\,(i\sigma(\theta))$$

(keeping θ fixed as $r \to \infty$).

Next we try to make the tangential component of $D\phi$ zero by setting

$$\frac{1}{r}\frac{\partial\phi}{\partial\theta} + ieA_\theta\phi = 0\;,$$

which implies for $r \to \infty$

$$A_\theta \sim \frac{1}{er}\frac{d\sigma}{d\theta}\;.$$

The radial component A_r can for the sake of simplicity be taken to be zero: $A_r = 0$. With such a choice of $\vec{A}$, ϕ can easily be chosen to make E finite.

The formula

$$\lim_{r \to \infty} \phi\,(r,\phi) = a\,\exp\,(i\sigma(\theta))$$

establishes a mapping $\theta \to \sigma(\theta)$ of a circle into a circle which must be smooth so that in particular for $n = \pm1, \pm2, \pm3, \dots$ $\sigma(2\pi n) = \sigma(0)$, i.e.

$$\int_0^{2\pi} \frac{d\sigma}{d\theta}\,d\theta = 2\pi n\;.$$

The flux

$$\iint \mathrm{curl}\,\vec{A}\;d\sigma$$

through the whole plane can be calculated by Stokes' theorem, obtaining

$$\iint \mathrm{curl}\,\vec{A}\;d\sigma = \lim_{r \to \infty} r \int_0^{2\pi} A_\theta(r,\theta)\,d\theta = \frac{1}{e}\int_0^{2\pi} \frac{d\sigma}{d\theta} = \frac{2\pi n}{e}\;.$$

Translating into geometric language, complex number (ϕ) plays the role of the tangent space, and parallel translation means setting the covariant gradient equal to zero. Curvature is the curl of the vector potential, which in our case can be identified with a scalar (it is actually a vector in the direction perpendicular to the 'physical' plane).

The last result,

$$\iint \mathrm{curl}\,\vec{A}\;d\sigma = \frac{2\pi n}{e}\;,$$

is a pale shadow of S. S. Chern's vast generalization of the Gauss–Bonnet theorem, to the effect that the integral of the Gaussian curvature over a smooth closed (orientable) surface is 2π times its Euler–Poincaré characteristic (see end of § 7.2).

A few final remarks are called for.

The question of existence of fields of finite energy E is reduced to the study of the mapping $\theta \to \sigma$ which also provides us with a classification (according to the integer n) of all such fields. The mapping $\theta \rightarrow \sigma$ is a mapping of a circle into a circle which is characterized by the winding number n. In more complicated situations, one is led to mappings of spheres into spheres and the underlying topological characterizations are, of course, more difficult.

In the example above I considered only *static* fields. By adding an appropriate kinetic energy term and considering the resulting Lagrangian, I could have written down field equations. Since total energy is conserved, a field of finite energy at $t = 0$ remains so for all times. At each t, since potential energy is *a fortiori* finite, we have a mapping $\theta \xrightarrow{t} \sigma$ which may be time-dependent. These mappings are, however, *homotopically* equivalent, i.e. they can be continuously deformed into each other (since the fields which define them change continuously in time) and have thus the same n. The winding number n is therefore a *constant* of *motion*, but it is different from the familiar constants of motion, e.g. energy. It is the simplest example of the much studied *topological constants* of *motion*.

7.4 A digression on catastrophe theory

The fact that mathematics plays an all-important, even crucial, role in formulating laws governing the physical universe and drawing valid conclusions from these laws, is too well known to require further comment. Nevertheless, the wonderment at the interactions between the universe and the mind never ceases. The desire to see creations of our imagination reflected in the order of the world around us is now, as it has always been, one of the strongest forces of intellectual progress. But such a desire, if not checked against cold reality of facts, will merely yield an illusion of accomplishment, or still worse, become an impediment to genuine progress. It was Fourier, I believe, who said that Nature is indifferent toward the difficulties it causes mathematicians, and mathematicians might do well to ponder this warning before attempting to mould Nature into an aprioristic design which they find aesthetically appealing.

This brings me to catastrophe theory. I was hoping to avoid this topic, especially since it is at the centre of a sharp controversy. However a chapter which is supposed to comment, albeit cursorily and superficially, on the present state of mathematics simply *cannot* ignore the only mathematical development of recent decades which found its way (in a garbled way to be sure) into the daily press (see, for example, the front page of the *New York Times* of 19 November 1977).

Let me first give a little background and terminology, restricting myself to a special case.[1]

Suppose that $f(x; \alpha, \beta)$ is a potential function depending on a *physical* (internal) or state variable and *control variables* (parameters) α, β.

---

[1] For a more complete presentation, as well as a detailed critique of applications of catastrophe theory, see the paper by Hector T. Sussman and Raphael S. Zahler: 'Catastrophe theory applied to the social and biological sciences: a critique' – to appear in *Synthèse*. For a more sympathetic presentation of the theory and its aspirations, see [4].

Equilibrium points are obtained from setting the derivative of f with respect to x equal to zero:

$$\frac{\partial f}{\partial x} = 0 \; .$$

In general, this equation determines a surface S in the (x, α, β) space. A point (x_0, α_0, β_0) of S is said to be *singular* if

$$\frac{\partial^2 f}{\partial x^2} = 0, x = x_0, \alpha = \alpha_0, \beta = \beta_0 \; .$$

If (α_0, β_0) is a point in the α, β control plane for which there is a x_0 such that $(x_0, \alpha_0 . \beta_0)$ is singular, we say that (α_0, β_0) is a *catastrophe point*.

To appreciate the terminology consider the example

$$f(x) = \frac{x^4}{4} - \frac{\alpha x^2}{2} - \beta x \; .$$

The equation of the surface S is

$$\frac{\partial f}{\partial x} = x^3 - \alpha x - \beta = 0 \; ,$$

and for a point (x_0, α_0, β_0) to be singular we must have in addition to

$$x_0{}^3 - \alpha x_0 - \beta = 0$$

also

$$3 x_0{}^2 - \alpha_0 = 0 \; .$$

Thus

$$\alpha_0 \geqslant 0 \; ,$$

and eliminating x_0 from the two equations above, we see that an x_0 can be found if and only if

$$\frac{\alpha_0{}^3}{27} - \frac{\beta_0{}^2}{4} = 0$$

or

$$\beta_0 = \pm \frac{2\alpha_0}{3} \sqrt{\left(\frac{\alpha_0}{3}\right)} \; , \alpha_0 > 0 \; .$$

What is 'catastrophic' about points (α_0, β_0) in the control plane lying on the curve

$$\frac{\alpha^3}{27} - \frac{\beta^2}{4} = 0 \; ?$$

The answer is that if

$$\frac{\beta^2}{2} > \frac{\alpha^3}{27}$$

the potential f has only one minimum, but if

$$\frac{\beta^2}{4} < \frac{\alpha^3}{27}$$

f has two minima (and one maximum), and thus as (α, β) crosses the curve

$$\frac{\alpha^3}{27} - \frac{\beta^2}{4} = 0$$

the appearance of f undergoes a dramatic change.

As long as the potential f has only one minimum it is *the* equilibrium point. Even after the catastrophe curve is crossed nothing will happen as the second minimum develops. Only after the second crossing when the original minimum disappears it happens that the minimum which was born, so to speak, at the first crossing becomes the sole equilibrium point, and a material point would actually jump discontinuously from where it stayed as long as it could to the new position of equilibrium. This is the catastrophe (or rather one of a set of so-called elementary catastrophies) which is called the *cusp catastrophe*, because the curve

$$\frac{\alpha^3}{27} - \frac{\beta^2}{4} = 0$$

has a cusp at the origin ($\alpha = 0, \beta = 0$).

Let now $f(\bar{x}; \bar{\alpha}, \bar{\beta})$ be an infinitely differentiable function and consider its critical set $\bar{S}$ defined by the equation

$$\frac{\partial f}{\partial \bar{x}} = 0$$

(in general this equation defines a surface in the $\bar{\alpha}, \bar{\beta}, \bar{x}$ - space). A point $(\bar{x}_0, \bar{\alpha}_0, \bar{\beta}_0)$ in $\bar{S}$ is called *ordinary* if, in the *neighbourhood* of this point, the equation of $\bar{S}$ can be written *uniquely* in the form

$$\bar{x} = g(\bar{\alpha}, \bar{\beta}) .$$

A point $(\bar{x}_0, \bar{\alpha}_0, \bar{\beta}_0)$ is called a *cusp singularity* if $\bar{S}$ in the neighbourhood of $(\bar{x}_0, \bar{\alpha}_0, \bar{\beta}_0)$ 'looks like' the critical set S of

$$\frac{x^4}{4} - \alpha x^2 - \beta x$$

and looks near the point $(0, 0, 0)$.

'Looks like' means that in a sufficiently small neighbourhood of $(0, 0, 0)$ we can find an *infinitely differentiable invertible* transformation

$$\bar{x} = h(x; \alpha, \beta) ;$$
$$\bar{\alpha} = j(\alpha, \beta) ;$$
$$\bar{\beta} = k(\alpha, \beta)$$

such that $h(0, 0, 0) = \bar{x}_0$, $j(0, 0) = \bar{\alpha}_0$, $k(0, 0) = \bar{\beta}_0$ and which maps the neighbourhood of $(0, 0, 0)$ on S into a corresponding neighbourhood of $(\bar{\alpha}_0, \bar{\beta}_0, \bar{x}_0)$ on $\bar{S}$. (Intuitively speaking, it means that by a very smooth reversible deformation we can transform a neighbourhood of $(0, 0, 0)$ on S into a neighbourhood of $(\bar{\alpha}_0, \bar{\beta}_0, \bar{x}_0)$ on $\bar{S}$.

Now comes the mathematically interesting (and non-trivial!) part in the form of a special case of René Thom's justly famous classification theorem:

'Most all' infinitely differentiable potentials $f(\bar{x}; \bar{\alpha}, \bar{\beta})$ are such that each point of their critical sets $\bar{S}$ is either ordinary or a cusp singularity or a fold singularity. (A fold singularity is one which looks like the singularity of $x^3/3 - \alpha x$ at $(0, 0)$.)

The term 'most' is a little too technical to explain and is one of the embarassments of catastrophe theory, since it must assume that nature provies us only with generic potentials and stays away from the exceptional ones.

Now let us see how all this is applied. I shall not comment on application to 'soft' sciences because there is no way to gauge the often exaggerated and almost always vague claims against anything sound. What I shall do is to examine the bearing of catastrophe theory on the theory of phase transitions.

Consider the equation of state of a gas

$$F(\rho; p, T) = 0 \ ,$$

where ρ is the density, which we treat as the state variable, and p, T, pressure and temperature respectively, as control variables.

If $F(\rho; p, T)$ can be derived from a potential function $f(\rho; p, T)$, i.e. the equation of state is equivalent to

$$\frac{\partial f}{\partial \rho} = 0 \ ,$$

and if f is *infinitely differentiable* and not 'exceptional', then just from the fact that there are only two control variables it already follows (from Thom's theorem) that we should expect a cusp (or fold) singularity if there is a singularity at all. This leads to the famous van der Waals law near the critical point (provided there is a critical point).

The argument is quite appealing[1] except that the weight of experimental evidence is against the critical point being 'classical', i.e. of the van der Waals type. In fact much effort has been directed in recent years toward understanding the non-classical nature of the critical point and the story is far from finished.

One should not, however, conclude that the van der Waals equation is devoid of all interest and importance. If molecules, were hard spheres with an attractive interaction potential of the form

$$- \gamma^3 \ \phi(\gamma r) \ ,$$

---

[1] It should be mentioned, however, that long before catastrophe theory, Landau, assuming only certain analytical properties of the equation of state, derived the 'classical' behaviour of the equation of state near the critical point (which has to be *assumed* to exist).

where r is the distance between molecules and $\phi(r)$ is non-negative and such that $r^2\phi$ is integrable, the equation of state would be, in the limit $\gamma \to 0$, of the van der Waals type, corrected below the critical temperature by Maxwell's rule to yield the horizontal parts of the isotherms. This is a *rigorous* consequence of statistical mechanics, and while one may regret that nature chose short-range attractive forces rather than the weak long-range ones, one should keep in mind that only in the latter case do we have a genuine derivation from first principles of the existence of critical point and of the Maxwell rule.

But even if the intermolecular attractive forces were of the weak long-range type so that the van der Waals equation would be valid, catastrophe theory would still be unable to even suggest what kind of discontinuous behaviour (catastrophe) should be expected. This is usually handled by an *ad hoc* rule and the most commonly accepted one in connection with the cusp catastrophe is the delay rule, i.e. the system will remain in a minimum until this minimum disappears. In the case of simple statics this rule is, of course, completely justified; but what to do in the case of the equation of state is left dangling.

D. H. Fowler in his article 'The Riemann–Hugeniot catastrophe and the van der Waals equation' writes:

Some procedure is required for specifying the choice between the two minima within the cusp region. The classical argument due to Maxwell is that, for thermo-dynamical reasons, the van der Waals loops should be cut off so as to make the shaded areas equal. This *almost* [italics mine] corresponds to saying that the lowest of the two minima is chosen, and it is this latter that Thom has called the *Maxwell convention*.... The construction described by Maxwell actually corresponds to another line.... [5]

What is involved is more subtle than is within reach of catastrophe theory, and it might help the reader if I summarize some results pertaining to a one-dimensional gas model whose equation of state is rigorously that of van der Waals.

The (one-dimensional) gas consists of hard rods of length δ, which attract each other according to the potential

$$\phi(r) = \alpha\gamma \exp(-\gamma r), \quad r > 0.$$

It can then be shown[1] that the equation of state of this gas is

$$\frac{1}{\rho} = v = \frac{\partial G}{\partial s},$$

where

$$s = \frac{p}{kT}, \quad \nu = \frac{\alpha}{kT},$$

and

$$G(s, \nu) = \min_{\eta} \log \left\{ -\eta\sqrt{2\nu} + \delta \left(s + \frac{\eta^2}{2} \right) + \log \left(s + \frac{\eta^2}{2} \right) \right\}.$$

[1] The result is due to M. Kac, G. E. Uhlenbeck, and P. C. Hemmer. For details see, e.g. [6].

The expression

$$- \eta \sqrt{(2v)} + \delta \left(s + \frac{\eta^2}{2}\right) + \log \left(s + \frac{\eta^2}{2}\right)$$

as a function of η has only one minimum if

$$\frac{2v}{\delta} < \frac{27}{8} \text{ (temperatures higher than the critical)}$$

and two minima if

$$\frac{2v}{\delta} > \frac{27}{8} \text{ (temperatures lower than the critical)}.$$

The rule for choosing between the minima is obvious: one *has to* choose the minimum minimorum, i.e. the lower of the two. The changeover from one to the other minimum occurs when the two are equal, i.e. according to what Professor Thom calls the Maxwell convention. It is easy to show that now this is *equivalent* to the Maxwell equal-area rule, i.e. the Maxwell construction.

If we set

$$f(v, s, v) = \frac{2v}{v} - \delta (s + v/v)^2 - \log (s + v/v^2),$$

then the critical set S is easily seen to be given by the equation

$$\frac{\partial f}{\partial v} = -\frac{2v}{v^2} + \frac{2v\delta}{v^3} - \frac{2v/v^3}{s + v/v^2} = 0 ,$$

which is equivalent to the van der Waals equation

$$\frac{p}{kT} = s = \frac{1}{v - \delta} - \frac{v}{v^2}$$

with the cusp singularity. However the correct 'convention' governing the choice of the appropriate minimum is determined by the function

$$\frac{2}{v} - \delta (s + v/v^2) - \log (s + v/v^2) ,$$

which is rooted in thermodynamics and *not* by the generic quartic of catastrophe theory on which Fowler's discussion is based. *Locally* near the critical point the two are equivalent (this, in fact, is the essence of Thom's theorm in the two parameter case), but there is simply *no way* of arriving at the correct form of the 'potential' knowing only the character of the singularity.

I should like to conclude this brief critique of a specific application of catastrophe theory by repeating a passage from an article I wrote some years ago, which seems rather fitting to the occasion:

When we propose to apply mathematics, we are stepping outiside our own realm, and such a venture is not without dangers. For having stepped out, we must be prepared to be judged by standards not of our own making and to play games whose rules have been laid down with little or no consultation with us.

Of course, we do not have to play, but if we do, we have to abide by the rules and above all not try to change them merely because we find them uncomfortable or restrictive. [7]

7.5 To whom does the computer belong?

The usefulness, even indispensability, of the computer as a tool of science and technology is now universally recognized and acknowledged. Imaginative uses of this remarkable instrument keep multiplying daily, covering an ever-widening spectrum of fields and disciplines. Even the humanities have not escaped — as witness concordances on which studies of stylistic peculiarities can be based. These can now be compiled and collated in a matter of weeks or even days, whereas they formerly required years of tedious and dull labour.

In a sense, the computer belongs to all of its users — even those who do not use it wisely or honestly. But the deeper question is: where does the computer belong in an intellectual sense? To put the question differently, what are the questions of general intellectual (as opposed to merely utilitarian) interest generated by the existence of the computer and to what, if any, discipline should they be assigned?

The question is not easy to answer because on the one hand the important (and certainly challenging) problems concerned with programming languages and related questions of 'software' do not easily fit into existing disciplines (if anywhere they belong somewhere between linguistics and logic), while on the other hand problems centring on the concept of algorithmic complexity,which unquestionably belong to mathematics, could have been formulated before computers, as we know them today, came into existence.

Since my concern is mainly with mathematics, let me say a few words about algorithmic complexity, especially since problems concerning this concept are, I believe, the only problem formulated in this century which are not traceable to the last one. It will facilitate discussion if I begin by stripping the computer of all its electronic gear and programming complexities, and return to its ancestor, namely the universal Türing machine.

The Türing machine is an infinite tape, subdivided into equal adjacent squares, each square being either black (denoted by *) or having a vertical bar (|) in it. There is also a scanning square which can be moved to the right (r) or left (l), one step (i.e. square) at a time. One can also replace (R) the symbol in the square by the other one, i.e. * by | and vice versa. Finally (and all-importantly) one can halt (h) the whole procedure.

A program is a (finite) set of numbered instructions of the form

$$12 : * R 8,$$

which is instruction 12 and reads: if the scanned square is blank print a vertical bar and look up instruction 8. This is all!

It has been proved that every algorithm (which is defined in terms of so-called recursive functions) can be programed on a Türing machine. The minimal length of a program is then the measure of the complexity of an algorithm. Thus we have a way of ranking algorithms by their complexities. Similarly we can define the complexity of a task as the minimum number of steps it takes to perform it on a Türing machine. One can then ask, for example, whether adding two n-digit numbers is of comparable complexity as multiplying such numbers — a question that is difficult and I believe still unanswered.

The subtle and elusive concept of randomness has also been coupled with that of complexity. A. N. Kolmogorov and, independently, Gregory J. Chaitin have proposed that a sequence of zeros and ones of length n be defined as 'random' if its complexity (i.e. the length of the shortest program needed to generate it) is of order n. This definition is tenable only in an asymptotic sense, i.e. in the limit as n approaches infinity, and is beset by a number of interesting logical difficulties [8]. Still, the idea is an interesting one, and for a while it received considerable attention.

Estimating complexities of algorithms or tasks is extremely difficult, and few general methods are available. The problem is also of considerable practical importance. Has the computer, however, been instrumental in solving an important purely mathematical problem?

The four-colour problem comes to mind, but here the major step in the solution was the reduction of the problem of colouring (with four colours) of *all* maps to colouring a *finite* number of specific ones. The latter task was relegated to the computer which came through with flying colours, completing the tedious and relatively uninteresting part of the proof.

Adherents of 'artificial intelligence' notwithstanding, it is difficult to see how the computer can be genuinely creative, although I admit that my scepticism may be vacuous, since nobody knows what 'genuinely creative' really means.

Computer experimentation may on the other hand be very illuminating and lead the mathematician to new and interesting fields to explore. Recently I found a particularly convincing illustration. It starts from a problem suggested by ecology, and what follows is based on two papers by R. M. May [9, 10].

For certain species population growth takes place at discrete time-intervals, and generations are therefore non-overlapping. Assuming that the population N_{t+1} during the time-interval $(t+1)$ – st depends only on the population N_t during the preceding time-interval, we are led to a difference equation

$$N_{t+1} = F(N_t),$$

where the function F is chosen to conform with biological intuition and knowledge.

The simplest case is the linear one

$$N_{t+1} = \exp(r)N_t,$$

where r is the growth rate. The solution is

$$N_t = \exp(tr)N_0,$$

which for $r > 0$ leads to the Malthusian explosion and for $r < 0$ to extinction. If the environment has carrying capacity k (i.e. as soon as the population exceeds k it must decrease), the linear model must be abandoned, and the following simple non-linear model is one of a number that have been proposed

$$N_{t+1} = N_t \exp \left[r \left(1 - \frac{N_t}{k} \right) \right] .$$

There is now no simple formula for N_t and the situation is already extremely complicated. If $0 < r < 2$, N_t approaches as t approaches infinity the carrying capacity, k, no matter what the initial population N_0 is. As soon as r exceeds 2, the population size N_t (again for t approaching infinity) shows a marked change in behaviour, for it now oscillates between two numbers exhibiting what is called a two-point cycle. Then comes the next critical value of r, namely 2·656 ... above which a four-point cycle is observed, then above 2·685 ... an eight-point cycle, etc. The critical numbers seem to approach 2·692 ..., and as soon as r exceeds this value, N_t fails to exhibit any kind of regularity and becomes reminiscent of sample traces of discrete stochastic processes. This 'chaotic' regime is not yet very well understood, but the phenomenon is striking and it poses problems of great mathematical interest. Much progress has however been made in providing a rigorous mathematical theory of the prechaotic behaviour, but even here the full story is far from finished.

Although I firmly believe that the computer's intellectual home is in mathematics, it is not a belief that is now widely shared within the mathematical community. One detects though, here and there, signs of awakening tolerance, and one can only hope that success stories like that of population-growth models will contribute to full acceptance of the computer as a member of the family. Mathematics could use a healthy dose of experimentation, be it only of the computer variety.

7.6 Where from and where to

I have already stated in the Introduction that to give an honest and even moderately complete account of the present state of mathematics is an impossible task. So much is happening on all fronts, so much of it is internal to this or that field or sub-field, and so much is a reflection of passing vogues that even if a review of acceptable accuracy were possible, it would be likely out of date in a year's time. Still, the enormous variety of preoccupations, which by some unstatable criteria are classed as mathematics, is holding together.

Is it possible to single out major developments or trends (as opposed to conquests of problems of long standing) which have had more than a local impact on this or that branch of mathematics?

Fifty, sixty years ago the theory of linear spaces would have been considered as such a development. This was the great period of geometrization of mathematics, and the aftershocks are still with us. Then came a period of algebraization of mathematics, and unification was sought by fitting as much mathematics as possible into algebraic structures: groups, rings, ideals, categories, etc. This trend is still very much with us, and the algebraic appearance of much of today's mathematics is perhaps its most notable characteristic.

This in a sense is an *organizational* trend — a trend to organize mathematics along formal algebraic lines. It has certainly been successful up till now, and it will certainly leave its mark on the mathematics of the future. And yet there is in such organizational successes an element of Monsieur Jourdain's surprise that one has been speaking prose all one's life.

Except for such generalities it is difficult to detect all encompassing trends. No doubt it is because an overall unifying direction is missing that one is sensitized to confluences of ideas coming from diverse sources, especially if one of the sources happens to lie outside of mathematics. Thus the excitement about theorems like the index theorem (linking geometry and differential equations) or the discovery of topological aspects of gauge fields, which brought fibre bundles to the peripheries of physics.

Let me deal briefly with yet another confluence, confessing at the outset to a strong personal prejudice. I have in mind the gradual infusion of probabilistic ideas and techniques into different parts of mathematics.

Probability theory had an unusual history. After a rather inauspicious beginning in problems related to games of chance, it evolved haphazardly until the early nineteenth century when it emerged, almost fully grown, as the magnificent *Doctrine de Chance* of Laplace. It was then almost immediately lost in the ruthless march toward absolute rigour, which had begun at about that time. By the time our century came along, probability theory was not even considered to be part of mathematics. It was readmitted into the fold in the 1930s only after A. N. Kolmogorov provided a complete axiomatization and made the subject 'respectable'.

Probabilistic ideas entered the more traditional branches of mathematics slowly, and their impact, though streadily on the increase, has not been dramatic. It is mainly felt in the theory of parabolic and elliptic differential equations, and the intervention of probabilistic concepts comes through utilization of the idea of integration in function spaces.

Integration in function spaces was first introduced by Norbert Wiener in a series of papers written around 1923, in which he developed a wholly novel approach to the theory of Brownian motion of a free particle. The standard theory tells us how to calculate the probability that a Brownian particle moving along the x-axis and having started at $t = 0$ from $x = 0$ will be found at times $t_1, t_2, ..., t_n$ ($t_1 < t_2 < ... < t_n$) in prescribed intervals $(a_1, b_1), (a_2, b_2), ..., (a_n, b_n)$. If a large number N of identical Brownian particles are released at $t = 0$ from $x = 0$ then the probability in question will approximate well the proportion of particles which at the prescribed times will pass through the preassigned 'gates'. Wiener proposed to interpret the probabilities as that above not in terms of *statistics of particles* as I have just done, but in terms of *statistics of paths*. In other words, he considered all possible paths (starting from $x = 0$ at $t = 0$) which a Brownian particle can choose, and interpreted probabilities as *measures* in the space of these paths.

Now, how does one go about constructing a measure theory? For the sake of simplicity let us think of the plane. We start by choosing a family of 'elementary sets' whose measures are postulated. In the case of the plane, the elementary sets could be rectangles. The measures assigned to them would be their areas as defined in elementary geometry. We now introduce the axiom of additivity to the effect that if *non-overlapping* sets $A_1, A_2, \ldots$ are 'measurable' (i.e. somehow a non-negative measure can be assigned to each of them), then the union of the sets is also measurable and its measure is the sum of measures of the As:

$$m(A_1 U A_2 U A_3 U \ldots) = m(A_1) + m(A_2) + \ldots$$

We also need the axiom of complementarity to the effect that if A and B are measurable and A is included in B, then B-A (the set of elements in B which are not in A) is measurable.

With these two axioms (rules) and with rectangles whose measures are given we can assign measures to extremely complicated sets. Once the measure is constructed, we can introduce the concept of integration.

We can now try to imitate this procedure in the space of paths. The rectangles are replaced by sets of paths which pass through preassigned intervals at prescribed times and the areas by probabilities assigned to these sets by the standard theory of Brownian motion. The axioms of additivity and complementarity are, of course, retained.

After overcoming a number of technical difficulties, one ends up with a theory of measure and integration. Remarkably enough this seemingly bizarre extension of the ordinary theory of areas and volumes has found many applications to problems of classical analysis and physics. For example, for a region R in Euclidean three-space, the measure (in the sense of Wiener) of the sets of paths emanating from a point p outside R, which at some time will hit R, is equal to the electrostatic potential at p corresponding to a charge distribution for which the potential on the boundary of R is normalized to unity.

Wiener's ideas were dramatically rediscovered by Richard Feynman in 1948 (actually in his Princeton doctoral dissertation in 1942, which was not published until 1948) in the context of non-relativistic quantum mechanics. Considering (for the sake of simplicity) a system of one degree of freedom (i.e. a particle) with the Lagrangian

$$L[x(\tau)] = \frac{m}{2}\left(\frac{dx}{d\tau}\right)^2 - V(x(\tau)).$$

Feynman defines the quantum mechanical propagator

$$K(x_0 \mid x; t)$$

as the integral

$$\int d(\text{path}) \exp\left\{\frac{i}{\hbar} S[x(\tau)]\right\}$$

over all paths which originate from x_0 at $\tau = 0$ and end at x at $\tau = t$. S is the action

$$S[x(\tau)] = \int_0^t L[x(\tau)]\, d\tau$$

and $\hbar$ the Planck constant divided by 2π.

The definition of the integral over the paths is now much trickier owing to the imaginary i in the exponent, and much mathematical work remains to be done.

Formally, the theory is in many respects analogous to a theory based on the Wiener integral.

Function space integrals (or path integrals, as they are often referred to) are at the centre of much of contemporary work, both in mathematics and in physics. I see no slackening of activity in this area in the near or even not-so-near future.

I should like to mention one more example of the influence of probabilistic ideas, this time on one of the 'purest' branches of mathematics, namely number theory. In 1939 Paul Erdös and I proved that the density of integers n whose number of prime divisors $\nu(n)$ satisfies the inequality

$$\log \log n + \alpha \sqrt{(\log \log n)} < \nu(n) < \log \log n + \beta \sqrt{(\log \log n)}$$

is equal to the error integral

$$\frac{1}{\sqrt{2\pi}} \int_{\alpha}^{\beta} \exp(-x^2/2) \, dx .$$

I regret that I am referring to my own work, but it happens to be a good illustration of a point I am about to make. The difficult part of the proof (which was supplied by Erdös) involved purely number-theoretical arguments. But it was an argument based on probability theory which suggested the theorem.[1] If probability theory was at that time still 'out of bounds' to mathematicians (and to many it was) there would have been no way of connecting prime divisors with the law of errors. On the other hand if in 1916 the Poisson distribution was not thought about as merely being empirically applicable to such things as the number of mule-kicks received by Prussian soldiers, then Hardy and Ramanujan could have discovered the theorem. As it is they came very close to it.

Brief, incomplete, and superficial as this account is, it must include at least a mention of combinatorial theory. Traditionally this part of mathematics, whose origins are buried in antiquity, has been concerned with the art and science of counting, enumerating, and classifying. It gradually annexed more and more ground, and so it has now become the study of all discrete and finite structures. Because of the enormity of its scope, embracing as it does the Ising model, the four-colour problem, and finite projective geometries, as well as more familiar questions of counting (e.g. how many different necklaces of n beads of k colours are there?), it lacked unified methodology and clear-cut criteria whereby puzzles could be distinguished from questions of genuine mathematical or scientific interest. It thus stood until recently on the periphery of mathematics, attracting few practitioners and frightening many by the prohibitive difficulty of its problems.

In the past twenty years we have witnessed a profound change. A search for unifying principles as well as an attempt to move combinatorial theory into the mainstream of mathematics had begun and strides have already been made. An article by G. C. Rota [12] (one of the principal figures in the combinatorial renaissance) written in 1969 paints an excellent picture of the subject and prophesizes an explosive future — a prophecy which to a large extent has been fulfilled.

---

[1] For a popular discussion of probabilistic ideas in number theory see, e.g. [11].

Of the example discussed by Rota, I have chosen one because it helps to illustrate the mysterious ways in which mathematics sometimes moves.

The example concerns the so-called travelling salesman problem, which can be stated as follows: suppose that we have n cities and the cost of travelling from city i to city j is $c(i, j)$; how should one plan the route through all the cities which will minimize the total cost?

The problem is of a type encountered in economics under the general heading of allocation of resources. It seems totally lacking in mathematical interest, since there is only a finite number of possible routes and, correspondingly, a finite number of possible costs to compare. But now comes the crux of the matter, since what is at stake is how long would it take to reach the solution, and this is clearly the problem of estimating the algorithmic complexity of the task at hand.

It has been shown that this complexity is less than

$$c\, n^2\, 2^n\ ,$$

where c is a certain constant, but it is not known whether this is the right order of magnitude. If it were, the problem would have to be considered as unsolvable, since the amount of computing time needed to find the solution, even for moderate n, would become prohibitive. I conclude with a quotation from Rota's article:

Attempts to solve the travelling salesman problem and related problems of discrete minimization have led to a revival and a great development of the theory of poly-hedra in spaces of n dimensions, which lay practically untouched − except for isolated results − since Archimedes. Recent work has created a field of unsuspected beauty and power, which is far from being exhausted. Strangely, the combinatorial study of polyhedra turns out to have a close connection with topology, which is not yet understood. It is related also to the theory behind linear programming and similar methods widely used in business and economics.

The idea we have sketched, of considering a problem $S(n)$ depending on an integer n as unsolvable if $S(n)$ grows too fast, occurs in much the same way in an entirely different context, namely number theory. Current work on Hilbert's tenth problem (solving Diophantine equations in integers) relies on the same principle and uses similar techniques.

As we have stated earlier (p. 197) Hilbert's tenth problem has in the meantime been solved.

I must end by apologizing for having left out large areas of mathematics which are vigorously pursued and are held in higher esteem by the community than the developments I have selected as being in some way indicative of the present state (good or bad) or possibly holding seeds of future concerns.

Certainly algebraic geometry, now much in vogue (two of the ten 'peaks' selected by Halmos *et al.* are solutions of outstanding problems in this field), is much broader in scope and clearly of greater intrinsic value to mathematics than catastrophe theory. But algebraic geometry, like several other important branches which I have failed even to mention, is to a great extent *internal* to mathematics, and I have tried, to the best of my ability and judgement, to single out for emphasis themes and trends which may contribute to lessening of the isolation of mathematics.

To the many omissions I must add one more. In recent years there has been much progress in the field of non-linear differential equations. Results about singularities of solutions of field equations of general relativity are surely among the most striking advances of the past few years. The recently discovered method of solving a class of important non-linear equations by the inverse scattering method is a new chapter of classical analysis of great beauty and appeal. It also lifted solitons from hydrodynamic obscurity into objects of intensive study in both mathematics and physics (the fields ϕ of finite energy discussed briefly in § 7.3 are solitons). Last, but perhaps not least, the remarkable non-linear phenomena in so-called dissipative structures, though still in their mathematical infancy, will surely inspire much further work. All these topics deserve much more than being included in an apology for their omission.

Mathematics today is a vital, vibrant discipline composed of many parts which in mysterious ways influence and enrich each other (*im Dunkeln befruchten* to borrow a Faustian phrase from Hermann Weyl[1]). It is beginning to emerge from a self-imposed isolation and listen with attention to the voices of nature. There is only one danger one must guard against, and that is that some zealots will, on either aesthetic or on counter-aesthetic grounds, try to redefine mathematics, so as to exclude this or that of its parts. Culture like nature has its ecological aspects, and the price for interfering with established equilibria may be catastrophically high.

References

[1] Stone, M. The revolution in mathematics. *American Mathematical Monthly* **68**, 715–34 (1961).

[2] Ewing, J. H., Gustafson, W. H., Halmos, P. R., Moolgavkar, S. H., Wheeler, W. H., and Ziemer, W. P. American mathematics from 1940 to the day before yesterday. *American Mathematical Monthly* **83**, 503–16 (1976).

[3] McKean, H. P. and Singer, I. M. Curvature and the eigenvalues of the Laplacian. *Journal of Differential Geometry* **1**, 43–69 (1967).

[4] Deakin, M. A. B. Catastrophe theory and its applications. *Mathematical Scientist* **2**, 73–94 (1977).

[5] Fowler, D. H. The Riemann–Hugoniot catastrophe theory and the van der Waals equation. In *Towards a theoretical biology*, Vol. 4, pp. 1–7. (Ed. V. H. Waddington.) Edinburgh University Press.

[6] Kac, M. *Quelques problèmes mathématiques en physique statistique.* Les Presses de l'Université de Montréal (1974).

[7] Kac, M. On applying mathematics: reflections and examples. *Quarterly Journal of Applied Mathematics* **30**, 17–29 (1972).

[8] Chaitin, G. J. Randomness and mathematical proof. *Scientific American* **232** (5), 47–52 (1975).

[9] May, R. M. Simple mathematical models with very complicated dynamics. *Nature* **261**, 459–67 (1970).

[10] May, R. M. Biological population nonoverlapping generations: stable points, stable cycles, and chaos. *Science* **186**, 645–7 (1974).

[11] Linnik, Y. V. Les nombres entries se pretent-ils aux jeux hasard? *Atoms* **245**, 441–6 (1967).

[12] Rota, G. C. Combinatorial analysis. In *The mathematical sciences: a collection of essays*. MIT Press, Cambridge, Mass. (1961).

[1] See his *Vorwort* to his classic *Gruppentheorie und Quantummechanik*.

8

Systems science

C. West Churchman

8.1 'System'

It is safe to say of 'systems science' that its investigators do not agree among themselves as to its meaning or the criteria of an important result. Nevertheless, it will be helpful at the outset to try to bound the meaning and suggest possible criteria of importance.

The Random House Dictionary (1967) defines a system 'an assemblage or combination of things or parts forming a complex or unitary whole'. This, of course, is much too general for the topic of this paper because all science studies systems in this sense. In order to narrow the definition, one needs to introduce the stipulation that the unifying principle for systems science is *teleological*, i.e., deals with means–ends relationships. Thus a system is a combination of parts which are organized to accomplish certain goals. Even so, this definition is again too broad, because it sweeps in most of technology. To narrow it further, I shall probably leave the domain of complete agreement. In this chapter I shall assume that the investigator not only studies teleological systems, but that he also attempts to be sufficiently comprehensive to estimate the worth or importance of the goals for humanity; that is, systems science has an ethical dimension. Some investigators would want to take issue with me for having chosen humanity as the ethical basis; after all, there is the open question as to whether the human species is really a most unfortunate aberration of living forms, and the sooner it disappears, the better for life on earth. Thus some investigators study 'ecosystems', which means systems where 'habitat' is the key concept. Human beings tend to disturb the natural interplay of the species in an ecosystem, e.g., a cubic foot of soil in a forest. But here there is no serious disagreement; the phrase 'for humanity' can be replaced by 'for living species'. However, in this chapter I shall be concentrating on human systems.

It may be helpful, too, to illustrate a system in the sense just given, by introducing a system which can be described in mathematical language. Suppose, for the moment, that we have been lucky enough to unitfy the system by means of a 'measure of performance' that adequately encompasses the goals. Examples of such a measure are net-profit per annum in the private sector, incidence of disease in a population (here the smaller the better), average annual income of a nation, and so forth. To suggest the ultimate mystery and ethical character of this unifying measure, we label it z. Suppose too that we have been lucky enough to identify a set of activities, each of which influences z in a specific manner; the set is exhaustive in the sense that no other activities influence z one way or the other. Next, our luck enables us to measure the amount of an activity, e.g., in terms of man-hours, amount of funding, etc. These amounts are represented by the variables $x_1, x_2,$ ..., x_n. Finally, we find that the manner in which the ith activity influences z can be represented by a linear relationship so that

$$z = a_1 x_1 + \dots + a_n x_n$$

If we assume that 'the more of z the better', then our systems scientists would like to maximize z by setting the x_is at the appropriate levels. This looks trivial, since common sense says that if the a_is are positive then we ought to make each x_i 'grow to infinity'. I should note in passing the 'systems significance' of the coefficients, a_i. If you are running activity x_i, and your coefficient is zero or negative, the implication for systems management is clear enough: fire the lot. So the a_is should at least be positive in 'the more the better' situation.

The next consideration is that the enlargement of all the x_is is not feasible. For example, if the x_is are man-hours or dollars, then their sum must be limited by a budget. There may be all sorts of other limitations brought about by pollution restraints, expertise limitations, etc. But our luck holds, so that these considerations can also be written down in our mathematical language as a set of constraint equations, e.g.,

$$x_1 + x_2 \dots + x_n \leqslant B .$$

What emerges is a classical mathematical problem, namely to maximize a linear function subject to a set of constraint equations (which we hope are linear). The trouble is that if the number of activities, n, is large, it takes a very long time for a human to solve the problem manually. In the late 1940s Dantzig found an algorithm for simplifying the number of computational steps, and, in the 1950s and subsequently, large computers were designed to eliminate the need for the human being and his error-prone behaviour.

The mathematics I have just described is called 'linear programming' (LP). Today there exists an LP with two million variables and thirty-five thousand constraint equations. For those with a mathematical curiosity, the LP describes an oil company; organizations like oil companies tend to 'decompose' rather readily, which means that the LP matrix has lots of 'blank spaces' in it.

Now an LP alone does not represent what I have called 'systems science', because we need to know whether z is ethically justified. For example, z might be the net take of a criminal organization, the number of deaths in a concentration camp, or sales of a tobacco company. Furthermore the constraint constants in the constraint equation may be ethically suspect; e.g., too big or too small a budget. But it can be shown that if z is the appropriate measure of performance, then the appropriate constraint constants can be deduced.

Hence, for LP and similar system descriptions, the basic question is the appropriate measure of performance. We have come to what might be called the stratetic question of systems science. The dilemma is this: if we take a comprehensive view of the term 'system', so that it includes all that we deem relevant, then we have yet to find a measure of performance which can stand up against critical arguments. (I should point out that some of my economist friends would object at this point because they believe that economic benefit minus cost is an adequate measure of performance; I shall return to this claim later on.) On the other hand, if we believe that systems must be studied by first subdividing them into subsystems, then plausible measures of performance seem possible.

The dilemma is easy enough to illustrate. Consider energy production and consumption as a subsystem. A plausible measure of performance might be total units of energy (measured in physical terms) divided by demand over some suitable time period. But viewed from a larger perspective, we would need to know the cost of maximizing such a measure, as well as the way in which the energy is to be distributed; the 'cost' would include considerations of other programmes (e.g., in health and education) which we would have to abandon in order to meet energy needs. Similarly, a military measure of performance might be the probability of deterrence over a time period; a broader perspective might call for a world decrease in armaments, and certainly would raise questions about costs.

8.2 A map of systems science

The dilemma just described is very old, and all races, cultures, and nations have had to face it. The question is whether 'systems' are coherent segments of human life, or whether there exists a total system which gives meaning to each of the segments. The segment-assumption is called pluralism, while the total-system assumption is monism.

Hence a report on systems science needs to consider both a pluralistic and monistic systems science. With some notable exceptions, most of the investigation in systems science today is pluralistic. That is, segments of the human social system are 'carved off', and the segment is regarded to be separate from the rest of human living. For example, most governments operate through a pluralistic philosophy. Health, education, defence, housing, international affairs, etc. are seen as separable systems, each with its own implicit or explicit measure of performance. Of course, overlapping is inevitable and is handled by negotiation between government agencies; and often issues 'fall between the cracks' and are never addressed. Furthermore, the governmental power charged with creating an overall budget allocation does try to use some sort of monistic judgment.

Monistic systems science is based on the reasoning that the attempt to solve problems separately ignores the fact that all our human problems are closely interconnected and that the basic global issue is to determine the nature of these interconnections and the global destiny of the human species.

Fig. 8.1 is an attempt to display the varieties of systems science. I apologize for it in both the explanatory and 'I'm sorry' senses. As a philosopher, I have used pluralism and monism as the basic categories, but a lot of systems science could of course be either, especially if it deals with techniques and methods, which I have put in an 'in-between' category. Furthermore, I have satisfied the exhaustive requirement by the bottom label, 'none of the above', to which I shall turn at the end of this chapter. Next, I apologize to the many authors I have omitted; it is impossible

PLURALISTIC

Maximization models
1. Queuing
2. Inventory
3. Statistical quality control
4. Statistical methods
5. Linear programming
6. Mathematical programming
7. Game theory, etc.

Non-Mathematical
1. Accounting
2. Time-motion
3. Social sciences
4. Political science
5. Planning

IN-BETWEEN

1. PPBS
2. C–B analysis
3. PERT
4. Decision trees
5. Fuzzy sets
6. Diagrams and flow-charts

NONE OF THE ABOVE

MONISTIC

Maximization models
1. Model planning
2. Forrester–Meadows
3. Maserovitch
4. etc.

Non-mathematical
1. General systems theory
2. (a) Biological base
   (b) Cybernetic base, etc.
3. Management
4. Traditional

Fig. 8.1. A map of systems science.

to cover the immense literature that could justifiably have been included. Finally, I have subdivided both pluralistic and monistic systems science. By 'maximization' models I mean models that implicitly or explicitly involve maximizing a mathematical function, whereas non-maximization methods usually do not involve maximization techniques. The distinction tends to be very important to the investigators and only of minor importance to systems managers, but since we are here concerned with the investigators, I have shaped the map to reflect their interests. When the non-maximizing investigators complain about the 'irrelevance of models', they are not complaining about models as such, because all thought uses models of some sort; rather, they are complaining about such things as differential calculus and abstract algebra.

8.3 Pluralism

The earliest use of the differential calculus for managerial decisions of which I am aware is A. A. Cournot's *Research on the mathematical principles of the theory of wealth* (1838).

A very simple piece of research is to assume that quantity sold, q, is linear in price, p:

$$q = -ap + b \ (a > 0) \ ,$$

where a is the market demand coefficient and b is a constant, and that one wants to maximize profit, π, where

$$\pi = pq - cq \ .$$

c being the unit cost. The problem is to maximize π by setting the correct price:

$$\pi = pq - cq = (-ap + b)(p - c) = -ap^2 + (ac + b) - bc \ ,$$

so that the maximum value of π occurs at $(ac + b)/2a$. Of course Cornot's investigations went far beyond this very simple model, which, nonetheless, does illustrate how the calculus can be used in systems investigations. It also shows the inevitable weakness of such models, e.g., the determination of a and c, the market demand coefficient and the unit cost, both of which tend to vary and to be quite difficult to estimate given the complexity of both the market and the manufacturing firm. Furthermore, the model assumes there is no competition, a matter which Cournot did study.

As in the case of Cournot, so in pluralistic systems science in general, the tendency has been to carve off segments by means of problem types, which are not specific to any domain like health or education, but can occur in a number of domains.

In the nineteenth century, and subsequently up to the present, many economists investigated economic problems in the style of Cournot. But all business firms and most government agencies can be studied by examining not only their economics but also their operations, i.e., by 'operations research' (OR).[1] The earliest example I have found of OR is Erlang's work on waiting lines for a telephone company, in about 1912. Assuming that the calls into a central office occurred in a random fashion, he was able to deduce the probability distribution of waiting times (for the operator to answer), length of waiting lines, etc., for one or more operators. But waiting lines occur wherever there are regular service operations. Thus there is a 'problem type' which can occur in various places and times. One 'carves off' the waiting line aspect from the rest of the operations of the system, e.g., from such issues as personnel policies, profitability of the telephone company, etc. Actually, Erlang did not use the calculus to maximize a measure of performance to balance waiting times of customers versus idle times of operators, and his method, called queuing theory, still lacks such measures even today. But there is the implicit assumption that some maximization technique, usually involving integer solutions, is possible.

Wilson, who in 1916 developed the first inventory management-model, had a measure of performance: the total cost. One is to select that amount to order into inventory (stock) which minimizes total cost. (It is to be noted again that sometimes the measure of performance is like a golf score: the smaller the better.) Inventory is also a problem type, and an inventory model usually does not address such questions as to whether one should sell the product at all, or the suitable price and advertising policies.

In the 1920s Shewhart developed a statistical model for controlling production, probably the first 'cybernetic' model for management. It was called 'statistical quality control' (SQC). I believe he had ambitions towards monism, i.e., a general management model, but all the real applications were to the type of problem where one is trying to control production of items with respect to some physical measurement (weight, size, etc.).

Statistical theory, e.g., in the hands of Neyman and Pearson (1930), was a model for testing hypotheses, which also had potential managerial implications, but was specifically applied to problems of inspection of material.

The Second World War saw the development of 'search theory', a model to help solve problems of the type involving looking for something. After the war 'game theory' was studied in order to help solve problems of conflict under rules. It also saw the development of a general theory of control, called cybernetics, as well as LP.

In recent history, other problem-oriented models have been discovered: replacement models, 'dynamic programming' (a sequential problem-solving method), investment models, etc.

As far as I am aware, very few tried to develop a general typology of problems for pluralistic systems science, perhaps because the pluralism would be threatened.

Finally, I should mention that many models are 'non-committal', in the sense that they can be applied to a wide variety of problems. LP, for example, is sometimes

---

[1] The label was first used just prior to the Second World War, for research into military operations, but the earliest examples fit clearly the same pattern.

thought to be a description of allocation problems (e.g., allocation of funds, man-hours, etc.), but can in fact be used to provide routes for a fleet of trucks and proper mixes for nut canners; it has even been applied to cellular biology. Some enthusiasts would argue that LP, or the more general non-LP (maximization of a non-linear function subject to constraint equations), or else cybernetics, can be applied to *all* managerial problems. In fact, none of the models described above needs be confined to problems with its label; an inventory model does not know it is an inventory model. As a symbolic logician would say, any sufficiently rich formal model can be given an extraordinary number of interpretations.

The non-maximization models for systems science are also quite plentiful and varied, and some of them are quite old. For example, accounting methods, which date back to the earliest civilizations that used numbers and writing, are 'systems descriptions' dealing with means and ends. But the accountant generally describes historical costs and profits. If the accountant reports that the business is losing money, then the manager may conclude that he has to change something, though in general the accounting figures do not tell him what. In more recent years accountants have tried to bring accounting closer to management decisions through 'managerial accounting'. Some argue that this is not possible because the true 'costs' in decision-making are 'opportunity' costs (the cost of *not* doing something else; e.g., of putting the money for a conference into another programme), and such costs never appear in the accounting figures.

In the late nineteenth century, Taylor suggested the need to measure more carefully the movements of workers, in order to improve worker performance. Some of his followers, Fayol in particular, tried to generalize on Taylor's theme in terms of the larger system. This has resulted in an 'industrial engineering' approach to managed systems.

Finally, there are a number of sociologists, social psychologists, and anthropologists who have tried to apply their disciplinary approach to the study of social systems, sometimes but certainly not always, with an intent of improving them. Indeed, there has been a running debate in Western social science as to whether the primary task of the social scientist is to describe or prescribe social processes. The influence of positivism on the social sciences led many investigators to claim that their conclusions were 'value-free', and' I suppose, 'prescription-free'. But applied anthropologists, consumer researchers, nutritional scientists in the field, industrial psychologists, all have to face situations in which their advice is sought by administrations and policy-makers.

In recent years, political science has witnessed a shift of emphasis from pure theory to application. A number of political scientists have taken an interest in systems science and planning, and have argued against monistic systems science on the grounds that it is politically naïve. Social systems, they say, cannot be changed on any global basis, because politics operates through highly specific issues, and change in social systems takes place through politics. They have introduced the term 'incrementalism' to describe their piece-by-piece approach. They have, in effect, attacked monistic systems science for its failure to take into account a strategy of implementation. I shall return to this subject after discussing the monistic approach.

In the 'map' (Fig. 8.1) I have included 'in-between' methods of systems science, which could be used in either a pluralistic or monistic approach. PPBS refers to 'Program planning and budget system', a label first used by the US Department of Defense, which suggests that budgets should be geared, not to bureaucratic departments, but to the programmes or missions of the system. In this regard, its philosophy is exactly the same as that which underlies LP; it is often the case that the activity of a single department of an organization cannot be related to the measure of performance of the whole organization. But PPBS in practice is more of a discussion-bargaining technique than a fully fledged logical approach.

'C–B' refers to 'cost–benefit' analysis. In its government applications it is an attempt to measure the value of government programmes, e.g., a health clinic programme for a ghetto, experimental schools for children, safety-belt programmes, etc. The idea is to try to estimate the total benefit of the programme, usually in economic terms, and to relate this to the total cost. If C–B is to be used to compare the effectiveness of programmes, then the comparison must be in terms of benefit-minus-cost, because the ratio, benefit divided by cost, can be highly deceptive. A great deal of controversy has been created over C–B analysis, primarily because when the analysts faced such issues as loss of life or limb, or the 'value' of an old person, or a biological species, they tended to leave these matters out of their calculations because they had no idea what to do with them. Some have suggested adding 'social indicators' to C–B, but the methodology of doing so is not clear and investigators disagree markedly.

PERT is a technique, usually applied to large-scale projects which require some months or years to complete, which tries to coordinate the various phases. The US National Aereonautics and Space Administration's *Apollo* programme of the 1960s is an excellent example; one should not have a well-trained crew on hand several years before a booster is available to send them up to the moon.

The remaining 'in-between' ideas can be briefly described. 'Decision trees' are maps indicating the nature of choices that must be made in a series of decision steps, and criteria for making the choices. 'Diagrams and flow charts' describe complicated processes and their interrelationships. 'Fuzzy set theory' is an extension of classical class logic, where the membership rules are not unambiguous ('*a* is a good worker' may not have precise rules of membership). The idea is that managers tend to deal with fuzzy sets rather than precise sets, though there is question whether they are even precise about their fuzziness.

8.4 Monism

Monistic systems science is based on the assumption that we humans can use our intellects to put the pieces together in an approximately accurate fashion. The word 'approximately' is important because no one has the hubris to claim he knows it all. Furthermore, 'putting the pieces together' may not mean putting everything together, but rather enough for the manager's purposes. Nor is the 'putting together' necessarily by means of a mathematical model of the type described above.

But I choose as my first example a model-builder. In the late 1950s, Forrester wrote *Industrial dynamics*, in which he attempted to put the pieces together for an industrial organization. The inventory models created in the early 1950s were often quite naive, because they failed to consider pricing as a decision variable, and used past demand as a basis for setting inventory policy. Forrester carried this criticism further by arguing that not only do pricing and inventory interact, but so do the other components of the firm: manpower, investment, etc. Forrester then applied his interaction model to cities in *Urban dynamics*.

In the late 1960s, a group of people met in Rome to discuss the seriousness of the world 'crisis'. One of them, Hasan Ozbekhan, wrote a position paper in which he introduced the term 'problematique', to describe the interconnection of the world problems (politics, malnourishment, war, production, etc.). Forrester was then able to persuade the 'Club of Rome' that his method could be applied to the world over a long period of time. The result was *The limits to growth* (Meadows *et al.*) which is unquestionably the most widely read systems science text to date. The book's conclusions were gloomy indeed, inferring from the last seven decades that if mankind proceeds as it has in the past then serious global starvation is in store for us. Its publication created a storm of protest and approval among the systems scientists, a storm that is still in progress. Some disagreed with details of the world model; others claimed that it left out important considerations (e.g. politics and ethics).

But to me the important result of *Limits to growth* was its raising to collective consciousness the question whether the policy-makers of the world did or did not need a monistic systems science to aid them in their decisions and, if they did, how we might find a suitable one.

The variety of monistic systems inquiry is large, and often the investigator does not use an optimization model. A good example is 'general systems theory', which seems to have been 'discovered' by Bertalannfy and Boulding when they worked together for a year at the Behavioral Science Center at Stanford. I have never been able to determine whether the adjective 'general' modifies 'system' or 'theory': are we talking about a general theory of systems or a theory of the general (i.e., comprehensive) system? I suspect that it is the former, as it clearly is in the writings of James Miller, who identifies seven levels of living forms and nineteen functions at each level. (The appearance of prime numbers in general systems theory is quite striking. The Batelle Memorial Institute initiated a programme to study the interconnection of the 'basic' world problems and found that there were 47 of them. When the Stanford Research Institute repeated the exercise, they came up with 41 problems.)

It is difficult to set boundaries on general systems theory. A recent (computerized) bibliography by Klir *et al.* leaves out modelling articles in such journals as *Management Science* and *Operations Research*, which leads to the speculation that general systems theory is more of a qualitative philosophy than an analytic approach.

There is also another genre of social systems literature which deals directly with the human system, its future, its precarious nature, and its destiny. Examples are:
(1) R. L. Ackoff's *Redesigning the future*, in which he dstinguishes among three types of management: 'reactive' (day-to-day coping), 'preactive' (decision-making

based on forecasts), and 'interactive' (designing the future as we want it to be); obviously his message is that we humans need more interactive management. (2) S. Beer's *Decision and control* and *Platform for change*, which are primarily based on Wiener's and Ashby's works, i.e., on a cybernetic philosophy, but which contain a great deal of the author's own contribution (e.g., to what extent can we apply our knowledge of the human brain to the solution of the question of control in human society?). (3) E. Jantsch's *Design for evolution*, which applies a vast literature of science to the understanding of designing evolution, especially a hierarchy of approaches to understanding changes, process, and evolution. (4) Vicker's *Freedom in a rocking boat* and other works, in which one theme is the weakening of a viable culture and often a cultural vacuum in Western civilization, and the pressing need to try to do something about it. (5) F. Schumacher's *Small is beautiful*, which develops the theme that humanity has become overwhelmed by large technologies, e.g., in transportation and energy, and needs to redesign technologies that are more appropriate to the human condition. (6) H. Linstone and W. Simond's (ed.) *Futures research*, a collection of essays dealing with ways of regarding the future segment of the human social system. I need to emphasize that these are only examples of the plethora of books now being published on social systems.

Since my own books belong to this genre, I feel pressed to add my own bias, namely, that the latter part of the twentieth century is not the first time that humans have tried to assess the human condition and to suggest pathways towards its improvement. Many have come to assume that a new label makes a new game, but this is clearly not the case. The oldest systems science text goes back probably as far as 2000 B.C. It is the *I Ching*, or book of changes. In modern systems language, it is a set of dynamic models that depict the present and future situations and give rather general advice to the person faced with a decision. Perhaps modern systems scientists would object to the apparently random element in the process of deciding which situation obtains, but they miss the main point, namely, than an expert is needed for the 'systems science' of the *I Ching* to work, a principle that we still cling to today.

The tradition of monistic systems science is very old, but also more or less continuous in all cultures throughout the ages: the Upanishads, the Bhagavad Gita, the pre-Socratics, Plato, Aristotle, the Stoics, St. Augustine, Thomas, Hobbes, Rousseau, Kant, Bentham, and Marx are but a few names associated with its history. So-called primitive cultures often worked out a vision of the human condition and its relation to reality, sometimes, as in the case of the Aztecs, for example, a very complicated system. What is helpful about studying the tradition is that today's systems scientists can find their meaning in an historical process; they tend to think of themselves as something brand-new.

8.5 Implementation

The history of systems science also enables us to perceive a mystery: namely, that its results have rarely been transformed into reality, or, in systems language, 'implemented'. I do not know the history of implementation of the *I Ching*, except that during Mao Tse-Tung's time the Chinese were forbidden to use it, and perhaps still are. Plato tried to implement his design of a Republic, and nearly got killed in the process. Kant turned bitter about mankind's lack of morality as he grew older. There is good reason to suspect that the Marx–Lenin systems approach was never implemented; certainly the 'state' has not withered away. The Jefferson–Madison dream of an equitable democracy is at best very crudely approximated in the USA today. The members of the Club of Rome are deeply disappointed by the fact that, though *Limits to growth* was very widely read, it has had little impact on social policy.

Indeed, if we take a 'systems view' of the human world today, the picture is one of irrationality to the point of insanity. A huge amount, the size unknown to most of us, is spent on armaments which have no value other than mutual deterrence, while something like a half a billion humans are grossly malnourished, a similar number or more are grossly ill-educated, an unknown but large number made to live lives they would rather not live, and so on. Furthermore, decisions are being made by fragmented, day-by-day crisis management. Why? Given that we humans have been given the gift of intelligence by God, evolution, whatever, how does it happen that the products of intelligence are so badly used?

I should hasten to point out that this mystery of systems science is probably not an item most systems scientists would put high on their list of unsolved problems. I suspect that the hours spent in trying to find better algorithms for solving problems of models are thousands of times greater than the hours spent on understanding implementation failure.

Of course the pluralist systems scientist will argue that there is no 'whole system' and that the failure of monistic systems science arises from the fact that it is basically unrealistic. One should rather consider the problems as they arise, in the terms the manager can understand, so that communication can take place, and hence implementation.

Unfortunately, this strategy usually implies that the pluralistic systems scientists tackle the unimportant issues of management. Solving an inventory problem, or even advertising problem, for a product that is inferior simply solves the wrong problem.

Furthermore, the vast majority of pluralistic systems studies are never implemented, especially if they run into political opposition.

There can be no doubt that the political approach to human affairs is the most powerful approach operating in the world today and that in most respects its decisions are contrary to either pluralistic or monistic systems science. For the purposes of this discussion, politics is a behavioural phenomenon in which people gather together over a specific issue, identify what needs to be done, identify their opponents, and seek to increase their power through numbers, persuasion, and occasionally violence. Politics includes, but is far more encompassing than, the 'party politics' of governments.

But there is a paradox in all that has just been said because, if politics is such a force, then it is real and hence a part of the social system. If it cannot be changed, then the systems scientist should recognize its unchangeable reality and include it in his calculations. If it can be changed, e.g., reduced in power, then by what right does the systems scientist try to destroy a human value?

I shall try to illustrate this paradox by discussing the word 'science' in the title of this chapter. Some (snobbish) wag once said that if an area of study uses the word 'science' in its name (e.g., nutritional science, political science, management science), you can be sure it isn't a science. Is systems science really a science or just a coordinated set of opinions? Now if one examines the traditional disciplines like logic, mathematics, physics, chemistry, and biology, one finds that they are made up of people who gather together over specific issues, identify the appropriate methods for addressing these issues, identify inadequate research methods, seek to increase the funding or excellent research through advertising the value of truth, and seek to stop the funding of inadequately designed research proposals. If we look back at the definition of politics, it is clear that the classical disciplines satisfy all the conditions of political behaviour: they are political in nature. A systems scientist who studies research and development policies asks how our limited resources should be spread among the areas of inquiry: which contribute most to the improvement of the human condition? To use 'truth-finding' as a standard of performance is much too limiting: all disciplines try to find, or approximate to, the truth; and is a gram of truth worth more or less than a gram of nutrient for a starving child?

What emerges is that systems science is the science which seeks to put 'truth-finding' science into the proper place (e.g., in terms of funding) in the whole human system. Of course, this kind of investigation is common in private companies, which must conduct some research and development to remain competitive, as well as in government agencies with practical missions. Cost–benefit is usually the technique that is employed for this purpose today. But should we conduct similar studies on, say, cosmology, high-energy physics, and basic microbiology? Most of the arguments used by the pure scientists are highly suspect to the systems scientists. There is, for example, at best only very weak evidence that all the funds spent on 'basic' research have really 'paid off' compared to other uses that might have been made of these resources.

But the question of rights is there: by what right should systems science judge the quality of investigation in all other areas of science? Seemingly it can do so only by claiming that rationality is the most important issue in the management of human affairs; it must press its point by publishing books like *Limits to growth* in order to persuade (or scare) people into accepting this doctrine; it must identify those who act irrationally (selfishly, ignorantly, etc.); it must identify what needs to be done, identify the opponents of systems science thinking, and seek to increase the power of systems science in the formulation of social policies. It must, in other words, turn political, which is exactly what the Club of Rome has sought to do.

Paradox is the philosopher's delight, and he does not seek to 'solve' it. Indeed, were it solvable in some reasonable way, it would cease to be paradox. The paradox I have just described has existed throughout history. It is one thing for Marx to be sitting harmlessly in the British Museum composing the pages of *Das Kapital*; it is quite another for him to have the hubris to write the *Communist manifesto*. How dare the monistic systems scientist tell the human world how to conduct its affairs? Of course, I should not fail to mention that the origin of this right to tell us how to live may be religious: the response is that the message is sacred, coming from a superior or supreme being.

8.6 None of the above

This brings me to the last item on my map, 'none of the above'. When one reflects a little, it is astonishing how much the map leaves out; a great many of the omissions have been rather thoroughly studied in history.

For example, there is the whole issue of the appropriate goals to be pursued by mankind, and a long history of ethics to guide us. A reflective person reading this history cannot help but be struck by the amazing agreements across time and cultures. Yet today's monistic systems scientists tend to ignore ethical issues, e.g., by the ploy of using survival of the human species as the main ethical goal. But what is the point of our surviving if our lives, as well as those of our fellow-species, turn out to be ugly, dangerous, or meaningless?

As another example, there is the plausible hypothesis that reality was designed and is controlled by a superior or supreme being. It would seem important for systems science to investigate this hypothesis, because if it is valid the central task of monistic systems science should be to determine the intentions of such a being. This is precisely what the great systems scientists of the seventeenth century did: Descartes, Spinoza, and Leibniz. Today's monistic systems science largely ignore the hypothesis, perhaps because it does not know how to investigate it, perhaps because, for political reasons, it would lose its right to be called a 'science'.

But there is another version of the theological hypothesis that systems science, as well as all sciences, cannot ignore. I take it that Descartes was right in one respect: all scientific investigations are based in part on the human intellect's ability to doubt and to criticize, and were this ability to be removed, e.g., by political edict, or drugs, or surgical operation, science as we know it would disappear. But the awesome image is that science expects the ability to be eternal, i.e., that every future generation will be able to doubt and criticize. The grand adventure of science otherwise becomes nothing but a small episode in human history. But what guarantees the endless pursuit of truth? We do not know, but it seems to me that this is one of the most important questions science can pose today, given the political forces that would destroy the ability if they could. Suppose we say that the appropriate 'destiny' of science is the preservation within the human species of the ability to doubt, criticize, and investigate freely. Then the question becomes: what is the Guarantor of Destiny? The acronym may or may not be helpful; but if it is, then it is noteworthy that the theological issue is not necessarily the proof of God's existence, as it was for Descartes and Leibniz, but perhaps is instead a design question: how can those who worship the freedom of inquiry design a guarantor of its immortality?

Finally, but briefly, the map leaves out the inner world of each and every human being, a world perhaps as immense as the universe. Most models of social systems must treat people collectively, e.g., by putting them in classes defined by age, occupation, income, etc. But the classifier does not really know what he is classifying. Economists are wont to regard each of us as a utility function operating over a set of commodities. But the all-shatterer, Freud, shattered that myth long ago: what is it that holds the utility function, the id, ego, or super ego? By the time Jung had set forth his foundations of psychology, our inner world became a vast complex of types and archetypes. Parhaps the failure of social systems science to implement arises because its investigators do not know very much about people. Jungians define psychology as the study of the soul. Systems scientists today do not understand, or do not believe in, the soul.

8.7 The future

If I can return to Ackoff's three types of management, I can make some forecasts. I think that pluralistic systems science will remain reactive. It will be sensitive to the waves of crisis that the human world exhibits. Today the crisis is energy. Tomorrow (and perhaps today) it will be information. Perhaps within the decade it will be malnourishment. All of these crises are rich in problems to be 'solved' by modelling. Monistic systems science will be preactive, trying to better the forecasts of *Limits to growth*, in order to determine, for example, whether humanity will in fact face a population crisis within the coming century.

As St. Paul so astutely remarked, hope is not based on knowledge; for if I know some event is bound to occur, it is meaningless to say that I hope it will or will not. I hope, but certainly do not know, that systems science will turn its attention to those matters it has so far ignored: the ethics of social systems, the theology of social systems, and the relationship between the inner life of humans and the outer social life, in order to help design mankind's future.

References

I will simply cite some books which themselves cite a large segment of the literature on social systems science, both pluralistic and monistic.

Churchman, C.W. (1969). *The systems approach,* Delacorte Press. The last section contains some historical references which I think are relevant.
Klir, G. and Rogers, G. (1977). *Basic and applied general systems research*. The University Center at Binghamton, New York. A bibliography.
Miles, R. F. (ed.) (1973). *Systems concepts*. John Wiley, New York. Lectures on contemporary approaches to systems.
Wagner, H. (1970). *Operations research,* Prentis Hall, New York. A compendium (of 1000 pages) of OR models.

9

L'électronique

Pierre Aigrain

9.1 L'avenir de l'électronique

L'électronique est un domaine extrêmement vaste, et afin de simplifier dans la mesure du possible l'étude de son évolution et de son avenir probable il est commode d'y distinguer quatre secteurs.

Cette classification, et cela peut sembler étonnant, sera basée sur la nature des clients du matériel correspondant. Mais cette distinction basée sur la clientèle recouvre en fait des différences profondes quant à la nature même des *matériels* concernés. Aussi traiterons-nous successivement :

(1) *L'électronique professionnelle*, dont la clientèle est constituée par les États, les grandes administrations, ou les très grandes sociétés d'exploitation.

(2) *L'électronique industrielle*, dont la clientèle est constituée, comme son nom l'indique, par d'autres entreprises industrielles.

(3) *L'électronique grand public*, dont la clientèle est constituée par l'utilisateur final individuel.

(4) *Les composants*, dont la clientèle est constituée par les entreprises d'électroniques elles-mêmes.

On peut considérer que l'informatique fait partie de l'électronique dont elle utilise les techniques. D'autre part, la séparation entre l'informatique et l'électronique n'est pas toujours très claire. Toutefois, compte tenu que d'autres études dans ce même volume ont traité le problème de l'informatique *stricto sensu*, c'est à dire de l'emploi de calculateurs dans la gestion, le calcul scientifique etc. lorsqu'ils sont employés seuls et non pas intégrés dans un système électronique plus complexe, nous n'aborderons pas ce problème ici.

9.2 L'électronique professionnelle

On peut y distinguer un domaine traditionnel, qui est d'ailleurs celui de l'électronique à ses débuts dans les années 1910. Il comporte, pour mémoire, les radiocommunications, la radiodiffusion et la télédiffusion au niveau des émetteurs et des studios, le radar et ses applications, les systèmes d'armes électroniques tels que les dispositifs de contrôle et de codage des missiles, les télécommunications etc.

Lorsqu'on analyse le développement de ce domaine depuis la création du Comité Scientifique de l'OTAN, il y a une vingtaine d'années, on est frappé par plusieurs caractéristiques :

(1) C'est un domaine en croissance quantitative considérable de l'ordre de 15% par an sur le plan mondial. C'est qu'en effet les besoins tant militaires que surtout civils en matière de dispositifs de communications et de contrôle ne cesseant de croître pour l'équipement des pays tant développés qu'en voie de développement.

(2) C'est probablement le secteur industriel où la part de la recherche et développement dans les prix de revient est la plus importante (25 % et plus).

(3) Corrélativement, c'est un secteur qui a connu des évolutions qualitatives extrêmement rapides, largement liées d'ailleurs aux possibilités nouvelles offertes par les composants.

Cependant, il est frappant de constater que les principes théoriques des matériels et des systèmes utilisés aujourd'hui, et même de ceux qui sont encore en cours de développement, étaient déjà connus il y a 20 ans, les améliorations qualitatives, si considérables qu'elles aient été, n'ont donc pas tellement porté sur l'invention de principes nouveaux, mais sur une mise en oeuvre plus évoluée de ces principes.

Quelles sont ces évolutions?

On trouve d'une part une tendance constante à la montée en fréquence atteignant aujourd'hui le domaine des ondes millimétriques et même de l'infra-rouge. La raison en est facile à comprendre : la multiplication même des dispositifs et des services crée un incontestable problème d'encombrement de l'éther que l'accroissement des fréquences permet de résoudre, d'une part par élargissement de la bande utilisée, et d'autre part par l'amélioration des directivités des aériens liées, à dimension constante, à la réduction de la longueur d'ondes. Il faut ajouter que cet accroissement de directivité peut présenter, dans certaines applications, un intérêt propre.

Une deuxième évolution porte sur le traitement des signaux en vue de leur transmission ou de leur exploitation. Avec le temps, ce traitement est devenu de plus en plus perfectionné et complexe, et les méthodes numériques sont de plus en plus utilisées, car la réalisation de traitements complexes par des méthodes analogiques se heurterait rapidement à un problème de bruit et de précision. A titre d'exemple, si le principe du radar Doppler à haute fréquence de répétition et ambiguité de distances est connu depuis 40 ans, la réalisation pratique de radars Doppler aéroportés n'est rendue possible que par les méthodes de traitement numérique des signaux.

Une troisième évolution porte sur la miniaturisation et la réduction concomitantes de poids des équipements à fonction constante rendues possibles, comme le traitement numérique, par l'évolution des composants; cette miniaturisation est souvent plus réelle qu'apparente, car elle a surtout servi à augmenter les performances et le nombre des fonctions remplies par les équipements plutôt qu'à en réduire efficacement la taille, sauf bien entendu en ce qui concerne les équipements portables.

Une quatrième évolution, qui est à la fois une conséquence et une condition nécessaire aux évolutions précédentes, porte sur la réduction des consommations énergétiques, particulièrement utile pour les équipements portables autonomes.

Enfin, la dernière évolution porte sur la baisse constante à fonction constante des coûts des matériels.

On peut dire que, bon an mal an, on gagne sur tous les points précédents, un facteur deux tous les deux ans (certains disent tous les ans). Lorsqu'elle atteint une telle rapidité, une évolution continue devient une révolution.

Mais il faut ajouter que ces toutes dernières années, sont apparues des techniques qui constituent de véritables percées se superposant aux évolutions rapides mentionnées ci-dessus. A titre d'exemple : les communications sur *fibres optiques*. Là aussi le principe de guidage d'une onde lumineuse par une fibre de matériau transparent est connu depuis longtemps, mais la possibilité de réaliser des fibres de verre ou de silice, dont les pertes, par unité de longueur soient suffisamment faibles, n'a, elle, été réalisée que très récemment. Il est intéressant de signaler que les fibres optiques pour télécommunications ont théoriquement des coefficients d'observations mille à dix mille fois plus faibles que les bons verres d'optique disponibles il y a cinq ans. Encore fallait-il, pour utiliser des fibres optiques dans un système de télécommunications, disposer aussi de l'émetteur de lumière modulable — et le laser à semiconducteur est arrivé à point à un degré de développement suffisant — et de détecteur suffisamment sensible et rapide, basé sur des jonctions semiconductrices.

C'est cette coïncidence de trois développements concomitants, et bien sûr stimulés l'un par l'autre, qui fait de l'emploi de fibres optiques comme support de télécommunication une réalité expérimentale aujourd'hui, et une probabilité industrielle énorme pour le futur immédiat. En effet, le nombre des communications qu'il est possible d'envisager de passer par unité de surface dans une fibre optique est plus de dix mille fois supérieur à ce que permettrait tout autre support de guidage d'ondes. De surcroît, les fibres optiques n'utilisent aucune matière première rare à l'échelon mondial, même si elles font intervenir dans leur fabrication une part considérable de connaissances technologiques.

Un deuxième exemple pourrait être fourni par les possibilités de stockage d'informations offertes par les techniques de vidéo-disques optiques. Comparé aux disques mangétiques traditionnels de l'informatique, le vidéo-disque optique permet en effet d'atteindre des densités en bit par cm^2 cent fois supérieures à des coûts inférieurs et avec des lecteurs beaucoup moins chers et plus performants. Par contre, la possibilité d'enregistrer en temps réel, et de réutiliser un grand nombre de fois chaque portion du disque pour réenregistrer de nouvelles informations semble exclue, dans le cas du vidéo-disque. C'est dire que cette technologie ne se pose pas en concurrente immédiate des mesures de masse classiques de l'informatique, mais constituera bien plus un moyen nouveau dont il faudra trouver le meilleur emploi dans les systèmes électroniques du futur.

Que pouvons-nous prévoir dans les 20 ans à venir?
Tout indique d'abord que les évolutions observées dans le passé vont continuer à peu près au même rythme pendant la plus grande partie de cette période. Il est probable que d'ici 20 ans, le traitement du signal sera entièrement numérique et que la part des techniques opto-électroniques et du moyen de transmission fibres optiques y sera devenue très largement prépondérante. Mais dans le courant de ces dernières années, l'électronique professionnelle a vu s'ouvrir un nouveau domaine d'application. où en peu de temps elle s'est substituée aux technologies traditionnelles. Il s'agit de

la commutation téléphonique dont les principes étaient demeurés assez largement inchangés, depuis son invention par Strowger en 1889 jusqu'à l'avénement de la commutation électronique. Il est probable que la commutation téléphonique électronique, qui fait usage en combinaison des possibilités de gestion de systèmes complexes des calculateurs électroniques et du traitement numérique du signal à transmettre, constituera sous peu une part importante du chiffre d'affaire des entreprises d'électronique professionnelle. Ainsi, même dans ce secteur, nous observons les trois modes de développement de l'électronique : l'amélioration des composant utilisés permettant la croissance des performances, l'introduction de technologies entièrement nouvelles, enfin la pénétration de l'électronique dans des domaines dont elle était jusque-là largement exclue.

9.3 L'électronique industrielle

Il est probable que, malgré les chiffres d'affaire déjà importants réalisés dans ce domaine, on est encore tout au début de l'introduction de l'électronique dans le contrôle des processus industriels, les machines-outils, les robots etc. C'est vraiment le secteur où l'électronique se substitue dans une large mesure à la partie la plus fastidieuse du travail des hommes. Elle lui apporte de surcroît la précision, la constance des résultats, et, par conséquent, la qualité.

Le problème que pose cette introduction de l'électronique dans ce secteur très lié à des secteurs autres que l'électronique elle-même est celui de la banalisation d'une technique réservée jusqu'à tout récemment à des entreprises spécialisées. Il est probable qu'une partie considérable des produits de l'électronique industrielle seront réalisés par les fabricants de machines plutôt que par les entreprises d'électroniques *stricto sensu*. Cela posera incontestablement le problème de la capacité de ces entreprises souvent plus traditionnelles à prendre ce virage : celles qui n'y réussirent pas seront impitoyablement éliminées par la concurrence internationale.

Cela pose également un problème sur lequel nous reviendrons au Chapître 14 et qui est celui de la disponibilité, pour ces entreprises non électroniques, de composants adaptés.

9.4 Le domaine du grand public

Pendant longtemps, l'électronique a été perçue par le consommateur individuel uniquement à travers la radio et la télévision. Là aussi, des évolutions de performance et de prix moins spectaculaires sans doute qu'en matière d'électronique professionnelle, mais cependant fort sensibles, se sont produites et sont liées au progrès des composants, à l'introduction des circuits intégrés, et pour une part encore faible, mais appelée à croître, à l'introduction des techniques numériques. Cette évolution continuera, mais il est probable que le secteur de l'électronique grand public va connaître dans le courant des 20 années à venir des bouleversements beaucoup plus profonds.

C'est une liste rapide de ce qui nous paraît le plus important : L'introduction de nouveaux services dont certains ne seront peut-être que des gadgets plus ou moins ludiques (mais la radio et la télévision ne sont-elles pas déjà par nature des activités de distraction, donc ludiques), mais dont d'autres pourraient avoir pour le consommateur une utilité pratique ou économique immédiate.

Citons par exemple l'accès sur un récepteur de télévision familial ou sur un autre écran de même nature à des banques de données variées, dont le consommateur ne fait pas usage aujourd'hui faute d'accessibilité.

D'autre part, de réelles percées technologiques apparaissent comme probables dans les 20 ans à venir.

Le grand écran plat de télévision, sur lequel tant de travaux ont déjà été faits jusqu'à présent sans grand succès, est probablement maintenant juste de l'autre côté du tournant. Il changera la nature même de l'image télévisée et probablement aura une influence sur l'aménagement de nos foyers. Nous voyons apparaître, de même, la possibilité déjà expérimentée dans une certaine mesure, du son super haute fidélité, en particulier grâce aux techniques d'enregistrement numérique à très haute densité.

Sans doute l'évolution la plus spectaculaire dans le domaine de l'électronique grand public est-elle et sera-t-elle encore plus dans le futur constituée par l'introduction de l'électronique dans de nouveaux domaines. Déjà, et en fort peu de temps, on a vu la montre électronique prendre une part croissante, et sans doute demain dominante, du marché de la montre jusque-là mécanique. Les calculettes ont fait plus que se substituer à la règle à calcul dont l'emploi était limité à des spécialistes: toute maîtresse de maison les utilise couramment, et si le calculateur domestique reste encore aujourd'hui un appareil de distraction, il est probable qu'il jouera demain un rôle important surtout en liaison avec l'accès aux banques de données dont nous avons parlé plus haut. Même des évolutions peut-être moins immédiatement sensibles à l'utilisateur, au moins dans leur aspect électronique, mais en fait plus importantes encore, vont se produire. Tout ce qui est contrôle de dispositif électro-ménager sera prochainement entièrement électronique avec des conséquences sur la précision et la fiabilité dont il est difficile de mesurer l'ampleur.

Une autre application de l'électronique chez l'usager sera celle du courrier électronique, c'est à dire du remplacement du transport physique des plis et lettres par des dispositifs de fac-similé transmettant sur le réseau téléphonique commuté une copie du document.

Techniquement la télé fac-similé existe dès aujourd'hui mais des baisses de prix considérables sont nécessaires pour qu'elle puisse véritablement se substituer à la poste classique. Les problèmes de discrétion, des communications se posent également, non pas tant en cours de transmission où des solutions dérivées de techniques cryptographiques seront concevables, qu'au sein même des familles où les différents membres d'un même foyer peuvent désirer que leurs correspondances ne soient pas accessibles à tous les autres. Pour ces raisons et par suite des modifications profondes dans le secteur même de l'activité de service considérable qu'est devenue la poste dans tous les pays, l'introduction du courrier électronique ne se fera probablement que progressivement, et en commençant probablement par les utilisateurs professionnels. Cependant, à échéance de 20 ans, il serait étonnant que l'on ne profite pas pleinement de ces avantages des transmissions électroniques par rapport au transport physique de documents, que sont la rapidité, la réduction des consommations d'énergie, la réduction des tâches fastidieuses pour certains travailleurs.

Au-delà du foyer, l'électronique va prendre une part croissante dans l'automobile. Son introduction doit en fait permettre d'améliorer à la fois la consommation, l'anti-pollution et la sécurité, tout en assurant des fonctions de détection automatique des pannes permettant ainsi de simplifier la maintenance dont le coût et la difficulté, dans les pays développés en particulier, ne cessent de croître. C'est pourquoi on peut penser que le foyer, les objets personnels de demain seront, en dehors peut-être de l'habillement, constitués largement par des produits électroniques. Là aussi l'influence sur les secteurs industriels pourrait être considérable et il est difficile aujourd'hui de la prévoir pleinement.

9.5 Les composants

Nous avons déjà mentionné plusieurs fois ci-dessus que la plupart des évolutions de l'électronique, tant dans le passé que dans le futur proche, sont liées aux évolutions des composants électroniques. Cette évolution a commencé dans l'immédiate après-guerre avec l'invention du transistor bi-polaire puis une dizaine d'années plus tard par la mise au point de transistor dit M.O.S. (métal oxyde semi conducteur). Tel qu'il était, le transistor apportait beaucoup à l'électronique en miniaturisation, en fiabilité et en réduction de consommation d'énergie, mais c'est il y a une vingtaine d'années, la possibilité d'intégrer sur un même substrat de nombreux transistors et composants annexes (diodes, résistances, condensateurs) qui a provoqué une véritable révolution. C'est qu'en effet le coût de circuit intégré ne croît que comme le nombre de ses connections externes, de ses 'pattes' comme il est usuel de le dire. Ceci tient à ce que la réalisation d'un circuit intégré complexe se fait par des procédés collectifs dont la difficulté technique croît certes avec le niveau d'intégration, c'est à dire le nombre de composants sur la même puce de silicium, mais dont le coût est de toute façon marginal par rapport à celui de la mise en boîtier et de la réalisation des connections externes. Or le nombre des connections externes ne croît que lentement (plus ou moins logarithmiquement) avec le nombre des composants. A titre d'exemple, le transistor le plus simple comporte trois pattes, un micro-processeur évolué, représentant une dizaine de milliers de composants individuels, en comporte 48. Il est évident que le prix, par fonction, aussi bien que l'encombrement ont été réduits d'une manière spectaculaire par l'introduction des circuits intégrés.

Tout laisse prévoir que dans les dix années qui viennent, ce niveau d'intégration continuera à progresser à la cadence déjà mentionnée d'un facteur 2 ou 4 tous les deux ans. Cette progression vers un niveau d'intégration élevé peut évidemment se faire de deux manières — accroître la surface des puces ou réduire la dimension des composants individuels. On montre facilement que les rendements de fabrication décroissent exponentiellement avec la surface de puce, toutes choses égales d'ailleurs.

C'est donc plutôt vers la réduction des dimensions des composants individuels que se fait et se fera l'évolution. Les dimensions les plus faibles des parties critiques des composants sont actuellement de quelques microns. De nouvelles techniques de micro-lithographic les amèneront progressivement à quelques dixièmes de microns, multipliant par 100 le nombre des composants par unité de surface. On peut sans grand risque, prévoir que dans une dizaine d'années, une bonne partie des circuits intégrés utilisés dans les applications électroniques contiendront de l'ordre d'un million de composants individuels, sans que le nombre des pattes ait notablement augmenté. A plus long terme, mais peut-être dans la période de 20 ans sur laquelle porte notre prospective il n'est pas inconcevable que l'on réalise des dispositifs où les composants individuels seront interconnectés en trois dimensions, avec la perspective d'atteindre le milliard de composants individuels par bloc de volume inférieur au cm^3. Il s'agit là d'une complexité qui devient comparable à celle du cerveau humain. Ces évolutions mettent en oeuvre, au-delà des méthodes de micro-lithographies optiques, des techniques nouvelles basées sur le masquage par faisceaux d'électrons ou par rayons X. Nous retrouvons là une autre caractéristique de l'électronique qui est elle-même largement génératrice de technologies lui permettant de progresser.

Cela étant, la puissance même des circuits intégrés n'est pas sans poser de problèmes, du point de vue de l'évolution de l'électronique. Tant qu'il s'agissait de composants discrets, c'est à dire jusqu'à il y a une quinzaine d'années, le métier de l'ingénieur électronicien s'apparentait à l'utilisation d'un jeu de mécano; un nombre relativement réduit de pièces élémentaires lui permettait, par un assemblage astucieux, de réaliser une énorme variété de fonctions. Le circuit intégré réintroduit dans cette activité une certaine rigidité qui pourrait constituer un frein pour l'évolution rapide des applications. C'est à ce problème que prétendent répondre deux approches non point concurrentes, mais largement complémentaires:

Les circuits intégrés programmables plus fréquemment appelés *micro-processeurs*. Au prix d'une légère augmentation de la complexité et de certains sacrifices sur les performances en particulier la rapidité, on peut en effet réaliser des circuits intégrés dont la fonction peut être programmée sur une mémoire morte qui peut ou non être intégrée sur la même puce.

Une autre manière de décrire cette approche est de dire que l'on réalise sur une ou un faible nombre de puces de silicium un véritable calculateur auquel par programmation en mémoire morte, on fait simuler la fonction à remplir. Il s'agit là d'une approche extrêmement simple, limitée bien entendu au traitement numérique des signaux. Mais, nous disons que le traitement numérique est probablement destiné à prendre à terme la plus grande place dans l'évolution de l'électronique. D'autre part, malgré l'inévitable réduction des performances qu'implique la programmabilité, par opposition à un circuit intégré spécial qui ne ferait que la fonction désirée, les économies d'échelle au niveau de la production font que l'approche micro-processeur peut dans de nombreux cas se révéler la plus économique. Par contre elle n'est jamais ni la plus performante, en particulier au point de vue rapidité, ni la plus compacte, ni celle qui consomme le moins d'énergie.

Le circuit intégré à la demande. Lorsque l'on a besoin des performances les plus poussées ou lorsque l'application envisagée justifie, à elle seule, une production de grande série, une autre approche du problème devient préférable: c'est celle du circuit intégré à la demande spécialisé, conçu pour remplir une fonction et une seule. Mais, dans ce cas, ce circuit doit être spécialement conçu puis réalisé par le fabricant de semi-conducteurs. Pour retrouver la souplesse d'utilisation au niveau de l'ingénieur qui étudie le matériel, en particulier lorsque celui-ci n'est pas dans une entreprise d'électronique — ce qui, comme nous l'avons vu précédemment, sera de plus en plus fréquemment le cas — on peut alors faire intervenir les possibilités de la conception assistée par ordinateur. Ce n'est plus le fabricant de semi-conducteurs qui conçoit le composant. Il met à la disposition de l'utilisateur final une bibliothèque de données, d'ailleurs fort coûteuse à constituer et dont l'amortissement pose d'incomparables problèmes financiers. Il donne d'autre part accès à l'utilisateur d'un programme de plus en plus élaboré, permettent à celui-ci, à partir d'une conception analogue à celle qu'il avait du temps de l'electronique mécano, de faire concevoir par un calculateur non seulement le circuit final, mais l'ensemble des masques qui seront nécessaires pour le construire, et il sera bientôt possible de passer de la bande magnétique de sortie de cet ordinateur à la réalisation physique de ces masques, sans intervention humaine.

Sans doute, compte tenu des inévitables erreurs qui se glissent dans un tel procédé, les circuits correspondants ne peuvent-ils être disponibles que plusieurs mois après les premiers travaux de l'utilisateur. Cependant le temps de développement de ce matériel électronique est généralement suffisant pour que ce délai soit tolérable.

Il est difficile aujourd'hui de définir la part relative qu'aura dans l'électronique du futur le circuit intégré programmable et le circuit intégré à la demande. Nous pensons que ces deux types de produits auront chacun leur application.

A signaler que dans les deux cas des problèmes de formation du personnel utilisateur, et dans le cas du circuit à la demande des problèmes de financement, se posent d'une manière difficile. L'importance de l'enjeu fait penser qu'ils seront résolus.

9.6 Les composants speciaux

Si les circuits intégrés constituent l'évolution la plus spectaculaire dans le domaine de l'électronique, il est bon cependant de ne pas oublier que les évolutions des performances et les applications de l'électronique à des domaines nouveaux passent souvent par le développement des composants en général discrets spéciaux. Citons à titre d'exemple:

Tous les capteurs suceptibles de transformer une grandeur physique (intensité lumineuse, composition chimique d'un gaz ou d'un liquide, force ou déplacement etc.) en signal électrique susceptible d'être ensuite traité par des moyens électroniques.

Le cas typique est celui de l'application de l'électronique à l'automobile, dont nous avons parlé plus haut. Elle passe nécessairement, non seulement par un traitement de signal, des dispositifs de contrôle et de régulation, mais aussi par des **capteurs variés susceptibles de fournir des données d'entrée.** Il s'agit en l'occurence de capteurs de position, de température, de composition du gaz d'échappement etc.

Les actuateurs. De méme, à la sortie du dispositif électronique, il est en général nécessaire d'agir souvent par des moyens mécaniques, ou parfois par des moyens optiques. C'est le rôle des actuateurs qui n'ont pas bénéficié, au même degré que le reste de l'électronique, des progrès spectaculaires dans le courant des dernières années.

Il est probable que l'utilisation de matériaux nouveaux, tels que les électrets piezo-électriques, ou les nouveaux matériaux magnétiques à haute énergie (terres rares – cobalt) vont dans les années qui viennent conduire à de nouveaux types d'actuateurs beaucoup plus performants.

On pourrait rattacher au problème précédent celui des dispositifs de visualisation, en dehors du traditionnel tube cathodique ou du grand écran plat du futur dont nous avons déjà parlé, les diodes electro-luminescentes, les critaux liquides, les panneaux à plasma ont déjà fait leur percée et sont susceptibles de progrès considérables.

Les composants actifs à trés haute fréquence, dont la puissance, le facteur de bruit et la fréquence de fonctionnement évoluent à des cadences comparables à celle de l'intégration des circuits intégrés et qui, dans le courant des 20 ans qui viennent, remplaceront probablement par des dispositifs à l'état solide les tubes à vide dans la plupart de leurs applications, sauf probablement pour les puissances extrêmement élevées.

Les lasers à semi-conducteurs sont certes une réalité dès-aujourd'hui, mais si leur fibailité, c'est à dire leur temps moyen entre pannes, est d'ores-et-déjà satisfaisant, des progrès sont encore nécessaires en ce qui concerne leur durée de vie et les gammes de longueur d'onde couvertes.

La percée des communications sur fibres optiques passe par la résolution de ces problèmes qui impliquent des études parfois très fondamentales sur les matériaux et les phénomènes physiques dont ils sont le siège.

Enfin *les composants passifs* constituent et constitueront toujours une part importante de la valeur des matériels électroniques. Leur évolution, sauf sur le plan de la fiabilité, a été plus longue que celle des circuits actifs. Aussi, si des efforts technologiques suffisants n'étaient pas entrepris, risqueraient-ils de devenir rapidement le facteur limitant de l'évolution de la discipline.

9.7 Conclusion

Malgré la rapidité de l'évolution dans ces 20 dernières années, la révolution électronique n'en est encore qu'à ses débuts. Il se trouve que les problèmes que peuvent aider à résoudre les techniques de l'électronique sont précisemment parmi les plus fondamentaux des économies des pays tant développés qu'en voie de développement (économie d'énergie, développement des activités de service, etc.). C'est dire que le développement de l'électronique correspond à un besoin, et à un besoin largement solvable.

Pour que cette révolution se déroule à la vitesse nécessaire pour qu'elle puisse aider à résoudre les problèmes de l'humanité, la maintien à un niveau très élevé des efforts de recherche et développement dans les pays industrialisés est nécessaire, et ceci, sur un front très large qui va des études les plus fondamentales sur les matériaux, jusqu'au développement technique des matériels en passant par la conception assistée par ordinateur et les études de grands systèmes.

Mais cette révolution s'accompagnera probablement de profondes modifications dans le secteur même de l'activité électronique. D'une part l'électronique se banalise. Elle constituait et constitue encore largement une branche d'activité industrielle; elle sera demain une méthode utilisée dans toutes les branches.

D'autre part le rôle des composants électroniques restera, pendant les vingt ans qui viennent, avec les études de systèmes émetteurs, l'essentiel de cette révolution.

Ces deux considérations vont inévitablement poser, dans les activités industrielles ou les services, un problème de formation des hommes à ces techniques, problème qui nous semble aujourd'hui insuffisamment abordé dans la plupart des pays.

10

The NATO science
programme

Sir Sam Edwards

In the survey of the scientific scene in 1978 sponsored by the NATO Science Committee on its twentieth anniversary, it is appropriate to have a brief review of what the Committee has sponsored, with some personal speculations on future possibilities.

10.1 The Alliance nations

The Allies are the most free and most equitable nations of the world, and contain many of the richest of nations. This strength does not stem from the size of their population, the size of their land, or the raw materials found on their soil. It stems from their efficiency. This efficiency stems from the native ability of their population and from the plurality of their organization, a plurality which comes from the free exchange of ideas, capital, and people within nations, and across national boundaries.

10.2 The Science Committee

From the beginning, the NATO Alliance was conceived as being broader than a purely military alliance, the original declaration of 1949 in Article II contained the following words:

The parties will contribute toward the further development of peaceful and friendly international relations by strengthening their free institutions, by bringing about a better understanding of the principles upon which these institutions are founded, and by promoting conditions of stability and well-being. They will seek to eliminate conflict in their international economic policies and will encourage economic collaboration between any or all of them.

In 1957 it was felt that explicit steps should be taken to maintain the scientific strength of the Alliance, and the Allies agreed to fund a scientific programme in these terms, taken from the *Report of the Committee of Three on Non-military Co-operation in NATO*, 13 December 1956:

The strengthening of political consultation and economic co-operation, the development of resources, progress in education and public understanding, all these can be as important, or even more important, for the protection of the security of a nation, or an alliance, as the building of a battleship or the equipping of an army.

It must be emphasized that 'science' is here used in the broadest sense of the word, that is the use of man's intelligence to solve problems by the scientific method; so technology and social science are fully included.

Individual nations spend large sums on research and there is no purpose in NATO spending money on research which is a marginal addition to national efforts. However, once one looks at international interactions an unsatisfactory picture emerges. Nations do not find it easy to interact freely in research work, which, unlike, for example, straight technical expertise, is not bought and sold. Research projects that transcend national boundaries are not easy to finance. The role of the NATO Science Committee is to make this kind of research possible and to act as a catalyst or lubricant which can enhance the mass of research.

The NATO Science Committee has now operated for twenty years and has worked out a variety of means of carrying out the task it was set. These means are under continuous test, since the development of science into new complex inter-disciplinary systems has made the original need of twenty years ago even more pressing today.

The Committee acts in what can be called 'self-organized' and 'centrally organized' ways which are now described.

10.3 The self-organized mode

Here individual enterprise is funded from NATO in response to requests. These can come from individuals, a few individuals, or groups. The three modes are: the Science Fellowships Programme, the Research Grants Programme, and the Advanced Study Institutes Programme.

The Fellowship scheme is one in which a research worker moves from one country to another to gain skill and experience. This programme is particularly esteemed by those members of the Alliance whose technology is not so highly developed as the leading technological nations. Over the twenty years more than 12 000 Fellowships have been held, and provide by their cementing of the Alliance one of the most striking successes of the NATO Science Programme.

The choice of fellows is left to the nations themselves, but broadly fits into the usual definition of 'fellowship' in the sense of higher education.

It is interesting to note that some of the special programmes (discussed below) have felt the need for fellowships at a much more practical level and, while acknowledging fully the value of the academically oriented fellowships, it is perhaps time for the Committee to see if there is a need for a broader fellowship scheme embracing an 'apprentice' or 'technician engineer' stream and perhaps a 'managerial' stream. The memorial to Winston Churchill in the U.K. is a fellowship scheme which is not based on academic values but solely on the value to the industrial base and the nation of the activity proposed by the applicant. Such a scheme among the Allies would be very much in line with the aims of the Alliance.

The Research Grant Scheme
Funds do not permit the NATO Science Committee to provide research grants as

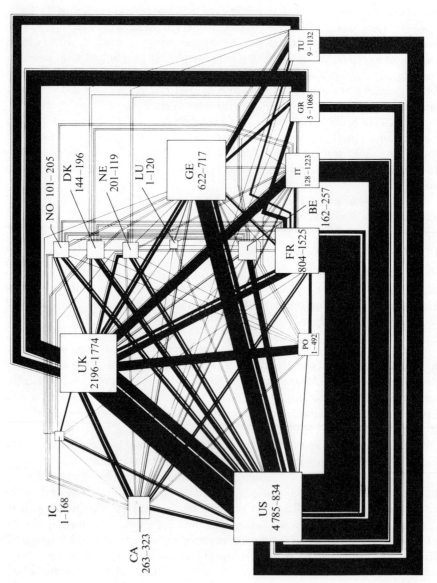

Fig. 10.1. NATO science fellows, 1964–76: the exchange of fellows between NATO countries. The first number in each box indicates the total number of fellows received and the second number indicates the total number of fellows

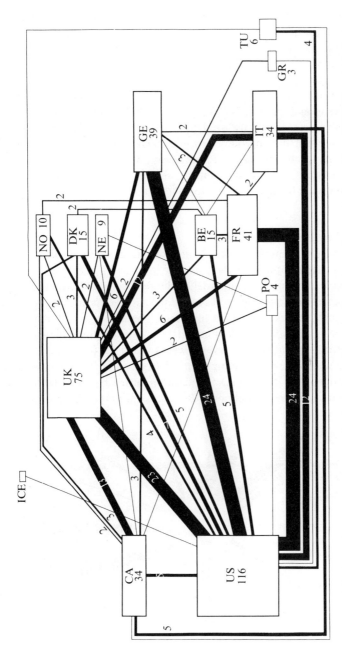

Fig. 10.2. Research grants programme: the frequency of co-operation between the member countries of NATO in a typical year. The figures appearing in the boxes show the number of projects in which scientists of that country are involved. The size of the box is proportional to the number of projects. The thickness of the connecting lines is proportional to the number of projects developed in collaboration by the scientists of the two connected countries. The figures on the connecting lines show the number of collaborative projects.

such; what this programme provides for is the interaction of established workers across national boundaries. This enables joint programmes to become a reality by providing the means for workers to visit one another's laboratories, to share in the exploitation of apparatus, and the development of computing schemes. There is no doubt that this programme, which turns away far more applications than it funds, could be greatly expanded even within the present coverage, while extensions of the programme cannot be contemplated at present.

The basic core of the research grants is to cover travel costs between the NATO nations. Rather than embark on a NATO system of full grants, which I feel would be expensive and would constantly overlap with national schemes, NATO could attempt to get national funding bodies together and act as a clearing house for international grants, each side paying its own way, with the NATO budget being limited to 'lubrication', i.e. matters which are difficult within national rules. Although all nations are for this in principle, there is no central body which can and does carry it out.

Advanced Study Institutes

The third of the 'self-organized modes' is the Advanced Study Institutes, where young research workers gather together, usually in some quiet retreat, to discuss their fields, led by lecturers who are internationally renowned authorities. The great success of this scheme prompts one to ask why it had never been done before the NATO Science Committee started these Institutes. The answer is that the Alliance (and for that matter any other group of nations) had no mechanism by which truly international institutes running for several weeks could be funded, but now an Institute usually succeeds in attracting support from national sources also. The Institutes are proposed to the NATO Science Committee by the scientific communit, and a representative panel chooses the successful proposals up to the limits

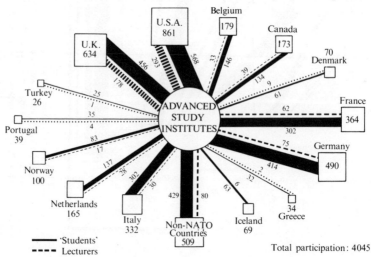

Fig. 10.3. Advanced Study Institutes programme: national distribution of participants in a typical year.

of its budget. The proceedings of the Institutes are published in the form of authoritative review texts, of which 700 have been published.

The pressure of applications had led to the NATO Science Committee to be content with this mode of operation, but there is no doubt that there are areas, particularly in the field of technology, where gaps can be seen; indeed, some of the special programme panels have acted to stimulate proposals in these gaps. It is perhaps time for a comprehensive study of the strengths and weaknesses of the Alliance by fields with a mind to improve the situation to be attempted with the aid of bodies like the OECD, the EEC, etc.

10.4 The centrally organized Special Programmes

Some research programmes, from the very geography of the Allies, cry out for international organization and stimulation. The Committee has responded to this situation by creating special programmes, which run for a number of years, which have a programme of experiments and conferences, and run cooperatively across the Alliance.

The current programmes of this type are: air–sea interaction, marine sciences, and eco-sciences. A programme just beginning under the auspices of the Air–Sea Interaction Programme is 'Remote sensing' i.e. the study by satellite and aircraft of the sea and land surfaces, which will coordinate national efforts in this rapidly developing field.

In addition the Committee has special programmes in areas of pressing importance where it is felt that an international stimulus is needed. Materials science is the newest of these, but it is still formulating its programme and so I will not report on it more fully. However, two other programmes are worth dwelling upon. These are (a) 'Systems science' and (b) 'Human factors'. Systems science is discussed in a powerful chapter in this volume, and so I need not discuss its content, but will remark that this subject, largely created by the pressure of war, has been transformed by systematic scientific study into a fully developed discipline whose power is not at all fully appreciated.

There has been a great advance in the control and organization of equipment and manufacture, from the level of trial-and-error experience to that of a major exact science. This subject now impinges upon all aspects of modern industrial life and hence is of direct consequence to the efficiency of the Alliance. This extends to the study of how industrial and social decisions are made. The NATO Science Committee programme has initiated research projects, Advanced Study Institutes, and inter-alliance conferences in this important, indeed crucial, area with great effort.

'Human factors' is the theme of the final special programme. This programme involves a systematic study of the interaction of man with instruments and operations in the factory or elsewhere. The interaction of psychology with the advances of physical and medical science has brought the whole issue of human efficiency and safety into the realm of science and this programme plays its part in the interchange of ideas, projects, and people to make work in the Alliance nations safer and more efficient.

In considering future special programmes, two suggestions appeal to me. One is that the air–sea interaction programme could be extended to the polar regions to become an air–sea–ice programme. I feel the time is ripe for an extensive polar study: the world-wide effects of the ice-cap, the potentiality of shallow seas, the great but ill-understood fish stocks, the energy potential of Greenland, and the purely scientific problems of the air–sea–ice interaction are ripe for study when so much data by satellite, air, land, and under water are becoming available.

I am impressed by the fact that the systems science and human factors programme are able to answer, by quite rigorous scientific study, questions which might otherwise be the subject of endless argument and speculation. I feel therefore that a joint extension of these two programmes could tackle problems such as urban and transport structure and renewal and energy and resource conservation. This should be done in a strictly numerical, experimental, and comparative way, and I strongly believe that when studied across the Alliance by an international body founded on conventional science and technology, many questions which seem difficult to answer on a purely national basis will be resolved.

10.5 The future

Professor Rabi has likened the work of the last twenty years as the creation of an invisible University, with a hundred thousand alumni spread through the Alliance nations. This achievement should, however, make one more concerned about the future. It would be a surprise, indeed a sensation, if a first comment about the future did not involve a plea for more money. The effect on inflation is difficult to assess in terms of figures in an international organization funded by many nations and currencies. But there does exist a real criterion: the work done: the man-months of fellowships, the number of visits to execute research, the number of Advanced Study Institutes, and the number of students at them. All these show a steady erosion of the programme. There is no part of the programme which can obviously be cut out; so the present activities are in difficulties and the various new ideas I have put forward are based solely on optimism. Nevertheless it is a good thing to take stock and make new suggestions in the hope that governments will like what we have done and ask for more.

Résumés des contributions

Chapitre 1 : Introduction : Analyse et bilan des progrès scientifiques John Maddox

Le progrès des sciences pures et appliquées depuis la seconde guerre mondiale a bouleversé les sociétés du monde industrialisé et ces changements ont été largement bénéfiques. Malgré cela, une certaine ambiguïté subsiste sur le rôle de la science — d'un côté on espère de nouveaux bénéfices du progrès scientifique et technologique mais, d'un autre côté une certaine méfiance s'exprime devant les changements sociaux imposés par le développement scientifique. Ce phénomène, bien qu'irrationnel est inséparable de la vision historique qu'ont les peuples de leur progrès.

Les années à venir recellent la promesse d'expériences aussi heureuses que celle des précédentes décennies — un approfondissement des connaissances des phénomènes naturels (comment l'univers est-il fait et de quelle matière ?) en même temps qu'un développement de la technologie (par exemple avec la diffusion de la micro-électronique).

Le texte procède à l'analyse des diverses contributions scientifiques au tome et souligne l'importance des progrès récents en biologie moléculaire à la fois comme moyen de compréhension des mécanismes de la vie et comme source de nouvelles techniques industrielles ; l'intérêt et l'importance de la recherche mathématique sont également soulignés. L'électronique et la science des matériaux sont à la base de futurs progrès technologiques importants, tandis que l'apport des sciences de l'environnement dans les décisions publiques sur le contrôle du progrès technologique et de l'analyse des systèmes appliqués aux problèmes sociologiques continue à susciter le débat.

Chapitre 2 : La science des matériaux M. F. Ashby

L'ingénieur chargé de concevoir un élément ou un ouvrage d'art appelé à supporter des charges, oriente immédiatement son choix vers les alliages. Ceci, parce que les métaux représentent un compromis particulier entre les différentes valeurs de leurs caractéristiques d'ensemble : modules relativement élevés, limites d'élasticité raisonnables, bonne résistance au choc et points de fusion relativement élevés. La mise au point d'alliages a connu un développement accéléré au cours des 100 dernières années et a débouché sur des variétés tout à fait originales d'aciers à haute résistance, d'alliages légers et d'alliages à haute température, dont l'apparition a stimulé ou permis, la création de nouvelles techniques et de nouvelles industries. Mais de nombreux métaux sont coûteux et en voie de raréfaction possible ; c'est pourquoi, au cours des dix dernières années, le rythme de mise au point de nouveaux alliages a diminué de façon spectaculaire.

Certaines des propriétés d'ensemble des céramiques et des polymères, dépassent de loin celles des métaux, mais dans la plupart des applications mécaniques, la possibilité de combinaison de ces propriétés est moins intéressante. Les céramiques, par exemple, ont des modules et des limites d'élasticité élevés, mais présentent une grande fragilité; les polymères sont potentiellement asses résistants, mais ont de faibles modules. Il existe désormais des solutions possibles à ces problèmes. Nous assisterons, au cours des vingt prochaines années, à des progrès rapides dans l'emploi des céramiques et des polymères, et à une évolution des matériaux non métalliques qui pourraient supplanter les métaux dans la plupart des domaines où ils dominent actuellement.

Chapitre 3: Sciences de l'environnement F. Kenneth Hare

Durant les années 1960 la prise de conscience en matière d'environnement s'est d'abord manifestée par l'inquiétude devant la pollution et les problèmes de 'qualité de la vie' puis s'est rapidement traduite en un effort visant à limiter le côte négatif des activités humaines sur le plan écologique. A l'heure actuelle, l'accent est mis sur la production de ressources, alimentaires et énergétiques notamment. Le document traite de l'impact de cette prise de conscience sur les sciences traditionnelles de l'environment — c'est-à-dire sur l'étude pure et simple de l'air, de l'eau, de la glace, du sol, des roches, de la faune, de la flore et de l'énergie — qui commencent lentement à s'intéresser également à l'activité de l'homme.

L'auteur présente en détail des études de cas de pollution stratosphérique, de désertification et de traitement des résidus nucléaires. Ces cas montrent la nécessité d'une action interdisciplinaire dans l'étude de tous les grands problèmes de l'environnement et font apparaître par ailleurs le rôle de la synthèse devant la disparité des questions.

L'auteur passe ensuite en revue les activités de surveillance et étudie les cycles biogéochimiques qui se sont révélés être les processus les plus centraux en matière d'environnement. Il appelle l'attention sur la nécessité de confier les problèmes majeurs à des équipes internationales et aussi sur le rôle spécifique de l'OTAN, du Comité scientifique et du Comité sur les défis de la société moderne.

L'analyse se termine sur une note optimiste. Il semble que dans nombre de pays occidentaux des mesures ad hoc aient permis d'acquérir une nette avance en matière d'environnement et que, sur le plan politique, l'opinion publique soit plus consciente que jamais de la gravité du problème. Dans le reste du monde, le principal souci reste la productivité de la nature — objectif primoridial des sciences de l'environnement pendant plusieurs dizaines d'années encore.

Chapitre 4: Biologie moleculaire Jean Brachet

Le but ultime de la Biologie moléculaire est d'expliquer les caractéristiques fondamentales (hérédité, reproduction) des êtres vivants, en fonction des propriétés chimiques et physiques des molécules qui les composent. Les plus importantes de ces molécules sont les acides nucléiques (ADN et APN) et les protéines.

La structure des bactéries (Procaryotes) et celle des cellules animales et végétales (Eurcaryotes) sont brièvement décrites. Les propriétés chimiques et physiques fondamentales des acides nucléiques et des protéines sont présentées. On montre ensuite que c'est l'ADN qui constitue le matériel génétique. La découverte, par Watson et Crick, de la structure en double hélice de l'ADN a permis d'expliquer sa réplication, sa transcription en ARN et le mécanisme des mutations génétiques. Le 'dogme' fondamental, universel, de la biologie moléculaire est le suivant :

$$\text{ADN} \xleftarrow{\text{Réplication}} \text{ADN} \xrightarrow{\text{Transcription}} \text{ARN} \xrightarrow{\text{Traduction}} \text{Protéines}$$

Un géne est une séquence spécifique dans une molécule géante d'ADN; elle code la synthèse d'une protéine qui lui correspond; un gène s'exprime lorsque cette protéine se synthétise. L'expression d'un gène se fait en deux étapes successives : sa transcription en un ARN messager, puis la traduction de ce messager en protéine. Le message génétique est inscrit, sous la forme d'un code, dans les molécules d'ADN présentes dans le noyau cellulaire; le code génétique est déchiffré dans le cytoplasme où la protéine spécifique correspondante est synthétisée par les polyribosomes. Les mécanismes de la réplication de l'ADN (qui est étroitement liée à la division cellulaire), de sa transcription (synthèse des ARN) et de la traduction (synthèse des protéines) sont décrits successivement.

La seconde partie de ce rapport a pour objet la biologie moléculaire des eucaryotes; l'accent est mis sur le problème de la différenciation cellulaire. On y montre que toutes les cellules d'un organisme possédent les mêmes gènes que l'oeuf fécondé; mais certains gènes ne sont actifs que dans un seul type cellulaire : par exemple, les gènes de l'hémoglobine ne sont exprimés que dans les cellules qui donneront naissance aux globules rouges. La différenciation cellulaire résulte donc de l'activation sélective de certains gènes; mais les mécanismes qui contrôlent l'activité génétique restent mal connus chez les eucaryotes. On montre que le problème de la différenciation cellulaire est étroitement lié à ceux du cancer, de la réponse immunitaire, du vieillissement, etc.

Enfin, on s'efforce d'imaginer l'avenir de la biologie moléculaire au cours des 20 prochaines années. L'accent est mis sur les potentialités des manipulations génétiques, qui permettent l'introduction et la multiplication dans des bactéries de gènes choisis par l'expérimentateur.

Chapitre 5: Biologie de l'ensemble de l'organisme: La machinerie immunitaire
G. Mathé

Si l'organisme est gouverné par son système nerveux via les nerfs et les hormones produites par les glandes endocrines, il se maintient dans son intégrité et dans la singularité que donnent à chaque organisme certains de ces constituants par sa machinerie immunitaire. Celle-ci reconnaît ce que pénètre en lui, qui lui est étranger en ce qui concerne les structures de singularité (antigènes), réagit contre ces composants vivants ou non qui possèdent ces structures étrangères, ces réactions aboutissant à leur rejet donc au maintien de l'intégrité de l'organisme. Ce rejet peut prévenir ou guérir certaines maladies, tout en étant responsable de certaines manifestations. Certaines de ces réactions sont dirigées contre des agents (microbes) directement 'nocifs' (infections), d'autres contre des agents qui ne seraient peut-être pas nocifs (allergènes) et qui ne le sont que via ces réactions (allergie). D'autres le sont contre des introductions thérapeutiques de cellules, tissus ou organes (graffes), et le thérapeute tente de les contrôler.

L'organisme reconnait aussi comme étrangers certaines cellules ou composants de cellules non étrangères puisque nées dans l'organisme, mais porteuses de composants étrangers qu'ont induits dans la ou les cellule(s) mère de la population cellulaire pathologique, des agents étrangers (virus, molécules chimiques, radiations). Ainsi reconnait-il les cellules cancéreuses (qui portent des antigènes dits 'associés aux tumeurs'). Ainsi admet-on que le machinerie immunitaire exerce une fonction de 'surveillance immunitaire' à l'égard des cancers.

Mais l'organisme peut aussi reconnaitre des cellules modifiées par des agents divers, non cancéreuses et non dangereuses directement, et réagir contre elles, déterminant, par ses réactions, des maladies d'expression trés variables pouvant affecter tous les organes ou presque (y compris les système nerveux et circulatoire) et appelés 'auto immunitaires'.

Tandis que nous avons créé l'immunothérapie des cancers pour obtenir de la machinerie immunitaire la destruction des cellules cancéreuses une fois la maladie établie, le médecin des maladies auto-immunitaires tente, comme le spécialists des greffes, de contrôler, voire de supprimer les réactions dirigées contre les cellules modifiées.

Ainsi l'immunité, avec son rôle en pathologie et sa possible action en thérapeutique dans les domaines des maladies aujourd'hui les plus fréquentes et qui constituent de grands fléaux comme le cancer, représente-t-elle un des sujets d'étude de la biologie à l'échelle de l'organisme entier qui ont le plus d'incidences surement prévisibles dns la connaissance fondamentale de la biologie, dans la meilleure compréhension de la plupart des maladies et dans la thérapeutique de beaucoup d'entre elles, directe par l'immunothérapie (ce qui est le cas du cancer et des maladies dites auto-immunes) et indirecte par les greffes.

Un résumé ne peut exposer tous les problèmes discutés dans le rapport, tels que la constitution de la machinerie immunitaire, ses principales fonctions et son rôle dans la surveillance immunitaire des cancers.

Chapitre 6: Astrophysique Bengt Strömgren

Au cours de la période 1958–78, la recherche en astrophysique a eu pour but de donner une description physique des objets célestes observés à notre époque, ainsi que de résoudre des problèmes posés par leur évolution au cours de millions et de milliards d'années. On examinera notamment les progrès accomplis dans la connaissance des structures stellaires et de leur évolution, de la matière interstellaire, de la croissance des étoiles et de leurs étapes successives de contraction, de développement principal, d'étoiles géantes aboutissant à l'étape finale de naines blanches, débris d'étoiles neutroniques. On soulignera l'apport de la physique dans l'analyse des propriétés de la matière et des densités extrêmement élevées dans lesquelles elle peut se présenter. On examinera également les changements intervenus dans notre galaxie par le formation d'étoiles et l'évolution de la masse stellaire. Les progrès accomplis dans l'analyse du système solaire sont abordés de façon indirect, de même que les questions de morphologie, les propriétés physique et la distribution spatiales des galaxies. La dernière section présente quelques indications sur les voies de recherche en astrophysique au cours des prochaines années.

Chapitre 7: **Mathematiques** Mark Kac

La recherche mathématique manquant de lignes directrices universellement acceptées, son évolution s'est poursuivie sur un certain nombre de fronts apparemment sans liens. Au fur et à mesure que notre siècle avançait elle s'est isolée et repliée sur elle-même. C'est assez récemment, toutefois, que quelques uns des développements les plus abstraits de la géométrie différentielle ont rejoint de manière inattendue certaines spéculations relatives aux champs de référence (gauge fields), qui émanent de la physique.

Mis à part § 7.2 'Escalade des sommets', dans laquelle sont décrites trois 'ascensions' victorieuses en mathématiques pures, le chapitre met l'accent sur ces développements de la recherche mathématique qui ont établi des contacts avec des domaines extérieurs aux mathématiques ou qui ont fait apparaître une convergence d'idées de différentes branches de la recherche mathématique elle-même. C'est ainsi que outre la rencontre récente de la géométrie différentielle et de la physique dont il est question à § 7.3, on trouve à § 7.4 une brève évocation de la théorie des catastrophes (seule partie des mathématiques à parvenir, bien que manière tronquée, au grand public par le biais de ses journaux quotidiens) et un aperçu, également bref, de l'ordinateur et de son rôle éventuel dans la problématique mathématique (§ 7.5).

La dernière Section ('D'où et vers où?') est consacrée principalement à la pénétration graduelle des idées probabilistes dans les mathématiques et au renouveau de la théorie combinatoire. On y trouve aussi, à titre d'essai, quelques aperçus sur le future.

Chapitre 8: **La science des systemes** C. West Churchman

La 'science des systèmes' se définit comme la démarche intellectuelle visant à une compréhension des éléments constitutifs d'un système social qui permette d'en améliorer le fonctionnement en fonction des objectifs qu'il s'est donné. Le document établit une distinction entre la science des systèmes pluraliste, qui étudie des systèmes spécifiques tels que l'énergie, les services de santé, l'enseignement, etc., et la science des systèmes moniste, qui ètudie le système global des activités humaines.

L'auteur tente de décrire les innombrables approches pluralistes, dont les unes utilisent des modèles mathématiques 'optimisation' alors que les autres n'y ont pas recours : étude des fils d'attente, contrôle, théorie de la recherche, programmation linéaire, ainsi qu'un certain nombre de sigles, tels que SQC, PERT et PPBS.

La formule moniste la plus récente est celle des 'modèles mondiaux' et des descriptions qualitatives de l'interconnexion des systèmes humains (et non humains). Il s'agit là du prolongement d'une longue tradition écrite remontant jusqu'à deux mille ans avant J.C., et cette tradition appelle quelques remarques.

Malgré les énormes moyens d'analyses qui sont apparus depuis le début de siècle, le système social comporte encore, pour les chercheurs, plusieurs grands mystères. (1) Pourquoi les recommandations concernant la science des systèmes ont-elles été aussi peu suivies d'effet? (2) Est-ce parce que les experts de la science des systèmes ont une certaine naïveté politique, ou (3) parce qu'ils perçoivent mal les vraies valeurs de l'humanité, ou (4) parce que le pluralisme néglige toujours des aspects importants ou (5) parce que le monisme est trop global pour être bien compris, ou (6) parce que la science des systèmes a, jusqu'ici, ignoré la nature profonde et spirituelle de l'homme?

Le document se termine à la fois sur une prévision et sur un espoir. Il prévoit que la science des systèmes cherchera de plus en plus à explorer les arcanea de l'analyse des systèmes, et qu'ainsi, elle s'intéressera davantage aux différents secteurs de la société, et moins à la destinée de l'espèce humaine. En même temps, l'auteur exprime l'espoir que cette prévision se révélera fausse, car la survie de l'espèce humaine et de sa fantastique 'connaissance' de la nature dépendra certainement de notre perception des moyens à employer pour que notre survie ait un sens.

Chapitre 9: L'électronique   Pierre Aigrain

L'électronique est une discipline technique très variée, puisque caractérisée par ses moyens et méthodes (l'utilisation du mouvement des électrons dans le vide, les gaz et les milieux condensés et des ondes électromagnétiques) plutôt que par ses domaines d'application qui sont fort nombreux, et dont le nombre croît constamment.

Pour simplifier il est utile de distinguer:

(1) Les domaines traditionnels − c'est-à-dire déjà développés il y a 20 ans − de l'électronique professionnelle (radar, transmissions professionnelles, systèmes de navigation, de guidage, ordinateurs, etc...).

(2) Les domaines traditionnels de l'électronique grand public (radio, télévision).

(3) Les domaines nouveaux de l'électronique professionnelle, c'est-à-dire ceux où les fonctions étaient réalisées sans l'utilisation de l'électronique mais où celle-ci s'impose progressivement : commutation téléphonique par exemple.

(4) Les domaines nouveaux grand public, fort nombreux et en naissance rapide : calculettes, montres à quartz, Fours à microonde, cuisinières à induction, jeux télévisés, mais aussi introduction de l'électronique dans l'électroménager, l'automobile, etc.

(5) Les composants et leur évolution.

Car c'est l'évolution extraordinairement rapide de ces derniers qui a permis et provoqué toutes les autres. En reprenant les domaines:

Les principes appliqués aujourd'hui étaient connus en principe il y a 20 ans. L'échelle et les complexités des systèmes a crû énormément grâce aux composants (prix, miniaturisation, fiabilité). Les communications sur fibre optique, et les ondes millimétriques constituent des moyens nouveaux eux-mêmes liés aux composants nouveaux (par exemple: Laser à semi-conducteurs).

L'évolution de l'électronique traditionnelle a sans doute été moins spectaculaire, mais là aussi les gains de prix, de fiabilité, de services annexes rendus possibles par les progrès des composants et des techniques d'assemblage auront à l'avenir un impact croissant.

D'ici 10 ans des domaines entiers, telle la commutation téléphonique, sera totalement modifiée par l'introduction déjà bien engagée de l'électronique. Cela dépassera la simple substitution de technique et l'introduction de nouveaux services et débouchera sur des modifications de la structure même des systèmes et des services qu'ils rendent.

Enfin l'introductin de l'électronique dans les produits existants (électroménager, montres, automobiles, etc.) ou entièrement nouveaux (calculateurs domestiques, télécopie remplaçant la poste, etc.) n'en est qu'à ses débuts. Elle connaîtra dans les 20 prochaines années des progrès spectaculaires.

Toute cette évolution est liée à celles des composants que l'on peut diviser en deux catégories:

(a) Les composants numériques à haut niveau d'intégration, en particulier les microprocesseurs dont la capacité à être 'programmée' redonnent au concepteur la souplesse que la nature 'Mécano' de l'électronique d'il y a 20 ans caractérisait, et que les débuts de l'intégration à grande échelle tendaient à faire disparaitre.

(b) Les composants spéciaux (lasers, tubes et semi-conducteurs hyperfréquences, Détecteurs d'infrarouge, senseurs variés, etc.– qui sont les composants clés de toute nouvelle application. Là aussi des progrès spectaculaires ont été accomplis et n'en sont qu'à leurs débuts.

En résumé, l'électronique a la réputation d'une technique en évolution rapide, et pourtant tout laisse penser que cette évolution va s'accélérer encore.

Chapitre 10: Le Programme Scientifique de l'OTAN Sir Sam Edwards

Le texte est une vue personnelle sur les résultats de deux décennies d'activité du Programme scientifique de l'OTAN, montrant comment le Comité scientifique a su répondre aux besoins de la communauté scientifique à travers ses programmes généraux et ses programmes spécifiques. Sont décrits les programmes des subventions à la recherche, des bourses, des cours d'été, et les programmes spéciaux; plusieurs suggestions sont faites sur les moyens d'améliorer et d'accroître leur impact, particulièrement dans la conjoncture économique présente.

Summaries of contributions

Chapter 1: Introduction: Retrospective and prospective reviews of scientific achievements John Maddox

The development of both fundamental and applied science since the Second World War has helped radically to change the societies of the industrialized world and, the argument goes, the changes have mostly been beneficial. Nevertheless, there persists a curious ambivalence on the role of science — the public looks forward to the benefits of science and technology but is disaffected from the social changes which scientific developments bring about. This phenomenon, irrational though it may be, is inseparable from people's perception of history.

The years ahead promise a continuation of the experience of recent decades — a deepening of understanding of natural phenomena (such as the question of what the universe is like or what matter is made of) and the development of technology (exemplified by the application of microelectronics to computing).

The chapter reviews the several scientific contributions to the volume and, in the process, underlines the importance of recent developments in molecular biology both in the understanding of life processes and as a source of new industrial techniques; and also the interest and importance of mathematical research. Electronics and materials science have an obvious potential for further beneficent technical change, but the value of environmental science as an aid to government decisions on the regulation of technological development and of systems analysis applied to sociological problems is more open to dispute.

Chapter 2: The science of engineering materials M. F. Ashby

Metallic alloys are the first choice of the engineer who is asked to design a component or structure which must carry load. This is because metals offer a particular compromise in the values of their important bulk properties: they have fairly high moduli, reasonable yield strengths, good toughness, and moderately high melting points. The development of metallic alloys accelerated rapidly over the 100 years between 1860 and 1960, resulting in entirely new ranges of high-strength steels, light alloys, and high temperature alloys, which in turn stimulated, or allowed the creation of, new technologies and industries. But many metals are expensive and potentially scarce; and in the last decade the rate of development of new alloys has slowed dramatically.

Certain of the bulk properties of ceramics and polymers surpass those of metals, but for most engineering applications, the combination of properties they offer is a less attractive one. Ceramics, for instance, have high moduli and yield strengths, but poor toughness; polymers are potentially tough enough, but have low moduli. Possible paths towards solving these problems now exist. The next 20 years should see rapid developments in engineering ceramics and polymers, and the evolution of non-metallic materials which could challenge metals in most of the areas in which they are now dominant.

Chapter 3: **The environmental sciences** F. Kenneth Hare

The environmental awareness of the 1960s began with anxiety about pollution and 'quality' issues, but was quickly transformed into an effort to render human action less ecologically damaging. At present the major emphasis is upon resource productivity, especially as regards food and energy. The chapter deals with the impact of this awareness on the old established environmental sciences – the pure study of air, water, ice, soil, rock, biota, and energy – which is slowly moving towards the inclusion of human action.

Detailed case studies are presented of stratospheric pollution, desertification and nuclear waste management. These illustrate the interdisciplinary response that is needed for the study of all broad environmental problems – and also the role of the individual synthesizer in holding disparate things together.

The text continues with a review of monitoring, and discusses the biogeochemical cycles, which have emerged as the most central environmental processes. Attention is given to the need for international teamwork in the proper treatment of major issues, and to the specific role played by NATO, the Science Committee, and the Committee on the Challenges of Modern Society.

The analysis closes on an optimistic note. In many Western countries, seemingly ad hoc measures have led to substantial advances in environmental practice, as political awareness of public concern has grown. For the rest of the world the major problem must continue to be concern for the productivity of nature – which will be the prime focus of the environmental sciences in the next several decades.

Chapter 4: **Molecular biology** Jean Brachet

The ultimate aim of molecular biology is to explain the main properties of living organisms (heredity and reproduction) by the physical and chemical properties of the molecules which build them up. The most important of these molecules are the nucleic acids (DNA and RNA) and the proteins.

The structure of bacteria (prokaryotes) and of animal and plant cells (eukaryotes) is briefly described. The main physical and chemical properties of the nucleic acids and the proteins are presented, as well as the evidence demonstrating that DNA is the genetic material. The elucidation by Watson and Crick of the double-helix structure of DNA has explained DNA replication, transcription, and mutation. The universal and fundamental 'dogma' of molecular biology is the following:

$$\text{DNA} \xleftarrow{\text{Replication}} \text{DNA} \xrightarrow{\text{Transcription}} \text{RNA} \xrightarrow{\text{Translation}} \text{Proteins}$$

The gene is a specific sequence in the giant DNA molecule coding for the synthesis of a specific protein. A gene is expressed when its corresponding protein is synthesized. Gene expression requires two successive steps: transcription of DNA into the corresponding messenger RNA and the translation of this messenger into the corresponding protein. The genetic message is encoded in the DNA molecules present in the cell nucleus and is deciphered in the cytoplasm, where the corresponding specific protein is synthesized by the 'polyribosomes'. The mechanisms of DNA replication (which is closely linked to cell division), of transcription (RNA synthesis) and of translation (protein synthesis) are presented.

The second part of the chapter deals mainly with the molecular biology of higher organisms (eukaryotes) with the emphasis placed on cell differentiation. It is shown that all cells contain the same genes as the fertilized egg, but that certain genes are active in given cells only: for instance, only the cells committed to differentiation into red blood cells will produce haemoglobin. Cell differentiation thus results from variable gene activity, but the control mechanisms of gene activity are not yet well known in eukaryotes. It is shown that the problem of cell differentiation is closely linked to those of cancer, immunological response, ageing, etc.

Finally, an attempt is made to evaluate the prospects of molecular biology in the next 20 years. Special attention is placed on the potentialities of genetic engineering, which allows the introduction and multiplication of selected genes in bacteria.

Chapter 5: Whole-organism biology: The immunological system G. Mathé

While the organism is governed by the nervous system, via the nerves and the hormones produced by the endocrine glands, it is the immunological system that maintains its integrity and the individuality that each organism owes to certain of its constituents. This system identifies any substance that penetrates it and is foreign to its differentiating structures (antigens)and it reacts against these living or non-living substances whose structures are alien; as a result, the foreign bodies are rejected and the integrity of the organism is maintained. This rejection may prevent or cure certain diseases, but it is also responsible for certain manifestations. Some of these reactions are directed against agents (microbes) that are directly harmful (infection), while other reactions are directed against agents that may possibly be innocuous (allergens) and that become harmful only as a result of these reactions (allergy). Other reactions, which the therapist tries to bring under control become harmful when cells, tissues or organs are introduced therapeutically (grafts).

The organism also recognizes as foreign bodies certain cells or cell components that are not foreign to the organism, since they were formed within it, but carry foreign components induced in the mother cell(s) of the pathological cell population by foreign agents (viruses, chemical molecules, radiation). It can thus identify cancerous cells (which carry 'tumour-associated' antigens). It is therefore generally agreed that the immunological system performs an 'immune surveillance' function against cancer.

However, the organism can also recognize cells that have been modified by various agents that are not cancerous and not directly dangerous and it can react against them; these reactions give rise to diseases, with very variable symptoms, that may affect all or almost all the organs (including the nervous and circulatory systems) and are known as the auto-immune diseases.

While we have created cancer immunotherapy in order to induce the immunological system to destroy cancerous cells once the disease has been diagnosed, the specialist in auto-immuno disease tries, like the graft specialist, to control if not supress the reactions against the modified cells.

Immunity, therefore, thanks to its role in pathology and its possible therapeutic action against the most widespread and dangerous diseases of our time, such as cancer, represents one of the areas of biological research, affecting the entire organism, with the most far-reaching foreseeable consequences as regards fundamental knowledge of biology, improved understanding of the majority of diseases and the treatment of many of the latter, both directly by immunotherapy (as in the case of cancer and the auto-immune diseases) and indirectly by means of grafting.

A summary cannot cover all the problems which are dealt with in the chapter such as the constitution of the immunization system, its main functions, and its role in immunization control of cancer.

Chapter 6: Astrophysics Bengt Strömgren

Research in astrophysics during the period 1958–78 has aimed at a description in physical terms of celestial bodies as observed at the present epoch, as well as at the solution of problems of their evolution over millions or billions of years. This chapter concerns itself with progress made over the last 20 years particularly in the areas of stellar structure and evolution, interstellar matter, and the structure and evolution of our galaxy. The processes of formation of stars out of interstellar matter, of the evolution of stars through contraction phase, main-sequence and giant star phases to the final stages of white dwarf or neutron star remnant are described. The role of the input from physics in dealing with the properties of matter over a range of densities that covers nearly forty orders of magnitude is emphasized. The evolutionary changes of our galaxy that are a consequence of star formation and stellar evolution are discussed.

Progress in the study of the solar system is referred to only very briefly. In sections on the morphology, physical properties, and spatial distribution of galaxies, and on cosmology, the discussion is brief, aiming at a specification of essential problems. In the last section some of the goals of astrophysical research during the next twenty years are mentioned.

Chapter 7: Mathematics Marc Kac

Since mathematics lacks central problems that are universally agreed upon, its evolution proceeds along a number of seemingly unrelated fronts. In our century it has also become progressively isolated and inward oriented. Quite recently, however, some of the more abstract developments in differential geometry came unexpectedly into contact with certain speculations concerning gauge fields, which have originated in physics.

Except for § 7.2 ('Scaling the heights'), in which three victorious 'ascents' in pure mathematics are described, the chapter emphasizes mathematical developments which have either made contact with areas outside of mathematics or which have shown a confluence of ideas coming from different branches of mathematics itself. Thus in addition to the already mentioned recent encounter of differential geometry and physics (treated in § 7.3) there is a brief digression in § 7.4 on catastrophe theory (the only piece of mathematics to reach, albeit in a garbled way, the general public via the daily press), and an equally brief discussion of the computer and its possible role in mathematical problematics (§ 7.5).

The last section ('Where from and where to?') is devoted mainly to the gradual incursion of probabilistic ideas into mathematics and to the renaissance of combinatorial theory. There are also a few tentative glimpses into the future.

Chapter 8: **Systems science** C. West Churchman

'Systems science' is defined as the intellectual attempt to understand enough about the components of a social system so as to improve its performance relative to human goals. The chapter makes a distinction between pluralistic systems science, which investigates specific systems such as energy, health services, education, etc., and monistic system science which investigates the global system of human affairs.

An attempt is made to describe the incredible variety of pluralistic approaches to systems, some using mathematical models ('optimization'), some not: queuing inventory, control, search theory, linear programming, as well as a number of acronyms, SQC, PERT, PPBS, etc.

The latest monistic systems science are 'world models', as well as qualitative descriptions of how human (and nonhuman) systems are interrelated. Since these follow a long tradition of writings dating back as far as 2000 B.C., some remarks on the tradition are in order.

Despite the enormous analytic skills that have been developed in this century, several critical mysteries remain for social system investigations. (1) Why has the implementation of systems science recommendations been so poor? (2) Is it because systems scientists are politically naive, or (3) because they fail to understand the real values of humanity, or (4) because pluralism always leaves out important facets, or (5) because monism is too global to be understood, or (6) because systems science has ignored the inner, spiritual nature of humans?

The contribution contains, at the end, a forecast and a hope. The forecast is that systems science will more and more try to fathom the intricacies of systems analysis, and thereby become more relevant to sectors of human society and less relevant to the destiny of the human species. The hope is that the forecast is wrong, because the survival of the human species, including its fantastic 'knowledge' of nature, must surely be dependent on our understanding of how we can survive in some meaningful way.

Chapter 9: Electronics Pierre Aigrain

Electronics is a very wide field of knowledge that is characterized more by its methodology (utilization of electron motion in vacuum, in gases and in 'condensed' media, and of electromagnetic waves), than by its areas of application, the already very large number of which is constantly increasing.

To simplify, it is useful to distinguish between:

(1) The traditional areas of industrial electronics, i.e. those already developed twenty years ago – radar, communications, navigation and guidance systems, computers etc.

(2) The traditional areas of electronics for the general public, such as radio and television.

(3) The new areas of industrial electronics, i.e. those where electronics was not previously used but is gradually taking over, for example telephone switching.

(4) The new 'general public' areas, of which there is a large and rapidly growing number: mini-calculators, quartz watches, microwave ovens, induction cookers, TV games, as well as the introduction of electronics into household appliances, etc.

(5) Components and their development.

It is the extraordinarily rapid developments in the field of components during the past few years that have been responsible for all the others. Returning to these areas, the principles being applied today were theoretically known twenty years ago. The scale and complexity of the systems have increased enormously, thanks to improved reliability, miniaturisation, and competitive pricing of the components. Optical-fibre communications and millimetric waves constitute new technologies that are themselves dependent on new components (for example, semiconductor lasers).

Developments in the traditional general-public field have no doubt been less spectacular but, here too, the improved prices, reliability and additional services made possible by progress in components and assembly techniques will have an increasing impact in the future.

Within ten years, whole areas such as telephone switching, will have been completely changed, thanks to the introduction of electronics which is already well in hand. This will be more extensive than the mere substitution of a technique and the introduction of new services, and will lead to changes in the very structure of the systems and services provided.

Finally, the introduction of electronics into existing products (household appliances, watches, cars, etc.), or entirely new ones (domestic computers, telecopying systems replacing postal services, etc.) is only just beginning. There will certainly be spectacular progress over the next twenty years.

All these developments are linked to the development of components, which may be divided into two categories:

(a) Digital components with a high level of integration, particularly microprocessors whose capacity to be programmed will give back to the designer the flexibility that was characteristic of the 'Meccano' period of electronics twenty years ago, but tended to disappear at the beginning of the large-scale scale integration stage.

(b) Special components (lasers, tubes and hyperfrequency semiconductors, infra-red detectors, various sensors, etc.), which are the key components for any new applications. Here, too, the spectacular progress that has already been achieved is hardly more than a beginning in an area of vast potential.

To sum up, although electronics already has the reputation of a rapidly-developing technology, there is no doubt that its development will speed up even more in the future.

Chapter 10: The NATO science programme Sir Sam Edwards

This chapter is essentially a personal review of the NATO science programme during the twenty years of its existence, showing how the Science Committee has operated by responding to the scientific community through self-organized modes and by developing explicit programmes. The research grants, science fellowships, advanced study institutes, and special programmes are described, and a number of suggestions are made that might improve and extend their influence, even in a time of financial stringency.

Index